GARY F. LINDGREN

Managing Industrial Hazardous Waste

A Practical Handbook

LEWIS PUBLISHERS

Library of Congress Cataloging-in-Publication Data

Lindgren, Gary F.
    Managing industrial hazardous waste.

    Rev. ed. of: Guide to managing industrial hazardous waste. cl983.
    Bibliography: p.
    Includes index.
    1. Hazardous wastes—United States—Management—Handbooks, manuals,
etc. 2. Factory and trade waste—United States—Handbooks, manuals,
etc. I. Lindgren, Gary F. Guide to managing industrial hazardous waste. II.
Title.
TD1040.L56 1989   363.7'28   88-27382
ISBN 0-87371-147-5

Cover designed by Eric M. Lindgren.
Photography by Neil M. Lindgren except where otherwise noted.

Fourth Printing 1990

Third Printing 1990

Second Printing 1989

LEWIS PUBLISHERS, INC.
121 South Main Street, Chelsea, Michigan 48118

PRINTED IN THE UNITED STATES OF AMERICA

To my lovely wife
Helen
and our two
wonderful children,
Elizabeth (age 4)
and Geoffrey (age 2)

Gary F. Lindgren is Regulatory Services Manager for Heritage Remediation/Engineering, Inc., the environmental consulting arm of Heritage Environmental Services. Mr. Lindgren has been responsible for a wide variety of hazardous waste management projects, including environmental audits, real estate assessments, enforcement negotiations, facility closures, RCRA compliance plans, training, regulatory interpretations, and contaminated site remediation.

He has prior professional experience as the environmental compliance director for a six-plant chemical company, and as a hazardous waste facility inspector for the State of Indiana. From 1983 to 1987, Mr. Lindgren also served as an Adjunct Professor in the School of Public and Environmental Affairs at Indiana University, teaching E510—*Hazardous Materials Regulation*.

Mr. Lindgren received his BS in Environmental Policy from Indiana University in 1978. He was then awarded a graduate fellowship for readings in Solid Waste Management. He received his MS in Environmental Policy from Indiana University. Mr. Lindgren is both a Certified Hazardous Materials Manager (CHMM) and a Certified Environmental Trainer (CET).

Preface

Industrial waste management has been revolutionized by Subtitle C of the federal Resource Conservation and Recovery Act of 1976 (RCRA), as amended by the Hazardous and Solid Waste Amendments of 1984 (HSWA). RCRA provides the statutory authority for the U.S. Environmental Protection Agency (EPA) to regulate and require the proper management of all wastes defined as hazardous. EPA issued the bulk of the implementing regulations in May of 1980. Since then, many amendments and additions to these regulations have been published in the *Federal Register*.

With these regulations, EPA provided a set of minimum standards for all who are involved in hazardous waste management—the generator, the transporter, and the owner/operator of treatment, storage, or disposal (TSD) facilities. These minimum standards apply nationwide, and many states acted to adopt similar standards and to develop state hazardous waste regulatory programs. Such state-level activity is allowed and encouraged by RCRA. States are allowed to have more stringent standards than those imposed by EPA. State hazardous waste regulatory programs that meet federal requirements receive what is known as "authorization" from EPA. This allows the state regulatory agency to have primary authority to regulate hazardous waste management within its borders. Certain HSWA regulatory initiatives, until adopted by the state agencies, will result in regulated entities having to satisfy the requirements of both state and federal governments.

The purpose of this book is to provide those responsible for waste management at manufacturing firms with (1) a framework to understand the complex web of regulatory requirements and (2) a philosophy to guide waste management decisionmaking within the regulatory context. Examples and practical applications of these considerations are offered throughout the text. This book should assist the practitioner in establishing or modifying the firm's regulatory compliance program. It focuses on those hazardous waste management activities most common at manufacturing firms: namely, generation of wastes and accumulation or storage onsite in containers or tanks, pending shipment offsite to commercial facilities for treatment or disposal. Certain onsite treatment activities, such as wastewater treatment and solvent distillation, are also discussed.

This book was written for those individuals at manufacturing firms with hazardous waste management responsibilities. The special concerns of practitioners, especially as they relate to making decisions and solving problems, are integrated into the text. Efforts have been made to organize and explain the necessary concepts and information in consideration of the practitioner's special requirements for relevance and utility.

Knowledge of the regulatory requirements is a necessary condition of proper hazardous waste management, but such knowledge is not in itself sufficient. The alternatives available, both onsite and offsite, can be overwhelming. Methods of and considerations in selecting between alternatives are offered, on both the conceptual and specific levels. The regulatory framework provided by the "three determinations" allows consideration of the philosophical alternatives offered by the Waste Management Hierarchy. This framework and philosophy are applied, for example, when selecting between various onsite and offsite technologies for those wastes determined to be hazardous.

Section I of this book provides an overview of the hazardous waste management system and outlines the regulatory definitions of solid and hazardous wastes. The three determinations necessary for potentially regulated manufacturing firms are described. These determinations are:

1. What are the wastes at the facility?
2. Are any of the wastes hazardous, under the existing regulatory definitions?
3. What are the applicable regulatory requirements, given the quantities of hazardous waste generated, and the types of hazardous waste management activities conducted at the facility?

Section II includes an in-depth explanation of the federal regulatory standards applicable to the different categories of hazardous waste generators. Areas where state hazardous waste regulations may differ from federal requirements are identified. Section III presents a philosophical basis for a corporate compliance program and then provides guidance and describes the actions and paperwork necessary for such a program. Key factors to be considered are specified.

Section IV concludes with some very practical information not commonly found in textbooks or regulations. Topics include selecting commercial treatment/disposal vendors, considerations in waste sampling and analysis, dealing with regulatory agency officials and consultants, legal liabilities, and examples of "best management practices."

Industrial hazardous waste management is an exciting and challenging endeavor, of great importance to the quality of the environment. Those involved in this rapidly developing field should be proud of numerous recent improvements. Though secondary to manufacturing functions, a good waste management program fulfills an important responsibility of the firm, and will lessen the potential environmental risks and liabilities facing manufacturing companies.

Acknowledgments

Looking back on the trail of events that led to the writing of this book, I realize that I have enjoyed the help of many people who have provided guidance, criticism, encouragement, assistance, and opportunities. I can name only a few of them here, but I want to thank Dick Young and Ed Lewis for recognizing my potential as an author, and Lee Langlotz and Guinn Doyle for allowing me to develop my capabilities while I was working for the State of Indiana. Ed Pitkin and Steven Hiatt of Ulrich Chemical and Tom Roberts of Heritage Remediation/Engineering have my gratitude for providing me with opportunities to put my ideas into practice. I want to express my deep appreciation to all of the contributing authors—Dick Reynolds, Freeman Cook, Chuck White, Steve Ball, John Kyle, and especially Jack Cornpropst. When I was an inexperienced state inspector, Jack taught me all I could understand about electroplating and industrial waste management. Jim Mabarak and Tim Kelley provided valuable insights during their review of the manuscript. I will always remember the patience and perseverance my wife Helen showed during the time this book was being written. Cindy Hartman has, without complaint, carefully typed the many drafts and the final version of the manuscript.

I would also like to express my appreciation to the members of the industrial community in Indiana. As a state hazardous waste inspector, a chemical company environmental director, and, most recently, an environmental consultant, I have had extensive interaction with many manufacturing firms. Their questions, problems, suggestions, complaints, and practices provided the basis for this book.

I am interested in reader comments on the usefulness of the material found in this book, and would appreciate receiving any suggestions for improving it in subsequent editions.

Contents

SECTION III
DEVELOPING THE CORPORATE
ENVIRONMENTAL MANAGEMENT PROGRAM

SECTION IV
SELECTED CONSIDERATIONS IN IMPLEMENTING
THE ENVIRONMENTAL MANAGEMENT PROGRAM

APPENDICES

While every effort has been made to present accurate information of general applicability, this book does not render legal or technical advice for specific applications. Competent professional assistance should be obtained for specific applications.

SECTION I

Basics

This section is a simple and straightforward explanation of the determinations necessary to see how a manufacturing firm's waste management activities fit into the hazardous waste management scheme. In essence, the following three determinations must be made:

1. whether the firm generates potentially regulated waste materials
2. whether those wastes are hazardous wastes according to the regulatory definitions
3. under which regulatory category the firm falls, given the quantities of hazardous waste generated and the types of hazardous waste management activities conducted onsite

For many manufacturing facilities, these determinations have not been made completely and correctly. Very commonly, the waste inventory necessary to the first determination was never performed, so the subsequent determinations are incomplete. In other instances, an assessment of the exclusions and conditional exemptions from the regulatory definitions of solid and hazardous waste was never made, resulting in over-regulation of certain waste materials. With respect to the third determination, precautionary permit applications have been filed, often for units exempt from regulation under RCRA. Other firms claim limited regulatory status as small quantity generators, without identifying and quantifying generation rates for all regulated waste materials.

The importance of these three determinations with respect to regulatory compliance cannot be overstated. While each level of determination is significant individually, it must be realized that each level builds on the information resulting from the previous level(s). For example, the waste inventory must be complete for the hazardous waste determination to be performed on all potentially regulated wastes. Identification of all hazardous wastes of the firm allows quantification of their generation rates and identification of the disposition of each waste necessary to determine which regulations are applicable. It is important to make these determinations correctly, and to document the rationale and information used in arriving at the conclusions.

CHAPTER 1

Introduction to the Hazardous Waste Management System

The purpose of this chapter is to give a general description of the overall workings of the federal hazardous waste management (HWM) system. Such an understanding will provide the reader with a framework to better grasp the specifics of the system, as found in the *Code of Federal Regulations* (CFR). New regulations, as well as modifications to existing regulations, are printed in the *Federal Register* (FR). The *Code of Federal Regulations* is modified annually to take into account changes promulgated in the *Federal Register*. The chapter begins with a description of the HWM system and concludes with an explanation of the FR/CFR system and other information sources on the regulations.

OVERVIEW

As it was conceived and as it has evolved, the hazardous waste management system is a "cradle-to-grave" system. In other words, the regulation of wastes defined as hazardous begins at the point of generation (e.g., a manufacturing facility), covers transportation, and continues to the waste's treatment or disposal (e.g., at an incinerator or secure chemical landfill). All participants in this system—generators, transporters, and owners/operators of treatment, storage, or disposal (TSD) facilities—are required to register with the U.S. Environmental Protection Agency (EPA). Each type of participant has a set of regulatory standards to abide by, and only those participants who have registered can perform hazardous waste management activities.

3

This registration, termed a "notification," was to have taken place only after completion of the three determinations required of potentially regulated parties under the Resource Conservation and Recovery Act (RCRA).

These determinations are:

1. What are the wastes at the facility? (termed the "waste inventory")
2. According to the regulatory definitions, are any of the wastes "hazardous"? (termed the "hazardous waste determination")
3. What is the regulatory status of the facility given the quantities of hazardous waste generated and the types of hazardous waste management activities conducted onsite?

All participants—"regulated parties," if you will—are required to notify EPA on EPA Form 8700-12. (See Appendix A.) Notifiers receive a site-specific 12-character alphanumeric identifier known as an EPA identification number. Owners/operators of TSD facilities must also submit permit applications for any existing facility engaged in TSD operations.

New TSD facilities are not to begin construction or operation without first obtaining a full RCRA permit. Generators accumulating hazardous wastes in containers or tanks are allowed to hold wastes onsite for up to 90 days without being subject to TSD storage permit requirements.

Existing TSD facilities that notified by August 18, 1980, and that submitted the first part ("Part A") of the RCRA permit application by November 19, 1980, qualified for what is known as "interim status." Interim status is a regulatory concept allowing such facilities to continue operation, pending final disposition of their full permit application ("Part B"). As additional wastes and waste management activities come under RCRA control, newly regulated facilities are allowed to obtain interim status.

By filing Part A of the federal consolidated permit application and abiding by what are termed "interim status standards" (40 CFR Part 265), the owner/operator of an existing TSD facility is allowed to continue HWM operations pending review of the Part B permit application. Timetables for termination of the interim status period were a part of the 1984 amendments to RCRA. Deadlines for Part B permit application submission and/or permit issuance were specified for the various types of TSD facilities.

The significance of EPA identification numbers and permits or interim status is that they delimit the firms and facilities that can be used to transport, treat, store, or dispose of hazardous wastes. A generator of hazardous waste can only use the services of a transporter having a valid EPA identification number, and the generator must have a valid EPA identification number in order to offer hazardous wastes to such a transporter. Further, hazardous wastes can only be sent to TSD facilities with a RCRA permit or under interim status (in addition to having an EPA identification number). This applies to facilities on the generator's property (onsite) as well as commercial (offsite) TSD facilities. There are, of course, certain limited exceptions to these requirements, such as the provisions allowing generators to accumulate their hazardous wastes onsite in containers or tanks for up to 90 days.

Manifest System

The transportation of hazardous wastes to offsite TSD facilities is to be accompanied by a document called a manifest (Figure 1.1). In its simplest form, a manifest is a four-part shipping document which also serves to track hazardous wastes to ensure delivery to the designated TSD facility. The manifest serves both shipping document and tracking document purposes, and contains the following information:

1. the generator's name, address, telephone number, and EPA identification number
2. the name and EPA identification number of the transporter
3. the name, address, and EPA identification number of the TSD facility to which the waste is to be transported
4. the description and quantity of the wastes and type and number of containers
5. a certification, by the generator, that the wastes are in proper condition for transportation, that a waste minimization program is in place, and that an appropriate treatment/disposal method has been selected for the waste. Spaces are provided for signatures from the generator, transporter, and owner/operator of the TSD facility.

The manifest system is fairly simple. The generator supplies all the required information on the manifest and signs and dates the manifest. The generator obtains the signature of the transporter and the date, acknowledging acceptance of the hazardous waste on that date. Each of the four parts of the manifest must contain this information, so carbon copies or carbonless forms are often used. The transporter leaves a signed copy with the generator and takes the remaining three copies of the manifest. The transporter then is required to take the waste to the TSD facility designated on the manifest. The TSD facility owner/operator signs and dates the manifest on arrival of the wastes, to certify that the hazardous waste shipment was received. Any significant discrepancies between the wastes as received and the manifest information is to be noted on the manifest. The transporter is then to receive a signed copy, and the facility retains a copy and sends a copy to the generator, acknowledging receipt.

If the generator does not receive a copy of the manifest with the signature of the TSD facility owner/operator within 35 days of the date of shipment, an investigation of the status and disposition of the wastes is to be initiated. If a copy of the manifest with the handwritten signature of the owner/operator of the designated TSD facility is not received within 45 days of the date of shipment, the generator is to file an "exception report" with EPA.

Some states have specific requirements regarding the manifest system, including the use of state forms and submittal of manifest copies to the state. The specifics

Please print or type.   (Form designed for use on elite (12-pitch) typewriter.)   Form Approved. OMB No. 2050-0039. Expires 9-30-91

UNIFORM HAZARDOUS WASTE MANIFEST	1. Generator's US EPA ID No.		Manifest Document No.	2. Page 1 of	Information in the shaded areas is not required by Federal law
3. Generator's Name and Mailing Address				A. State Manifest Document Number	
				B. State Generator's ID	
4. Generator's Phone ( )					
5. Transporter 1 Company Name	6.	US EPA ID Number		C. State Transporter's ID	
				D. Transporter's Phone	
7. Transporter 2 Company Name	8.	US EPA ID Number		E. State Transporter's ID	
				F. Transporter's Phone	
9. Designated Facility Name and Site Address	10.	US EPA ID Number		G. State Facility's ID	
				H. Facility's Phone	

	11. US DOT Description (Including Proper Shipping Name, Hazard Class, and ID Number)	12. Containers		13. Total Quantity	14. Unit Wt/Vol	I. Waste No.
		No.	Type			
G E N E R A T O R	a.					
	b.					
	c.					
	d.					

J. Additional Descriptions for Materials Listed Above	K. Handling Codes for Wastes Listed Above

15. Special Handling Instructions and Additional Information

16. **GENERATOR'S CERTIFICATION**: I hereby declare that the contents of this consignment are fully and accurately described above by proper shipping name and are classified, packed, marked, and labeled, and are in all respects in proper condition for transport by highway according to applicable international and national government regulations.

If I am a large quantity generator, I certify that I have a program in place to reduce the volume and toxicity of waste generated to the degree I have determined to be economically practicable and that I have selected the practicable method of treatment, storage, or disposal currently available to me which minimizes the present and future threat to human health and the environment; OR, if I am a small quantity generator, I have made a good faith effort to minimize my waste generation and select the best waste management method that is available to me and that I can afford.

Printed/Typed Name	Signature	Month Day Year

T R A N S P O R T E R	17. Transporter 1 Acknowledgement of Receipt of Materials		
	Printed/Typed Name	Signature	Month Day Year
	18. Transporter 2 Acknowledgement of Receipt of Materials		
	Printed/Typed Name	Signature	Month Day Year

F A C I L I T Y	19. Discrepancy Indication Space		
	20. Facility Owner or Operator: Certification of receipt of hazardous materials covered by this manifest except as noted in Item 19.		
	Printed/Typed Name	Signature	Month Day Year

EPA Form 8700-22 (Rev. 9-88)   Previous editions are obsolete.

Figure 1.1.   Uniform Hazardous Waste Manifest (EPA Form 8700-22).

of the regulations implementing the HWM system, as well as areas where certain states have chosen to more stringent than the EPA, are discussed in Section II.

Authorization

RCRA allows EPA to "authorize" states to regulate hazardous waste management within their borders. The authorized state regulatory program is said to act in lieu of the federal program. There are five requirements for a state to receive final authorization. The state program must:

1. be equivalent to the federal program
2. be consistent with the federal program and other state programs
3. be no less stringent than the federal program
4. provide for adequate enforcement
5. provide for public participation in the permitting process

As the federal program changes, authorized states are allowed one year to adopt regulatory changes, and up to two years to adopt changes requiring statutory modification. However, states are not required to adopt federal program changes that would lessen existing state requirements. States are allowed to have more stringent regulatory programs, with limited exceptions.

The 1984 amendments to RCRA (the Hazardous and Solid Waste Amendments or HSWA) changed this scheme somewhat. New requirements, prohibitions, and deadlines imposed by HSWA take effect in authorized states at the same time they take effect in nonauthorized states. EPA is directed to carry out these requirements and prohibitions, including the issuance of permits, until the various states are authorized specifically for the HSWA portions of the RCRA program. As a result of HSWA, the regulated community is forced to deal, temporarily, with dual state/federal regulatory programs, even in finally authorized states.

Even with the HSWA impact, the net effect of final authorization is that state-level regulatory agency officials administer and enforce the RCRA program, by:

- conducting inspections
- interpreting regulations
- reviewing permit applications
- issuing permits
- bringing enforcement actions
- negotiating fines and penalties

Primary regulatory contacts will be with state regulatory agency employees instead of EPA personnel. In some instances, however, there is the possibility of EPA involvement or joint state/EPA review.

HWM REGULATIONS

The relevant citations for the various parts of the federal hazardous waste management system are listed in Table 1.1, along with their subject matter. These citations show how the HWM program is organized. Note that each set of participants—generators, transporters, and TSD facility owners/operators—has its own set of regulatory requirements. However, the Part 262 requirements for generators accumulating onsite less than 90 days incorporate by reference many

Table 1.1.   Federal HWM System Regulations (40 CFR)

Part	Title and Brief Description	Discussed in Chapter(s)
260	**Hazardous Waste Management System: General** definitions, rulemaking, and delisting petitions	1,2,4
261	**Identification and Listing of Hazardous Waste** definitions of solid waste and hazardous waste; exclusions; requirements for conditionally exempt small quantity generators; characteristics of hazardous waste; lists of hazardous wastes	2,3,4,5,7,8,9
262	**Standards Applicable to Generators of Hazardous Waste** hazardous waste determination; EPA identification numbers; the manifest; pretransport requirements; 90-day accumulation requirements; requirements for 100–1000 kg/month generators; recordkeeping and reporting	4,5,7,8,9,10,12
263	**Standards Applicable to Transporters of Hazardous Waste** EPA identification numbers; compliance with the manifest system; recordkeeping; hazardous waste discharges during transportation	11
264	**Standards for Owners and Operators of Hazardous Waste Treatment, Storage, and Disposal Facilities** general status (permit) standards that will be used to evaluate Part B permit applications and that will apply to facilities possessing a full RCRA permit	
265	**Interim Status Standards for Owners and Operators of Hazardous Waste Treatment, Storage, and Disposal Facilities** operating requirements for existing HWM facilities under interim status, until final disposition of their permit application	5,9,10,11
266	**Standards for the Management of Specific Hazardous Wastes and Specific Types of Hazardous Waste Management Facilities** special (limited) requirements for certain types of wastes being reclaimed or used as fuel. Materials so regulated include hazardous wastes and used oil burned for energy recovery, as well as precious metals and lead-acid batteries being reclaimed	5,13
268	**Land Disposal Restrictions** prohibitions on the direct land disposal of untreated hazardous wastes; treatment standards established to determine the acceptability of land disposal; provisions for extensions, petitions, and variances. A schedule is specified for the establishment of treatment standards for all listed and characteristic hazardous wastes.	4,5,8,9,10
270	**EPA-Administered Permit Programs: The Hazardous Waste Permit Program** contents of the Part A and Part B permit applications; standard permit conditions; permit modifications; duration of permits; interim status	5,11
271	**Requirements for Authorization of State Hazardous Waste Programs** requirements for state programs to receive authorization to operate in lieu of EPA	1,11

Table 1.1.   cont'd.

Part	Title and Brief Description	Discussed in Chapter(s)
280	**Technical Standards and Corrective Action Requirements for Owners and Operators of Underground Storage Tanks (UST)** notification requirements for underground *product* (not hazardous waste) storage tanks; definitions; design, construction and installation; general operating requirements; release detection; release reporting, investigation, and confirmation; release response and corrective action; out-of-service UST systems and closure	5,14
281	**Approval of State Underground Storage Tank Programs** requirements for state programs to receive approval to operate in place of federal UST requirements	
124	**Procedures for Decisionmaking** procedures to be followed in making permit decisions; requirements for public participation in permit decisions	11
148	**Hazardous Waste Injection Restrictions** prohibitions and restrictions on the deep well injection of hazardous waste	

of the Part 265 requirements for interim-status TSD facilities. Most state hazardous waste regulations are similarly organized around subject matter.

These standards, as they relate to the activities of manufacturing firms, will be explained in detail in Section II. The chapters in Section II are organized to correlate with the determinations outlined in Section I. As a result, the regulatory standards applicable to each regulatory category are in separate chapters.

Federal Register and *Code of Federal Regulations*

Before delving into the regulations, it is important to note where they are found. All federal regulations, proposed regulations, notices, and other documents of general applicability or legal effect are published in the *Federal Register* (Figure 1.2). Printed every working day, the *Federal Register* serves as a source of information regarding the current activities of all federal agencies. The general and permanent rules published in the *Federal Register* during the preceding 12 months are codified annually according to subject matter in the *Code of Federal Regulations*, thus revising the existing regulations to include the new regulations and amendments to existing regulations. The CFR is divided into 50 titles covering the areas subject to federal regulation. Title 40 of the CFR (Figure 1.3) deals with protection of the environment. This is where EPA regulations are found. Title 29 of the CFR deals with labor, and is where Occupational Safety and Health Administration (OSHA) regulations are found. Title 49 of the CFR deals with transportation, including hazardous materials and hazardous waste transportation.

Each title is divided into chapters, parts, subparts, and sections; a part consists of a unified body of regulations devoted to a specific subject. Hazardous waste

Figure 1.2. The *Federal Register* of the United States Government is printed every working day.

Figure 1.3. The *Code of Federal Regulations*—Title 40: Protection of the Environment.

regulations are found in Parts 260–268 of Title 40 of the CFR. As referenced here and in other documents, the federal hazardous waste regulations are (Title) 40 CFR (Parts) 260–268. Sections of each part are referenced by part number and section number, separated by a period (e.g., 261.3). When referring to individual *Federal Register* issues, the reference is ordered by volume and page number. Volume 1 was published in 1936, so the years 1986, 1987, and 1988 are Volumes 51, 52, and 53, respectively. Occasionally the date of publication is included. An example would be (Volume) 45 FR (Page) 33066, which is the first page of Book 2 of the May 19, 1980, *Federal Register*. For those new to the management of hazardous waste, May 19, 1980 is a landmark date. On that date, the hazardous waste management regulations (40 CFR 260–265) were printed in the *Federal Register*. The hazardous waste regulations were printed separately from other regulations, and the result was approximately 525 pages. Many copies were distributed by the EPA to the regulated community. Copies of this *Federal Register* still contain the basis of the HWM system, but 40 CFR 260–265 has been amended so much since May 19, 1980 that the *Federal Register* of this date is unreliable as a source of regulatory responsibilities.

To have a complete and up-to-date set of federal HWM regulations, one of several alternatives must be pursued. The most obvious is to obtain the yearly edition of the *Code of Federal Regulations*, Title 40, Parts 190–399, and to subscribe to the *Federal Register*. Both are available from the Superintendent of Documents, U.S. Government Printing Office, Washington, DC 20402. In 1988, the cost of purchasing a copy of Title 40, Parts 190–399 was $29.00. A yearly subscription to the *Federal Register* was $340.00. One may also subscribe to one of several services existing for the purpose of providing up-to-date regulatory information.

There are other, less expensive options available. Many state chambers of commerce and manufacturers' associations are active in the field of environmental regulation. Summaries of federal and state regulatory activity in HWM may be available from them. In addition, many trade organizations also offer such services and may have waste-related articles of interest in their publications. Further, magazines in the pollution control field also cover the topic of HWM. Several such magazines, including *Pollution Engineering*, *Hazardous Materials Control*, and *Hazardous Waste Management*, publish summaries of *Federal Register* rules and notices, and have articles of interest to environmental compliance personnel.

Certain libraries are designated Government Depository Libraries and will have government publications, including the *Federal Register* and the *Code of Federal Regulations*, available for public viewing. In addition, they may also subscribe to the regulatory reporters mentioned earlier. Check with your city library or any nearby university or law school library for the availability of the *Federal Register* and CFR, as well as other related publications. These libraries may also have finding aids for the CFR and the *Federal Register*. The *List of Sections Affected* and the *Federal Register Index* may be of some value in staying current with regulatory developments (Figure 1.4).

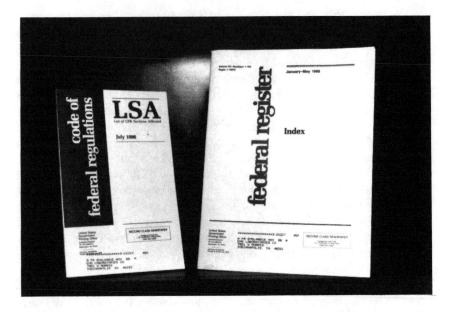

Figure 1.4. CFR—*List of Sections Affected* and *Federal Register Index.*

State Regulations

Not to be overlooked, due to their immediate relevance, are state hazardous waste management regulations. A state hazardous waste regulatory program must have regulations substantially equivalent to the federal regulations to receive authorization to operate a regulatory program in lieu of EPA. The various states have acted on this requirement by adopting the federal regulations by reference, or by reprinting them in their Administrative Codes using appropriate language and format.

As indicated earlier, a state's regulations may be more strict than the federal regulations, and the more stringent requirement will apply. Further, most state regulations relate to the federal regulations as they existed on a certain date. The impact of this is that certain non-HSWA amendments published in the *Federal Register* may not apply immediately at the state level, especially if the amendment reduces or eliminates existing requirements. Contact your state hazardous waste management agency for a copy of the state regulations and for information regarding differences between the federal and state programs and regulations. If necessary, obtain a copy of the federal regulations that the state regulations reference. Appendix B lists such state agencies. Appendix C lists all the regional EPA offices.

CONCLUSION

RCRA was designed to "close the circle" in earlier environmental legislation by regulating the management of all solid wastes. The Clean Air Act (CAA) and Clean Water Act (CWA) regulatory programs resulted in manufacturing process wastes and pollution control residuals being disposed of via land disposal, instead of direct discharges to air or water. RCRA was intended to minimize the public health and environmental quality threats from the management of solid and hazardous wastes.

RCRA, together with the Comprehensive Environmental Response, Compensation and Liability Act (CERCLA or "Superfund") imposes requirements upon generators of hazardous waste with respect to the disposition and environmental impact of such wastes. Generators of hazardous wastes are ultimately and even perpetually responsible for onsite and offsite damages to human health and the environment resulting from their wastes. This responsibility is sometimes termed "cradle-to-grave," although it really extends well beyond the grave.

The remaining chapters in Section I will examine the regulatory requirements promulgated in the *Federal Register/Code of Federal Regulations* system. Chapter 2 explains the regulatory definitions of solid and hazardous waste. Chapters 3 and 4 discuss the waste inventory and determinations of each waste's hazardous or nonhazardous nature. Finally, Chapter 5 serves as a guide in determining the regulatory categories potentially applicable to a manufacturing firm.

CHAPTER 2

Regulatory Definitions of Solid and Hazardous Waste

In environmental regulatory programs in general, and the hazardous waste program in particular, certain words and phrases have been given special meaning and significance. Often, the regulatory definition of these terms differs significantly from their common usage. Acronyms such as RQ, TSD, and RCRA are used extensively in regulatory language as a form of shorthand. Hazardous waste management, like other specialized areas of endeavor, has acquired its own language or jargon. The newcomer to this field quickly perceives that the use of acronyms, regulatory terms, and extensive cross-referencing makes reading EPA regulations analogous to translating a foreign language.

Learning this language is a very necessary endeavor for environmental compliance personnel. Knowledge of the meaning of key terms and acronyms becomes critical to regulatory compliance, as well as to effective communication with regulatory agency personnel, consultants, and attorneys. In order to understand the regulations, and appropriately apply them to particular situations, environmental managers must be aware of the existence of key terms, and interpret regulatory language in light of the special meanings given such terms.

Definitions of many key terms and acronyms, as they are used in the hazardous waste management regulations, are found at 40 CFR 260.10. These terms take on special meaning and significance within the context of this regulatory program. Other key terms are defined within the body of the regulations. This book will attempt to highlight the most significant regulatory terminology, and give its regulatory definition and interpretation in the text. A list of commonly used acronyms is provided in Appendix D. A glossary of selected key regulatory terms is found in Appendix E. Please note that a few terms, such as solid waste and hazardous waste, have both statutory and regulatory definitions.

SOLID WASTE

An understanding of the provisions of 40 CFR Part 261, "Identification and Listing of Hazardous Waste," is crucial to correct determinations of whether a manufacturing facility generates a solid waste, and if so, whether the solid waste is also a hazardous waste. These determinations are explained in detail in Chapters 3 and 4.

To be a hazardous waste, a material must first meet the definition of solid waste. Hazardous wastes are, by statutory and regulatory definition, a subset of solid wastes. Solid waste, therefore, is a term with a very special meaning within the RCRA regulatory program. Regulatory terms must always be associated with their regulatory definitions in order for the regulations themselves to make any sense, to determine applicable compliance measures, and to communicate with regulatory agency personnel.

A solid waste, according to the regulatory definition, need not be in a solid physical state. A solid waste may be in a liquid, semisolid, or even a contained gaseous physical state. The definition of solid waste is found at 40 CFR 261.2. A solid waste is any material not excluded from regulation which is:

- disposed of or abandoned in lieu of disposal
- burned or incinerated or recycled
- considered inherently waste-like

Please note that "inherently waste-like" is regulatory jargon for dioxin- and furan-contaminated wastes (see 261.2(d)).

The EPA deliberately made the definition of solid waste extremely broad and inclusive. This broad definition requires a variety of exclusions, conditional exemptions, and variances to avoid over-regulation of certain waste materials and certain recycling scenarios. The regulations specify at 40 CFR 261.4(a) ("Materials Which Are Not Solid Wastes") certain materials which are not considered solid wastes for purposes of solid and hazardous waste determinations. The excluded materials are:

1. domestic sewage and mixtures of domestic sewage and other wastes passing through a sewer system to a publicly owned treatment works for treatment. "Domestic sewage" means untreated sanitary wastes that pass through a sewer system.

   Comment: This exclusion, often termed the "domestic sewage exclusion," assumes that such discharges are adequately regulated under Clean Water Act requirements. It is assumed that such discharges are in compliance with the general pretreatment standards (40 CFR Part 401), any applicable categorical standards (40 CFR 405-469), and any local sewer ordinances for industrial discharges.

2. industrial point source wastewater discharge requiring a National Pollutant Discharge Elimination System (NPDES) permit under Section 402 of the Clean Water Act

   Comment: This exclusion applies only to the actual point source discharge. It does not exclude industrial wastewaters while they are being collected, stored, or treated before discharge, nor does it exclude sludges that are generated by industrial wastewater treatment.

3. irrigation return flows
4. radioactive materials regulated by the Nuclear Regulatory Commission
5. in-situ mining wastes not removed from the ground
6. pulping liquors (i.e., black liquor) that are reclaimed in a pulping liquor recovery furnace and then reused in the pulping process, unless they are accumulated speculatively as defined at 40 CFR 261.1(c)
7. spent sulfuric acid used to produce virgin sulfuric acid, unless it is accumulated speculatively as defined at 40 CFR 261.1(c)
8. secondary materials that are reclaimed and returned to the original process or processes in which they are generated (where they are reused in the production process), provided:
   a. Only tank storage is involved, and the entire process through completion of reclamation is closed by being entirely connected with pipes or other comparable enclosed means of conveyance.
   b. Reclamation does not involve controlled flame combustion (such as occurs in boilers, industrial furnaces, or incinerators).
   c. The secondary materials are never accumulated for over twelve months without being reclaimed.
   d. The reclaimed material is not used to produce a fuel, or used to produce products that are used in a manner constituting disposal.

If a material is not considered a solid waste, it cannot by definition be a hazardous waste. EPA excludes certain solid waste materials from being hazardous waste, at 40 CFR 261.4(b), "Solid Wastes Which Are Not Hazardous Wastes." These following solid wastes are not hazardous wastes for purposes of the hazardous waste determination:

1. household wastes
2. solid wastes from agricultural crops and animals returned to the soil as fertilizer
3. mining overburden returned to the mine site
4. fly ash waste, bottom ash waste, slag waste, and flue gas emission control waste generated primarily from the combustion of coal or other fossil fuels
5. drilling fluids, produced waters, and other wastes associated with the exploration, development, or production of crude oil, natural gas, or geothermal energy

6. wastes containing chromium in the trivalent state and failing the characteristic of EP (extraction procedure) toxicity solely because of chromium or that are listed solely because of chromium. The waste must be generated from an industrial process using trivalent chromium exclusively, the process must not generate hexavalent chromium, and the waste must be managed in nonoxidizing environments. Specific examples are given at 40 CFR 261.4(b)(6)(ii).
7. solid wastes from the extraction, beneficiation, and processing of ores and minerals. Specific examples and exceptions are given at 40 CFR 261.4(b)(7).
8. cement kiln dust waste
9. discarded wood or wood products generated by users of arsenical-treated wood or wood products

For the majority of manufacturing firms, only exclusions 4 and 6 would be of potential interest.

Recycling Exemptions

EPA eliminated most exemptions (''loopholes'') for hazardous wastes when recycled, in their so-called ''redefinition'' of solid waste (50 FR 614). Most waste recycling activities are completely regulated. Others are partially regulated at 40 CFR Part 266. A few recycling scenarios are conditionally exempt, leaving very few recycling scenarios totally exempt from regulation.

In order to determine whether a certain recycling scenario is totally exempt from regulation, it is necessary to know both what type of material is to be recycled and what type of recycling activity is to be utilized. In the EPA's universe, there are five types of materials and four types of recycling activities.

The five types of materials are as follows:

1. *Spent materials* are those which have been used and, as a result of contamination, can no longer serve the purpose for which they were produced without reprocessing.
2. *Sludges* are residues from treating air or wastewater, or other residues from pollution control. EPA distinguishes between sludges which are listed hazardous wastes and those which exhibit characteristic(s) of hazardous waste.
3. *By-products* are residual materials resulting from industrial, commercial, mining, and agricultural operations that are not primary products, are not produced intentionally or separately, and are not fit for a desired end use without further processing. EPA distinguishes between by-products which are listed hazardous wastes and those which exhibit characteristic(s) of hazardous waste.
4. *Commercial chemical products* are products and intermediates, off-specification variants, spill residues, and container residues listed in 40 CFR 261.33(e) and (f) (acutely hazardous P-wastes and toxic U-wastes).
5. *Scrap metal* is defined as bits or pieces of metal that are discarded after consumer use or that result from metal processing operations. Scrap metal is not held to include residues from smelting or refining operations (e.g., drosses, slags, and sludges), liquid wastes containing metals (e.g., spent plating bath

solution), liquid metal wastes (e.g., liquid mercury), or metal-containing wastes with a significant liquid component (e.g., spent lead-acid batteries).

The four types of recycling activities specified by EPA are:

1. *use constituting disposal*, which involves directly placing wastes or products that contain hazardous wastes as an ingredient onto the land. These practices are considered to be equivalent to land disposal.
2. *burning waste or waste fuels* for energy recovery or using wastes to produce a fuel
3. *reclamation of wastes*, which involves the regeneration of wastes or the recovery of material from wastes
4. *speculative accumulation*, which involves accumulating wastes that are potentially recyclable, but for which no feasible recycling market exists. It also involves recycling less than 75% of accumulated wastes during a one-year period.

Table 1 of 40 CFR 261.2 places the types of materials and the types of recycling activities in a matrix. Very few recycling scenarios are totally or conditionally exempt from regulation. Materials which are recycled in lieu of disposal need to be evaluated with respect to the matrix. Most combinations of materials and recycling activities are regulated. The only combinations not considered solid waste, and therefore not regulated, are:

1. reclamation of unlisted sludges exhibiting a characteristic of hazardous waste
2. reclamation of unlisted by-products exhibiting a characteristic of hazardous waste
3. reclamation of commercial chemical products
4. accumulation of commercial chemical products

Failure to recycle at least 75% of these materials within a one-year period may result in the revocation of the exemption, as the waste material would then become a solid waste. Such materials could become subject to the RCRA regulatory scheme by virtue of being characteristically hazardous (sludges or by-products) or by virtue of meeting a listing description (commercial chemical products). There are, however, provisions for obtaining one-year extensions under extenuating circumstances (40 CFR 260.31).

Please note that recycling of scrap metal is excluded from regulation at 40 CFR 261.6(a)(3)(iv).

As discussed above, few recycling scenarios are totally exempt. Some recycling scenarios are conditionally excluded from regulation. These scenarios involve use or reuse activities resembling ordinary production operations.

The redefinition of solid waste provided for conditional exemptions for materials which are used or reused. Use or reuse implicitly assumes that reclamation operations (e.g., solvent distillation) are not necessary for the material to be used or reused. Materials are not considered solid wastes subject to regulation when their use or reuse resembles ordinary production operations or where such materials are used/reused in lieu of commercial materials. Three specific situations are given as being conditionally exempt from regulation:

1. direct use/reuse of materials as an ingredient or feedstock in production processes (e.g., use of distillation bottoms from chemical A as feedstock in the manufacture of chemical B)
2. direct use/reuse of materials as an effective substitute for raw materials (e.g., use of spent pickle liquor as a wastewater conditioner, instead of virgin ferric chloride)
3. return of materials to the original production process without first reclaiming the materials (e.g., closed loop processes)

These situations are best described as conditional exemptions. Materials are not considered to be solid wastes when used or reused in the above-specified situations, with the following exceptions:

1. materials used in a manner constituting disposal, or used to produce products that are applied to the land
2. materials burned for energy recovery, used to produce a fuel, or contained in fuels
3. materials accumulated speculatively (less than 75% recycled in a calendar year)
4. inherently waste-like materials listed at 40 CFR 261.2(d)(1) (dioxin- and furan-contaminated wastes)

These exceptions will nullify the conditional exemption provided. Thus, the specifics of the entire scenario must be known to determine if the conditional exemptions are applicable. Note that EPA requires persons claiming that materials are not solid wastes to document such claims (40 CFR 261.2(f)). Several aspects of the recycling transaction should be documented, including the actual disposition of the materials, as well as the location and permit status of the reclamation facility. Contracts, shipping papers, purchase agreements, etc., would serve as documentation. The absence of records regarding the transaction would likely be viewed as sham recycling. The burden of proof is on the generator with respect to qualifying for any regulatory exemption or exclusion.

To allow for the possibility that other recycling scenarios might legitimately deserve to be exempt from regulation, or subject only to limited requirements, the regulations provide for petitions for variances from classification as solid waste (40 CFR 260.30–31). Remember that under the RCRA regulations, hazardous wastes are a subset of solid wastes. If a material is not a solid waste, it cannot be a hazardous waste. Regulatory authority is also available to regulate hazardous waste recycling activities on a case-by-case basis (40 CFR 260.40).

There are three recycling scenarios which, on a case-by-case basis, under specified conditions, can be determined by the EPA to not be solid wastes. These scenarios, specified at 40 CFR 260.30, are:

1. materials that are accumulated speculatively without sufficient amounts being recycled (as defined at 40 CFR 261.1(c)(8))
2. materials that are reclaimed and then reused within the original primary production process in which they were generated
3. materials that have been reclaimed but must be reclaimed further before the materials are completely recovered

In order to be granted a variance for any of these scenarios, an application must be made to the EPA Regional Administrator, according to procedures specified at 40 CFR 260.33.

The implication of these regulatory changes regarding use, reuse, recycling, and reclamation is that most waste materials being recycled in lieu of disposal must be shipped, under hazardous waste manifest, to a facility pursuing a RCRA permit for TSD activities. These regulatory changes, along with changes in the market value of certain secondary materials, are likely the cause of the decision by many reclamation operations to no longer accept regulated recyclable materials for reclamation. To continue to do so would require these secondary material reclamation facilities to pursue a RCRA hazardous waste TSD facility permit.

HAZARDOUS WASTE

When compared to the definition of solid waste, the definition of hazardous waste is a model of clarity. Basically, a solid waste can become a hazardous waste by virtue of any one of the following conditions:

1. being on one of four lists of hazardous wastes
2. being a mixture of a solid waste and a hazardous waste on one of the lists
3. being derived from the treatment, storage, or disposal of a hazardous waste on one of the lists
4. exhibiting one or more of the four characteristics of a hazardous waste

Hazardous Waste Listings

There are four lists of hazardous wastes. The first two of these lists (261.31 list and 261.32 list) deal with process wastes and sludges. The 261.31 list is entitled "Hazardous Waste from Non-Specific Sources." These wastes are generic industrial process wastes resulting from degreasing, solvent usage, electroplating, and heat treating, along with certain dioxin-contaminated wastes from the production of organic chemicals. The lists assign an EPA hazardous waste number to each listing. The 261.31 list is often referred to as the "F-list" because the hazardous waste number for all listings begins with the letter "F." The hazardous waste number is used in the notification process, on manifests, and in certain recordkeeping and reporting requirements. Each listing also has a hazard code, which tells why the waste was listed. The hazard codes used throughout the four lists are:

- I: ignitable waste
- C: corrosive waste
- R: reactive waste
- E: extraction procedure (EP) toxic waste
- H: acute hazardous waste
- T: toxic waste

Operations generating spent solvents, performing electroplating or metal heat treating, or producing specified acutely hazardous organic chemicals must review the 40 CFR 261.31 listings to see whether any of the waste materials generated meet the listing descriptions.

Manufacturing firms utilizing proprietary (trade name) solvents or solvent blends need to be aware of the potential application of the spent solvent listings to certain blends or mixtures. Material Safety Data Sheet (MSDS) information should be reviewed to identify, by chemical name, the individual components of the mixture as well as their concentrations. All spent solvent mixtures or blends containing, before use, a total of 10% or more (by volume) of one or more of the solvents listed in the F001, F002, F004, and F005 descriptions are regulated as listed hazardous waste. The mixture limitation does *not* apply when only F003 solvents are involved. The F003 listing applies only to technical grades of individual F003 solvents, when spent. Spent solvent mixtures from the F003 listing are considered listed only if they are combined with 10% of one or more spent solvents from the F001, F002, F004, and F005 groups. Table 2.1 identifies the spent solvents regulated under RCRA.

Manufacturing firms utilizing solvents should carefully review the F001–F005 listings, keeping in mind that the listings apply to *spent* solvents, no longer suitable to solibilize (dissolve) or mobilize other materials. The F001–F005 listings do not include manufacturing process wastes containing solvents (e.g., waste paint), unless the listed solvents were used for cleaning purposes, or unless listed spent solvents were mixed with the manufacturing process wastes. *Unused* obsolete, off-specification, or expired solvents do not fall under the F-list, but need to be evaluated against the 40 CFR 261.33 commercial chemical products lists.

The second list is found at 40 CFR 261.32 and is entitled "Hazardous Waste from Specific Sources." Like the 261.31 list, each listing has an EPA hazardous waste number and a hazard code. The 261.32 list is referred to as the "K-list," because each of the hazardous waste numbers begins with the letter "K." This list is organized by industry, unlike the 261.31 list. Thus, a manufacturing firm's

Table 2.1.   Spent Solvents Regulated Under RCRA

Solvents Listed at 40 CFR 261.31

Toxic Solvents (F001, F002, F004, F005)

Benzene
Carbon disulfide
Carbon tetrachloride
Chlorinated fluorocarbons (freon, genesolv)
Chlorobenzene
Cresols and cresylic acid
2-ethoxyethanol
Isobutanol
Methyl ethyl ketone
Methylene chloride
Nitrobenzene
2-nitropropane
Ortho-dichlorobenzene
Pyridine
Tetrachloroethylene (perchloroethylene)
Toluene
Trichloroethylene
Trichlorofluoromethane
1,1,1,-Trichloroethane
1,1,2-Trichloroethane
1,1,2-Trichloro-1,2,2,-trifluoroethane

Ignitable Solvents (F003)

Acetone
n-Butyl alcohol
Cyclohexanone
Ethyl acetate
Ethyl benzene
Ethyl ether
Methanol
Methyl isobutyl ketone
Xylene

operations must match the industry description before the firm's solid waste can be matched correctly against the 261.32 listing description. The industries covered by the 261.32 list are:

1. wood preservation
2. inorganic pigments
3. organic chemicals
4. inorganic chemicals
5. pesticides
6. explosives
7. petroleum refining
8. iron and steel

9. primary copper
10. primary lead
11. primary zinc
12. primary aluminum
13. ferroalloys
14. secondary lead
15. veterinary pharmaceuticals
16. ink formulation
17. coking

If the operations of a manufacturing facility do not come under one of the above-referenced industrial categories, the 261.32 list is not applicable. If the firm does fall into one of the specified industrial categories, the waste materials generated must be compared to the listing descriptions within the applicable industrial category. If the waste materials generated meet the listing description, the waste is a listed hazardous waste.

The second set of lists deals with commercial chemical products, manufacturing chemical intermediates, or off-specification species, *if and when* they are discarded or "intended to be discarded," applied to road surfaces or otherwise applied to the land in lieu of their original intended use, or used as a fuel. The phrase "intended to be discarded" includes the accumulation or storage before or in lieu of disposal. These listings could apply to off-specification chemical products, obsolete chemical inventory, or excess or surplus inventory, as well as expired chemical products. However, the listings do *not* apply to regulated wastewater discharges containing *de minimis* losses of commercial chemical products and chemical intermediates listed at 40 CFR 261.33(e) or (f), where such *minor* losses are from normal manufacturing operations where such materials are used as raw materials or are produced in the manufacturing process (40 CFR 261.3(a)(2)(iv)(D)). *De minimis* losses, in this context, include losses from normal material handling operations (e.g., spills from the unloading or transfer of materials, as well as leaks from pipes or valves or other devices used to transfer materials); minor leaks of process equipment, storage tanks, or containers; leaks from well-maintained pump packings and seals; sample purgings; relief device discharges; discharges from safety showers and rinsing and cleaning of personal safety equipment; and rinsate from empty containers or from containers that are rendered empty by that rinsing.

There is an "acute hazardous" list (261.33(e)) and a "toxic" list (261.33(f)). These lists consist of generic chemical names (not trade names), and their Chemical Abstracts Service Registry Number. Each entry has a unique hazardous waste number and a hazard code. The absence of a printed hazard code following the listed chemical indicates it was listed only for acute toxicity (261.33(e) list) or toxicity (261.33(f) list).

It must be noted that these lists were intended to designate the chemicals themselves as hazardous waste, if discarded or intended to be discarded. These lists do not apply to all wastes that might contain these chemical constituents. For example, the two commercial chemical product lists do *not* apply to spent materials or manufacturing process wastes consisting of or containing the listed chemicals.

The substances listed in the two commercial chemical products lists are considered hazardous only if the technical grade of the chemical, or the off-specification variants, are discarded or intended to be discarded. In the case of proprietary (trade name) chemicals or formulations, the listed chemical must be the sole active ingredient for the listing to apply. MSDS information must be consulted for proprietary (trade name) chemicals or formulations, in order to ascertain the applicability of either of these two listings, if the chemical is to be discarded.

For both commercial chemical products lists, the listing of each chemical applies not only to the commercial or technical grade of the chemical itself and obsolete, expired, surplus, or off-specification species when discarded, but also to any residue, contaminated soil, water, or other debris associated with the cleanup of a spill of the listed material. The limited exclusion for *de minimis* losses was discussed earlier.

Acutely hazardous commercial chemical products (261.33(e) list) have additional regulatory implications. First, the residues remaining in unrinsed "empty" containers and any inner liners that have not been triple rinsed (or cleaned by an equivalent method) are considered to meet the listing description and thus to be listed as acutely hazardous wastes.

Second, there are lower small quantity exclusion levels for generators of acutely hazardous wastes. The exclusion level for acute hazardous wastes listed at 261.33(e) is, and always has been, 1 kg. With respect to spill residues, contaminated soil, water, or other debris from a spill cleanup, the small quantity exclusion level is 100 kg. The provisions for 100–1000 kg/month generators are not available to generators of acutely hazardous waste exceeding the appropriate exclusion level.

To conclude the discussion of the four lists of hazardous wastes, it is necessary to emphasize that the listings always take precedence over the characteristics. A waste material must be evaluated first with respect to the listing descriptions before it is evaluated with respect to the four characteristics.

There are definite regulatory implications associated with listed hazardous wastes, primarily with respect to the purposeful or inadvertent mixing of listed hazardous wastes and other waste materials. Remember that a waste material (solid waste in the regulatory jargon) becomes a listed hazardous waste when it first meets the listing descriptions found in the regulations. In order to prevent listed wastes from escaping regulation by dilution with nonhazardous wastes, the so-called mixture rule was included in the regulations. The mixture rule, found at 40 CFR 261.3(a)(2)(iv), states that a solid waste becomes a hazardous waste when a listed hazardous waste is first added to the solid waste. With two very limited exceptions, there are no *de minimis* concentrations established that would prevent application of the mixture rule.

The first exception involves wastewater mixtures. Wastewaters with specified *de minimis* concentrations of certain solvents are not subject to the mixture rule provided that the wastewater discharge is subject to regulation under the Clean Water Act (40 CFR 261.3(a)(iv)(A), (B), and (E)). Wastewater mixtures containing heat exchanger bundle cleaning sludges from the petroleum industry (K050) and *de minimis* losses of commercial chemical products are similarly excluded from the application of the mixture rule (40 CFR 261.3(a)(2)(iv)(C) and (D)).

The second exception to the mixture rule, found at 40 CFR 261.3(a)(2)(iii), deals with mixtures of solid waste and hazardous wastes listed solely for one or more of the characteristics of hazardous waste (Hazard Codes I, C, R, and E). This exception to the mixture rule applies only when the resulting mixture no longer exhibits any of the characteristics of hazardous waste. The only listing of hazardous waste solely for the I, C, R, or E characteristics is the F003 spent solvent listing, which is listed solely for ignitability (I). Therefore, mixtures of F003 spent solvents and other wastes, such as paint booth filters or paint booth sludges, are not considered listed hazardous wastes as long as the mixture does not exhibit any of the characteristics of a hazardous waste. Such mixtures would not be regulated under RCRA. Otherwise, mixtures of solid waste and listed hazardous wastes can escape regulation only through the delisting process. The requirements for delisting are found at 40 CFR 260.20 and 260.22. Delisting is a waste- and facility-specific process.

In addition to the mixture rule, the regulations specify that any solid waste generated from the treatment, storage, or disposal of a listed hazardous waste remains a hazardous waste. Thus, treatment residues (sludges) and spill residues of listed hazardous wastes remain listed hazardous wastes until delisted. This regulatory concept, often termed the "derived-from" rule, is found at 40 CFR 261.3(c)(2). There are no *de minimis* concentrations established that would prevent application of the derived-from rule. Derived-from listed hazardous wastes can only escape regulation through the delisting process, with the exception of waste pickle liquor sludge generated by lime stabilization of spent pickle liquor from the iron and steel industry (40 CFR 261.3(c)(2)(ii)). This lime-stabilized waste pickle liquor sludge must, in addition, be characteristically nonhazardous to escape regulation.

Manufacturing operations generating listed hazardous wastes have to be aware of the potential implications and ramifications of the mixture and derived-from rules. For example, mixing distillation residues ("still bottoms") from listed degreasing solvents with waste oil will cause the waste oil to become a listed hazardous waste. Also, gun flushing with a listed solvent into filters or waters in dry filter or waterwall paint booths will cause the filters or sludge to become listed hazardous wastes, with the exception of F003 solvents, where the filters or sludges are not characteristically hazardous. Finally, spill residues, such as contaminated soils from the release of listed hazardous wastes, are considered to be listed hazardous wastes.

There is a definite incentive for manufacturing firms to be very diligent in segregating listed hazardous wastes from other wastes, as a hazardous waste minimization and cost-containment measure. By segregating, the potential feasibility of reuse, recycling, or treatment becomes greater. In addition, obtaining approvals from regulatory agencies and waste disposal companies to treat or dispose of the material may become easier. Treatment/disposal costs may be less on a per-unit basis.

Hazardous Waste Characteristics

Having discussed the listings of hazardous waste, it is now necessary to cover the characteristics of a hazardous waste. It is important to note that a solid waste not meeting any of the listing descriptions may still be a hazardous waste by virtue of possessing any one or more of the four characteristics. These characteristics are (1) ignitability, (2) corrosivity, (3) reactivity, and (4) EP toxicity. For characteristics 1, 2, and 4, specific analytical procedures and decision levels are used. These analytical procedures need to be specifically requested when having work done by a laboratory. Otherwise, the results may be inappropriate for regulatory determinations. Any analytical work performed to determine if a solid waste possesses any of the characteristics should only be performed by a laboratory with the proper equipment and qualified personnel well versed in these special procedures and protocols.

It is also important that the laboratory receive a representative sample to analyze. The regulatory definition of "representative sample" was purposely left very general. The definition of a representative sample is "a sample of a universe or whole (e.g., waste pile, lagoon, groundwater) which can be expected to exhibit the average properties of the universe or whole" (40 CFR 260.10). No procedures are specified as mandatory for obtaining representative samples. The generator is expected to meet the performance standard implicit in the definition. The EPA included Appendix I, "Representative Sampling Methods," in 40 CFR Part 261 to provide references to acceptable methods of obtaining a representative sample. Because no sampling methods are designated mandatory, it would behoove the generator to put some thought into meeting the performance standard. This would include both methods and locations for sampling waste streams. This thought process and conclusions should be documented. In this way, there will be written evidence (in addition to the laboratory results) should there be any question of the validity of a nonhazardous determination for a waste stream. Sampling procedures and analytical methods are further discussed in Chapter 24.

Ignitability

The characteristic of ignitability is defined at 40 CFR 261.21. An ignitable waste will present a fire hazard during routine management. A representative sample of a solid waste is considered ignitable if:

1. It is a liquid and has a flash point less than 140°F (60°C) as determined by a Pensky-Martens Closed Cup Tester using American Society for Testing and Materials (ASTM) standard D-93-79 or D-93-80 for a test method, or as determined in a Setaflash Closed Cup Tester using the test method specified in ASTM standard D-3278-78. Aqueous solutions with less than 24% alcohol by volume are excluded because they typically do not sustain combustion.

2. It is not a liquid and is capable, under standard temperature and pressure, of causing fire through friction, absorption of moisture, or spontaneous chemical changes and, when ignited, burns so vigorously and persistently that it creates a hazard (40 CFR 261.21(a)(2)). This is a vague definition without a testing protocol or decision level. The intent is to ''capture'' thermally unstable solids likely to cause fire. It is assumed that generators of such wastes are aware of this property from prior experiences with spontaneous combustion, autoignition, or vigorous chemical fires. While the flash point test protocol does not apply to nonliquids, some laboratories do report solids ignitability determinations for such materials.

3. The solid waste is an ignitable compressed gas or an oxidizer as defined by U.S. Department of Transportation (DOT) regulations.

Wastes meeting any of the above aspects of ignitability, and not already listed as a hazardous waste, are assigned the EPA hazardous waste number of D001.

Corrosivity

The characteristic of corrosivity is found at 261.22. A corrosive waste is able to deteriorate standard containers, damage human tissue, and/or dissolve toxic components of other wastes. A representative sample of a solid waste is considered corrosive if it is either:

1. aqueous (water-based), and has a pH less than or equal to 2 or greater than or equal to 12.5

2. a liquid (nonaqueous), corroding SAE 1020 steel at a rate greater than 0.25 inches per year at a temperature of 130°F

There is no provision for a corrosive solid in the definition. However, many state regulatory agencies require that the sample be placed in distilled water and the pH of the aqueous solution be measured. The test methods for both of the above-referenced corrosive properties are found in an EPA manual entitled, ''Test Methods for the Evaluation of Solid Waste, Physical/Chemical Methods,'' SW-846. A laboratory testing solid wastes for any of the characteristics should have a copy of this manual. A solid waste meeting the regulatory definition of corrosivity, and not already listed as a hazardous waste, is assigned the EPA hazardous waste number D002.

Reactivity

The characteristic of reactivity is described at 40 CFR 261.23. A reactive waste has a tendency to become chemically unstable under normal management conditions or react violently when exposed to air or mixed with water, or can generate toxic gases. Like the definition of solid ignitable wastes, the reactivity characteristic is a narrative definition without a mandatory testing protocol or specified decision levels. The intention is to identify the wastes that are extremely unstable, have the capability of liberating toxic gases, or have a tendency to react violently or explode. Again, generators of reactive wastes are assumed to know that their wastes possess this property and require special handling. This knowledge can be inferred by an examination of MSDS information on purchased components of the waste material, along with knowledge of the hazard characteristic of the waste in light of the materials or the process used.

A representative sample of a solid waste is said to exhibit the characteristic of reactivity if it has any of the following properties:

1. It is normally unstable and readily undergoes violent change without detonating.
2. It reacts violently with water.
3. It forms potentially explosive mixtures with water.
4. When mixed with water, it generates toxic gases, vapors, or fumes in a quantity sufficient to represent a danger to human health or the environment.
5. It is a cyanide- or sulfide-bearing waste which, when exposed to pH conditions between 2 and 12.5, can generate toxic gases, vapors, or fumes in a quantity sufficient to present a danger to human health or the environment.
6. It is capable of detonation or explosive reaction if it is subjected to a strong initiating source or if heated under confinement.
7. It is readily capable of detonation or explosive decomposition or reaction at standard temperature and pressure.
8. It is a forbidden explosive as defined in 49 CFR 173.51, a Class A explosive as defined in 49 CFR 173.53, or a Class B explosive as defined in 49 CFR 173.88.

A solid waste possessing any of the above properties is a reactive hazardous waste, and is assigned the EPA hazardous waste number of D003. Again, the waste must not already be listed to be classified properly under D003. Please note that many state regulatory agencies require special attention be given to cyanide- and sulfide-bearing wastes, when evaluating such wastes with respect to reactivity. Typically, regulatory agency guidelines are to be used by generators in making the determination with respect to reactivity. These guidelines usually include both a test method (e.g. total cyanide, total available cyanide, cyanide amenable to chlorination, reactive cyanide) and a regulatory threshold or decision level. Generators of cyanide- and sulfide-bearing wastes are advised to contact their state regulatory agency to determine the recommended test method(s) and the appropriate decision level. EPA, in SW-846, identifies the test methods of reactive cyanide (SW-846 9010) and reactive sulfide (SW-846 9030) as recommended methods. No regulatory thresholds are identified; however, reactive cyanide levels of 250 mg HCN/kg waste or above and reactive sulfide levels of 500 mg H_2S/kg waste or above are typically considered characteristically hazardous.

Table 2.2. Maximum Concentration of Contaminants for EP Toxicity Characteristic

EPA Hazardous Waste Number	Contaminant	Maximum Concentration (mg/L)
D004	Arsenic	5.0
D005	Barium	100.0
D006	Cadmium	1.0
D007	Chromium	5.0
D008	Lead	5.0
D009	Mercury	0.2
D010	Selenium	1.0
D011	Silver	5.0
D012	Endrin	0.02
D013	Lindane	0.4
D014	Methoxychlor	10.0
D015	Toxaphene	0.5
D016	2,4-D	10.0
D017	2,4,5-TP (Silvex)	1.0

Source: Table I, 40 CFR 261.

EP Toxicity

The fourth characteristic is EP toxicity, specified at 40 CFR 261.24. The extraction procedure is a special sample preparation step in which the sample of solid waste is exposed to conditions simulating improper land disposal. It was designed to identify wastes likely to leach relatively large concentrations of any of the heavy metals and pesticides identified in the EPA's Primary Drinking Water Standards (40 CFR Part 141). In essence, the solid waste is mixed with an acidic leaching medium (pH 5.0 ± 0.2) for a period of 24 hours. The liquid extract from this leaching procedure is then analyzed for the presence of any of the specified contaminants. Table 2.2 gives the contaminants of concern and the applicable regulatory threshold. Concentrations equal to or greater than the values in the table cause the waste to be hazardous by virtue of EP toxicity. Liquid wastes that, after filtration, are found to have had less than 0.5% solids by weight are analyzed directly for the specified constituents. The analytical procedures for these and other chemical constituents are specified in Appendix III to Part 261, "Chemical Analysis Test Methods." Appendix III largely references the appropriate method numbers, which are described in SW-846—"Test Methods for Evaluating Solid Waste, Physical/Chemical Methods."

The above description of the EP to prepare a waste for analysis is, of course, a simplification. The EP is described in Appendix II to Part 261 and is also included in SW-846 as method SW-846 1310. The purpose of this brief description of the EP is to indicate that levels of constituents obtained from the EP extract are not necessarily going to be the same or similar to the levels obtained through an analysis for the individual levels of those metals on a total metals basis. The levels should be less. Again, the importance of specifying the proper sample preparation and analytical procedure is evident.

If the EP extract of a representative sample of the waste contains any of the specified contaminants at a concentration equal to or greater than their respective regulatory threshold, the waste material is said to exhibit the characteristic of EP toxicity. EP toxic wastes that are not already listed as hazardous waste are assigned EPA hazardous waste number(s) (specified in Table 2.3) corresponding to the contaminant(s) causing the solid waste to be hazardous. Please note that analysis of the EP extract for the six pesticides is not typically performed, unless they are used or produced or would otherwise be expected to be present in the waste material.

The EPA hazardous waste numbers assigned to each characteristic of a hazardous waste, as well as to each listed waste, are important due to their use in notification, manifesting, biennial reporting, and other recordkeeping and reporting requirements.

The importance of representative sampling and of utilizing the specified SW-846 methods cannot be overemphasized. Procedures and considerations in sampling and analysis are discussed in Chapter 24.

Note that some state regulatory agencies require that the method of standard additions be used for quantification of contaminant concentration. The method of standard additions is described in SW-846. Commercial laboratories utilized for analyses for any of the characteristics of hazardous waste should be directed to use SW-846 methods in general, and the method of standard additions if required by the state regulatory agency. Otherwise, the resulting data will be suspect, and may not be considered acceptable for regulatory purposes.

Toxicity Characteristic

In June of 1986, EPA proposed to replace the Extraction Procedure utilized for the EP toxicity characteristic with a new leaching procedure, called the Toxicity Characteristic Leaching Procedure (TCLP). TCLP is a zero headspace extraction, analyzed by gas chromatography/mass spectrometry (GC/MS) for 38 organic toxic constituents. The eight metals, four insecticides, and two herbicides under the EP toxicity characteristic would also be analyzed and remain of concern under the new toxicity characteristic.

EPA is expected to take final action in this matter in March of 1990. The final rule would take effect six months after publication in the *Federal Register.*

Table 2.3.   Proposed Toxicity Characteristic Constituents and Regulatory Thresholds

EPA HW No.	Constituent	CAS RN	Regulatory Threshold (mg/L)
D018	Acrylonitrile	107-13-1	5.0
D004	Arsenic	7440-38-2	5.0
D005	Barium	7440-39-3	100.0
D019	Benzene	71-43-2	0.07
D020	Bis (2-chloroethyl) ether	111-44-4	0.05
D006	Cadmium	7440-43-9	1.0
D021	Carbon disulfide	75-15-0	14.4
D022	Carbon tetrachloride	56-23-5	0.07
D023	Chlordane	57-74-9	0.03
D024	Chlorobenzene	108-90-7	1.4
D025	Chloroform	67-66-3	0.07
D007	Chromium	1333-82-0	5.0
D026	o-Cresol	95-48-7	10.0
D027	m-Cresol	108-39-4	10.0
D028	p-Cresol	106-44-5	10.0
D016	2,4-D	94-75-7	1.4
D029	1,2-Dichlorobenzene	95-50-1	4.3
D030	1,4-Dichlorobenzene	106-46-7	10.8
D031	1,2-Dichloroethane	107-06-2	0.40
D032	1,1-Dichloroethylene	75-35-4	0.1
D033	2,4-Dinitrotoluene	121-14-2	0.13
D012	Endrin	72-20-8	0.003
D034	Heptachlor (and its hydroxide)	76-44-2	0.001
D035	Hexachlorobenzene	118-74-1	0.13
D036	Hexachlorobutadiene	87-68-3	0.72
D037	Hexachloroethane	67-72-1	4.3
D038	Isobutanol	78-83-1	36.0
D008	Lead	7439-92-1	5.0
D013	Lindane	58-89-9	0.06
D009	Mercury	7439-97-6	0.2
D014	Methoxychlor	72-43-5	1.4
D039	Methylene chloride	75-09-2	8.6
D040	Methyl ethyl ketone	78-93-3	7.2
D041	Nitrobenzene	98-95-3	0.13
D042	Pentachlorophenol	87-86-5	3.6
D043	Phenol	108-95-2	14.4
D044	Pyridine	110-86-1	5.0
D010	Selenium	7782-49-2	1.0
D011	Silver	7440-22-4	5.0
D045	1,1,1,2-Tetrachloroethane	630-20-6	10.0
D046	1,1,2,2-Tetrachloroethane	79-34-5	1.3
D047	Tetrachloroethylene	127-18-4	0.1
D048	2,3,4,6-Tetrachlorophenol	58-90-2	1.5
D049	Toluene	108-88-3	14.4
D015	Toxaphene	8001-35-2	0.07
D050	1,1,1-Trichloroethane	71-55-6	30.0
D051	1,1,2-Trichloroethane	79-00-5	1.2
D052	Trichloroethylene	79-01-6	0.07
D053	2,4,5-Trichlorophenol	95-95-4	5.8
D054	2,4,6-Trichlorophenol	88-06-2	0.30
D017	2,4,5-TP (Silvex)	93-76-5	0.14
D055	Vinyl chloride	75-01-4	0.05

Source: 51 FR 21647–21693 (June 13, 1986).

As indicated earlier, this would include the introduction of the new extraction procedure (TCLP), and the expansion of the constituents of concern to include an additional 38 organic toxicants. The proposed list of toxicity characteristic parameters and their respective regulatory thresholds is found in Table 2.3.

There are two issues of relevance to generators of hazardous waste. The first involves the cost of laboratory analyses by TCLP to determine whether a waste material possesses the toxicity characteristic. The TCLP extraction, with subsequent organic analyses by GC/MS and metal analyses by atomic absorption, will be more expensive than EP toxicity analyses. Preliminary pricing by commercial laboratories for TCLP testing is in the $1200–$2200 per sample range, depending upon whether pesticide and herbicide analyses are appropriate, among other factors. This compares to the $150–$500 per sample range for heavy metal analyses for the EP toxicity characteristic.

The second concern involves the expansion of the constituents of concern to include organic toxicants, including several common solvents. Solvents identified as organic toxicants include toluene, methyl ethyl ketone, methylene chloride, perchloroethylene, trichloroethylene, and 1,1,1-trichloroethane. The proposed regulatory thresholds specified for the solvents are rather low, ranging from 0.07 mg/L for trichloroethylene to 30 mg/L for 1,1,1-trichloroethane. The end result will be that many wastes currently considered nonhazardous could become hazardous wastes. This is especially likely for wastes containing trace organics, such as paint booth sludges. The effort and potential expense of redoing the hazardous waste determination for potentially affected wastes will also be necessary.

Please note that TCLP as an extraction method is specified in the Land Disposal Restrictions (40 CFR Part 268). The constituents of concern and the regulatory thresholds will differ between TCLP analyses for hazardous waste determination purposes and TCLP analyses for land disposal restriction purposes.

IDENTIFICATION OF HAZARDOUS WASTE

Now that the lists of hazardous wastes and the characteristics of hazardous waste have been discussed, it is appropriate to return to the definition of hazardous waste (40 CFR 261.3) for guidance on when a waste becomes hazardous and when a hazardous waste ceases to be a hazardous waste.

A solid waste not excluded from regulation under 40 CFR 261.4(b) becomes a hazardous waste if and when it meets any of the following criteria:

1. when it meets the listing description on any of the four lists of hazardous waste and has not been excluded (delisted) from the lists under 40 CFR 260.20 and .22
2. for mixtures of listed and solid wastes, when the listed waste is first added to the solid waste
3. in the case of any other waste (including mixtures), when the waste exhibits any one of the four characteristics of hazardous waste

In general, except for certain listed process residues and wastewater treatment sludges, the point in time when materials can first become hazardous wastes is when their intended use has ceased (e.g., off-specification or obsolete commercial chemical products) or is no longer possible (e.g., spent solvents), and the materials are segregated, accumulated, or stored for treatment or disposal, re-use, or reclamation. Remember, a material first must be a (solid) waste before it can become a hazardous waste. Further, manufacturers can have residues that would be considered hazardous wastes in a raw material or product storage tank or in a manufacturing process unit that are not subject to regulation until the wastes are removed from the tank or unit. However, such residues are subject to regulation if they remain in the product storage tank or manufacturing process unit more than 90 days after the unit ceases to be operated for manufacturing, storage, or transportation of product or raw materials (40 CFR 261.4(c)). Plating bath sludges and raw material tank bottoms or heels are examples of material potentially qualifying for this temporary exemption.

Samples of waste materials or of water, soil, or air collected for the purpose of monitoring or testing are excluded from regulation under RCRA at 40 CFR 261.4(d). There are certain requirements and limitations to the so-called "sample exclusion provision." Samples shipped to the laboratory for testing must be packaged so as to prevent any leakage, spillage, and vaporization, and must otherwise be shipped in conformance with DOT or U.S. Postal Service requirements, as applicable. If DOT or U.S. Postal Service requirements do not apply, then specified information must accompany the sample. This exclusion no longer applies when the laboratory determines that the waste is hazardous and no longer meets the conditions stated in 40 CFR 261.4(d)(1).

EPA expanded the scope of the sample exclusion provision to include waste samples used in small-scale treatability studies. In this final rule, published in the July 19, 1988 *Federal Register,* EPA conditionally exempts from regulation waste samples of up to 1000 kg of nonacute hazardous waste, 1 kg of acute hazardous waste, or 250 kg of soils, water, or debris contaminated with acute hazardous wastes. There are a variety of limitations and conditions to the applicability of the exclusion for treatability study samples, including recordkeeping requirements. A generator intending to utilize this exclusion when shipping quantities of hazardous waste offsite for treatability studies is well advised to carefully examine the regulatory provisions found at 40 CFR 261.4(e) and (f).

Certain conditional exemptions are given from the definition of hazardous waste for:

1. mixtures of solid waste and hazardous waste listed solely for one or more of the characteristics (Hazard Codes I, C, R, or E in the listings), where the resultant mixture no longer exhibits any characteristic of a hazardous waste (40 CFR 261.3(a)(2)(iii))

2. mixtures of wastewater subject to the Clean Water Act regulatory program, and certain listed wastes, where the concentration of such listed contaminants does not exceed trace levels (either 1 or 25 ppm, depending upon the listed contaminant) (40 CFR 261.3(a)(2)(iv)(A), (B), and (E))
3. mixtures of wastewater subject to the Clean Water Act regulatory program and *de minimis* losses of commercial chemical products from manufacturing operations (40 CFR 261.3(a)(2)(iv)(D))

The regulatory definition of hazardous waste also specifies when a hazardous waste ceases to be a hazardous waste (40 CFR 261.3(c) and (d)). In the case of solid wastes generated from the treatment of a characteristic hazardous waste, a hazardous waste remains a hazardous waste unless and until the solid waste does not exhibit any of the characteristics of a hazardous waste. For listed hazardous wastes, mixture rule listed hazardous waste, or wastes derived from a listed hazardous waste, the waste(s) must not exhibit any of the characteristics of a hazardous waste, and must be excluded (delisted) from regulation under the rulemaking procedures of 40 CFR 260.20 and 260.22.

CONCLUSION

It is important to note the limited nature of most exclusions or exemptions from the application of certain regulatory provisions. Most exclusions have definite limitations in terms of their nature, scope, and application. The otherwise-regulated party intending to take advantage of any exclusion or exemption is well advised to make absolutely certain that all the conditions necessary to qualify for the exclusion are met. In many cases it is also advised to document the factual circumstances and rationale used in arriving at the decision that the exclusion was applicable and legally available to the firm.

With this understanding of the regulatory definitions of solid and hazardous waste, those people assigned waste management duties at a manufacturing facility are now prepared to move on to the determinations necessary for corporate compliance. Determination of all the wastes generated at the facility is discussed in Chapter 3. Determination of which of the solid wastes generated are hazardous waste is covered in Chapter 4. Finally, the likely regulatory categories for manufacturing firms generating wastes are outlined in Chapter 5.

CHAPTER 3

Conducting a Waste Inventory: Procedures and Considerations

Perhaps the most logical place to begin compliance with federal and state HWM requirements is with preparation of an inventory of waste generated from the facility. In this manner, there can be no question as to how many different waste streams there are, and where they come from. The term "waste stream" is used to describe waste materials generated from a known point of origin. Common waste streams include spent solvents from vapor degreasers, wastewater treatment sludges from electroplating operations, and paint booth sludges from water wall paint booths. Once each waste stream is identified, determining whether each waste is hazardous can then be conducted in a straightforward and comprehensive manner. Quantifying the total amount of hazardous waste generated per month and evaluating how each waste is (or should be) managed allows the determination of which HWM requirements apply to the firm.

BLACK BOX APPROACH

One approach of particular usefulness in preparing an inventory is to look at the plant operations as a whole. This requires, initially, viewing the entire manufacturing facility as a black box, with inputs (raw materials, chemicals, energy, water, and air) and outputs (finished products, wastewater, air emissions, and other waste liquids and solids). It is important to keep in mind the regulatory definitions of solid and hazardous wastes when looking at facility inputs and outputs.

Chemical purchasing records should be reviewed, and the fate of each chemical should be determined by tracing its path through plant processes or operations.

The chemical may be utilized on or in the product, may be dissipated or consumed, and/or some portion may remain as a by-product or waste requiring proper disposal.

For manufacturing firms, the reason for existence is the production of a marketable item. However, in most factories the item to be sold is not the only thing produced. There are also manufacturing process wastes, pollution control residuals, obsolete chemical inventory, and a variety of other solid and liquid waste products. Although these waste materials are only incidental to the manufacture of a marketable item, they still must be handled in a proper and safe manner.

Inputs

In using the black box analogy, the search for waste outputs begins with examination of all inputs that could possibly become wastes. To discover all chemicals and materials used in facility operations, an examination can be made of all purchase orders for a period of time, say 12 months. Annual quantities, purpose in facility operations, price per unit, and usage per unit time should be noted. This can become a major endeavor at many manufacturing firms. However, a thorough examination can serve the purposes of three separate regulatory programs, as discussed later.

From this examination, a list of "suspect" inputs should be compiled. As a general rule, the following types of inputs, if used, should always be on such a list:

1. cleaning agents: acids, alkalis, solvents, and organic cleaning solutions
2. process chemicals
3. industrial oils: hydraulic oils, cutting oils, coolants, lube oils, and quench oils
4. water treatment chemicals: water softening agents, deoxgenators, biocides, and coagulants
5. surface finishing agents: rust inhibitors, primers, paints, coatings, and all chemicals involved with electroplating

Remember that spent solvents and commercial chemical products, if discarded, can be hazardous wastes. Even though such chemicals are often sold under a trade name, the chemical user must determine if such proprietary chemicals meet any of the listing descriptions, if discarded.

The evaluation of chemicals utilized at the facility may have been performed as part of the compliance program developed in response to the OSHA Hazard Communication Standard (29 CFR 1910.1200). This type of evaluation will also be necessary for other purposes, including compliance with various EPA regulations issued under the authority of the Emergency Planning and Community Right-to-Know Act (EPCRA), also known as Title III of the Superfund Amendments and Reauthorization Act of 1986 (SARA).

Material Safety Data Sheets

The vendors selling the "suspect" inputs to the facility should be contacted and asked to furnish MSDSs. The Material Safety Data Sheet provides much useful information. The information on these sheets is compiled by the chemical manufacturer and includes:

1. identity (name) of the product and appropriate chemical and common names
2. physical and chemical characteristics
3. physical hazards
4. health hazards
5. routes of entry (inhalation, ingestion, skin/eye contact)
6. exposure limits (OSHA Permissible Exposure Limit [PEL] and American Conference of Governmental Industrial Hygienists Threshold Limit Value [TLV®])
7. carcinogenicity
8. precautions for handling and use (protective measures, procedures for spill cleanup, appropriate hygienic practices)
9. control measures (engineering controls, work practices, or personal protective equipment)
10. emergency and first aid measures
11. date of preparation
12. name, address, and telephone number of the chemical manufacturer

The value of such information for the proper management of the chemicals (as products, process chemicals, or wastes) is obvious. It is important to note, however, that the information on MSDSs is oriented toward worker safety, as workers are exposed to the concentrated chemical product. The purpose in obtaining the MSDS at this point is to have some preliminary information as to the chemical composition of purchased components of facility waste streams. Such information is important when deciding whether the chemicals used meet any listing descriptions, and which chemical analyses to perform on each waste stream.

Outputs

Continuing the black box examination, the emphasis switches to the outputs. The inputs that leave the factory as products obviously are not of concern here. The inputs not accounted for in production, now known as wastes or residuals, are the focus of attention. Determination of these unwanted outputs is less straightforward than determination of inputs. Examination of any waste hauling records and hazardous waste manifests is a start, although they are typically utilized for waste materials already classified as hazardous wastes. The objective is to trace the path of the inputs to the manufacturing process, in order to determine where, how, and why portions of these inputs become waste materials. Here it is important to actually get on the factory floor and examine the production lines and talk to operating personnel. Detailed examinations of facility processes and operations should be performed using the black box methodology. Interviews with select personnel (equipment operators, maintenance and repair employees, and housekeeping staff) are also necessary. At this point, a listing should be prepared, detailing the outputs from each machine and activity at the facility. The information that should be obtained in this regard includes:

1. wastes associated with process startup and shutdown and process changeovers
2. wastes associated with normal production
3. wastes associated with equipment cleaning, maintenance, or repair
4. wastes associated with process upsets, including the use of off-specification inputs and production of any off-specification products
5. wastes associated with inventory (of inputs and outputs), including obsolete and expired chemicals, materials, and products
6. wastes associated with products not meeting customer specifications (returned goods, goods damaged in transit, etc.)

The information on inputs and outputs gathered allows the development of a crude (nonnumerical) materials balance diagram. This flow diagram correlates the various operations, processes, and support activities of the plant with the inputs and outputs discussed above. Evaluation and implementation of cost-effective waste minimization measures will require this effort.

The waste inventory will reveal several areas requiring additional effort:

1. The waste streams identified need further classification. The composition and quantity of each waste stream need to be determined. Quantity determinations must be made using historical data or statistically rigorous sampling methods. Startup and shutdown procedures, maintenance activities, and differences in production levels must be taken into account.
2. It is necessary to determine if any of the waste streams are regulated under federal and state HWM regulations. All necessary laboratory analyses should be performed on representative samples using the EPA SW-846 sampling and analysis protocol.
3. For the waste streams classified as hazardous, it is necessary to determine whether they are excluded from regulation at 40 CFR 261.4, or qualify for the conditional exemptions for certain use, reuse, recycling, or reclamation activities.
4. Quantities of regulated hazardous wastes must be totaled to determine which generator category the firm falls into.
5. Federal, state, and municipal wastewater and air pollution regulations must be examined to determine whether the facility's air and water effluents are subject to regulation. If so, it must be determined whether those effluents meet the applicable standards, and all necessary permits must be obtained.
6. The amount of wastes per unit of production remains unknown.

7. The effect of process and production variables on waste composition and quantity remains unknown.

8. The current storage practices and final disposition of all regulated waste materials need to be determined. The appropriateness of onsite storage practices and offsite treatment/disposal methods needs to be evaluated with respect to regulatory standards, the waste management hierarchy, and best management practices. The compliance status of waste haulers and offsite (commercial) HWM facilities also needs to be determined. These topics are addressed in Section IV of this book.

INTEGRATED APPROACH

It is obvious that the above concerns must be addressed. However, the complexity of the ever-growing body of environmental regulations and the high cost of compliance (or noncompliance) seem to indicate the advantages of some sort of integrated approach toward residuals management. Such an approach would include a strategy to effectively manage total short- and long-term costs of complying with all applicable requirements.

An integrated approach to solid and hazardous waste management at manufacturing facilities is the subject of Section III of this book. First, the wastes at a manufacturing facility must be uncovered, and determinations made as to each waste's hazardous or nonhazardous status. The firm's regulatory category is then ascertained and the applicable regulations are identified and understood. Only after these preliminary steps are completed can a workable environmental management program be implemented. An integrated approach toward the various requirements is a desirable element of any corporate environmental management program.

CHAPTER 4

Hazardous Waste Determinations: Procedures and Considerations

Federal and state HWM regulations require certain actions on the part of waste generators with respect to the nature of their wastes. Those who generate solid wastes are required to determine if their wastes are hazardous. Other considerations come into play once these determinations are made. It is the purpose of this chapter to aid in these determinations.

Chapter 2 dealt with the definitions of solid and hazardous waste, and explained the listings and the four characteristics of hazardous wastes. Chapter 3 outlined the procedures and considerations involved with a waste inventory at a manufacturing facility. This chapter builds on these previous chapters and delves into some related considerations.

IDENTIFICATION OF SOLID WASTE

The first step in any determination of the hazardous waste(s) at a facility is the application of the regulatory definition of solid waste to the waste streams uncovered during the waste inventory. As mentioned in Chapter 2, solid wastes need not be solid (i.e., they can be liquid, solid, semisolid, or contained gaseous material). Further, the regulatory definition of solid wastes includes any material disposed of or abandoned in lieu of disposal; burned, incinerated, or recycled; or considered "inherently waste-like." Finally, there are limited exclusions and exemptions from the definition of solid and hazardous waste.

This broad definition will likely result in the situation where spent materials, sludges, and manufacturing by-products (all uncovered during the waste inventory) are considered solid wastes, unless they are recycled or reused in a manner qualifying for the conditional exemptions (discussed in Chapter 2) or they qualify for one of the 40 CFR 261.4(a) exclusions. The exclusions of interest to the majority of manufacturing firms are those for:

1. domestic sewage and any mixture of domestic sewage and other wastes that passes through a sewer system to a publicly-owned treatment works for treatment
2. industrial wastewater discharges that are point source discharges subject to regulation under Section 402 of the Clean Water Act, as amended

Determination of whether a material is a solid waste is important because hazardous wastes are a subset of solid wastes; therefore, a material must be a solid waste before it can be considered a hazardous waste.

IDENTIFICATION OF HAZARDOUS WASTE

Once it is determined that a material meets the regulatory definition of solid waste, the next step is to determine whether the material also meets the regulatory definition of hazardous waste. As explained in Chapter 2, a solid waste is hazardous if it (1) meets the description of any item on any one of the four lists of hazardous waste; (2) is a mixture that contains a listed hazardous waste or is derived from the treatment, storage, or disposal of a listed hazardous waste; or (3) exhibits any one of the four characteristics of a hazardous waste.

Supplemental instructions for generators regarding hazardous waste determinations are given at 40 CFR 262.11. A generator of a solid waste is to first determine whether the waste is excluded from regulation under 40 CFR 261.4. At this point, the generator is to determine if the waste is listed, by virtue of meeting any of the listing descriptions of the four lists of hazardous waste. Please remember that waste materials can be considered listed hazardous wastes by virtue of the so-called mixture rule and derived-from rule. If the waste is not listed, it should be examined to see if it possesses any one or more of the four characteristics of hazardous waste. In determining if the waste is characteristically hazardous, the generator may either test the waste according to EPA-approved (SW-846) methods or apply knowledge of the hazard characteristic of the waste in light of the materials or processes used.

The regulations (40 CFR 262.11) do not require that a generator go to the effort and expense of chemical analysis for wastes thought to possess a characteristic of a hazardous waste. The generator has the option of declaring a waste stream to be a hazardous waste. Lack of laboratory analyses in regulatory determinations may result in enforcement problems, particularly where nonhazardous determinations are not well documented. Also, some laboratory analyses are typically required to make arrangements with commercial (offsite) treatment/disposal vendors. However, such analyses are not always the same as those required for hazardous waste determinations.

Sufficient information and data need to be compiled in order to make a defensible hazardous waste determination as well as to make offsite treatment/disposal arrangements. At a minimum, for each waste stream uncovered during the waste inventory, the components need to be identified. MSDSs should be obtained for each purchased component of each waste stream. For waste materials *not* meeting any of the listing descriptions, the determination with respect to the characteristics can be made by applying knowledge of the hazard characteristic of the waste in light of the materials or process used.

When MSDS information indicates that the waste material may possess one or more of the characteristics, then laboratory analyses are typically indicated. Analyses of representative samples for the appropriate parameters using EPA-specified (SW-846) methods become necessary to make the determination in a legally defensible manner. The burden of proof is on the generator to determine whether or not a waste material is hazardous (40 CFR 262.11), and to document and keep records of the determination (40 CFR 262.40(c)).

It is important that the hazardous waste determinations be made correctly, with an understanding of the regulatory definitions of solid and hazardous waste. It is also important that the determinations be made comprehensively, based on a thorough waste inventory. It is necessary to document the information and rationale used in arriving at a hazardous or nonhazardous determination for each waste stream. The need for thorough and comprehensive documentation is all the more critical when the waste material is thought to be nonhazardous, or to fall under one of the exclusions or conditional exemptions.

A few other points merit further discussion.

1. Manufacturing operations with limited quantities of hazardous wastes need to quantify and document the rate of hazardous waste generation in order to determine whether they qualify for one of the two regulatory categories with limited requirements. Conditionally exempt small quantity generators (SQGs) are granted an exclusion at 40 CFR 261.5 from most RCRA requirements. Further discussion of the nature of and conditions associated with this exclusion can be found in Chapter 5, and the regulatory standards for these generators can be found in Chapter 7.

2. Generators, other than conditionally exempt SQGs, must also determine whether any of their hazardous wastes are subject to the 40 CFR Part 268 Land Disposal Restrictions. This requirement is found at 40 CFR 268.7, and applies regardless of whether the restricted hazardous wastes are to be managed by direct land disposal or not. For example, notification of a restricted waste's status must be made with each shipment to offsite RCRA treatment facilities.

   EPA has promulgated a schedule for the restriction of land disposal and for the establishment of treatment standards for all listed and characteristic hazardous wastes (40 CFR Part 268 Subpart B). To date, the F001–F005 solvent listings, dioxin wastes, the so-called California List Wastes, and First Third listed wastes have been restricted from direct land disposal. The statutory language in RCRA, as amended, is such that failure of EPA to promulgate regulations (including treatment standards, variances, exemptions, and extensions) by the statutory deadlines can result in the complete and absolute prohibition on the land disposal of the hazardous wastes involved. EPA has until May 8, 1990 to complete the land disposal restriction process for all existing listed wastes and characteristics of hazardous wastes.

3. Procedures for delisting are specified at 40 CFR 260.20 and 40 CFR 260.22. Petitions to amend 40 CFR Part 261 to exclude a waste produced at a particular facility are allowed at 40 CFR 260.22, and 40 CFR 260.20 gives the general requirements for all petitions. These petitions are for the purpose of excluding a listed hazardous waste at a particular generating facility from the lists of hazardous waste in 40 CFR Part 261. Any exclusion granted only applies to the particular waste generated at the facility and will not apply to similar wastes from any other facility.

Petitioners must provide EPA with sufficient information to establish that (1) the waste is not hazardous based upon the criteria for which it was listed, and (2) no other hazardous constituents are present at levels of regulatory concern.

4. The final point to be discussed, but certainly not the least important, is that of the disposition of wastes determined to be nonhazardous. It must be realized that wastes not falling under the RCRA definition of hazardous wastes are still potentially regulated materials. Nonhazardous industrial wastes can still pose a threat to the environment if mismanaged. In other words, nonhazardous status for a waste stream does not relieve the generator of responsibility for proper management. Regulatory standards for the management of nonhazardous industrial waste are discussed in Chapter 6.

SUMMARY

The definition of solid waste, the four lists of hazardous wastes, and the four characteristics of hazardous wastes are to be used by generators to determine if their solid wastes are hazardous. The listings always take precedence over the characteristics. Wastes that meet the listing description of any of the lists are automatically hazardous. The generator has the responsibility for identifying a waste that is hazardous by virtue of possessing a characteristic of a hazardous waste. The generator can do so by testing the waste and/or by applying his knowledge of the hazard characteristics of the waste.

Chapter 5 concludes Section I by describing regulatory categories under which manufacturing operations most commonly fall. The waste management activities of some manufacturing firms can be described by more than one category. The regulatory standards for each category are described in Section II.

CHAPTER 5

Determining the Firm's Regulatory Category

Once an inventory of the wastes produced at the manufacturing facility is conducted, and the wastes have been determined to be either hazardous or nonhazardous, the next step necessary for compliance is to determine how the firm's hazardous waste management activities fit into the available regulatory categories. This will involve both a quantification of the total amount of all hazardous wastes generated per month and an assessment of the types of HWM activities conducted onsite. There are several categories, and the 1984 amendments to RCRA added two peripheral categories, dealing with used oil and underground product storage tanks. As a result, many manufacturing firms will fall into more than one category. The categories of regulated parties under RCRA that potentially apply to manufacturing firms are:

1. solid waste/nonhazardous industrial waste generator
2. conditionally exempt small quantity generator
3. 100–1000 kg/month small quantity generator
4. generator who accumulates wastes onsite less than 90 days
5. generator who ships wastes offsite for treatment, storage, or disposal
6. generator who stores wastes onsite under RCRA permit or interim status
7. generator who treats hazardous waste by an exempted method or in an exempted manner
8. generator who markets or burns used oil fuel
9. underground product storage tank owner/operator
10. generator who treats hazardous wastes onsite under RCRA permit or interim status
11. generator who disposes of hazardous wastes onsite under RCRA permit or interim status

Please note that categories 1–4 are mutually exclusive.

The regulatory definition of generator, found at 40 CFR 260.10, is "any person, by site, whose act or process produces a hazardous waste identified or listed in Part 261 of this Chapter or whose act first causes a hazardous waste to become subject to regulation." This definition is very broad and makes no reference to the quantities of hazardous waste generated. The regulations themselves, however, do differentiate between classes of generators in terms of the quantities of hazardous wastes generated in a calendar month. Reduced requirements are applicable to generators of less than 100 kg of hazardous waste per calendar month, as well as to generators of between 100 and 1000 kg of hazardous waste per calendar month.

Regardless of the quantities of hazardous waste produced, generators need to perform the three determinations of any potentially regulated party under RCRA. The importance of formal, written determinations based on thorough documentation becomes critical when the firm qualifies for and intends to take advantage of any reduced requirements, such as those granted to conditionally exempt small quantity generators. If there is any doubt on the part of the regulatory agency, the burden of proof with respect to the applicability of any reduced requirements will be placed solely and squarely on the generator.

This book deals with the various categories of generators of hazardous waste. It does not cover treatment or disposal operations of manufacturing facilities falling into categories 10 or 11, nor does it cover storage of hazardous wastes in surface impoundments or waste piles. Inclusion of such operations would complicate matters unnecessarily, since most small- and medium-sized manufacturing firms fall into categories 1–9. This chapter provides guidance in the determination of which category or group of categories best match the HWM activities of the manufacturing firm. Section II of this book covers the regulatory standards for categories 1–9 in separate chapters.

SOLID WASTE/NONHAZARDOUS INDUSTRIAL WASTE

If, after an extensive examination of the manufacturing process wastes, pollution control residuals, and commercial chemical products sometimes discarded, it is determined that none of the solid wastes produced at the facility are hazardous (or that the wastes fall under 40 CFR 261.4(a) or (b) exclusions or recycling/reuse exemptions), the HWM regulations issued under Subtitle C of RCRA do not apply. More likely is the finding that many, but not all, of the solid wastes generated are nonhazardous.

It is important to note the existence of state regulations for proper management of solid waste including nonhazardous industrial wastes. These regulations include requirements for any onsite disposal of such wastes. Such wastes, if managed by land disposal, are typically regulated under sanitary landfill requirements. These requirements will be discussed in Chapter 6.

It is also important to document the information and logic used in the determinations regarding each nonhazardous industrial waste. The Material Safety Data Sheets obtained for purchased components of such wastes, and any laboratory analyses performed on representative samples of each waste, provide important evidence in the determination. A short narrative to accompany these documents should be prepared for each waste stream. Such a narrative would, if applicable, explain why the waste is not a solid waste. If the material is a solid waste, the narrative would state (1) that the waste meets one of the 40 CFR 261.4(b) exclusions, if applicable; (2) that the waste does not meet any of the listing descriptions of process wastes under 40 CFR 261.31 and 40 CFR 261.32 and that the waste does not meet the 40 CFR 261.33(e) or (f) commercial chemical product listings; and (3) why the waste does not possess any of the four characteristics. This narrative can be based on knowledge of the hazard characteristic of the waste in light of the materials or the processes used. More common, however, is the use of laboratory analyses, using SW-846 methods, for waste materials with the potential to be characteristically hazardous.

This is not an onerous task, and will provide ready documentation should there be any question on the part of federal or state hazardous waste inspectors, landfill operators, insurance companies, local emergency planning committees, fire departments, or corporate management. The burden of proof with respect to a nonhazardous determination will be placed squarely on the generator of the waste material.

Remember, however, that the firm's status as a nonhazardous industrial waste generator is not static or permanent. Any one of the following events could alter the firm's status, or the regulatory status of a particular waste material, and should be investigated as necessary for regulatory impacts.

1. a change in production processes or waste treatment operations
2. trial use or change in process chemicals
3. spills, leaks, or contamination of listed commercial chemical products not normally discarded, or of process chemicals normally nonhazardous
4. infrequent cleaning, maintenance, or construction/demolition operations (e.g., cleaning of product storage tanks, manufacturing process units, pollution control equipment, and exhaust ventilation ductwork; or removal of production equipment, chemical pipelines, or contaminated concrete or soil)
5. discovery of surplus, obsolete, off-specification, or expired commercial chemical products on the 40 CFR 261.33(e) or (f) lists
6. changes in the regulatory definitions of hazardous waste at either the federal or state levels

Should any of these events result in the generation of a hazardous waste, the firm's waste management operations need to be evaluated once again. Quantities of hazardous waste generated must be calculated and proposed waste management methods must be evaluated as to their compatibility with the applicable regulatory standards. Transportation and disposal arrangements utilized for industrial nonhazardous wastes are typically unsuitable for newly generated hazardous waste. Special arrangements will be necessary with commercial HWM firms for the offsite transportation and treatment/disposal of such hazardous wastes.

CONDITIONALLY EXEMPT SMALL QUANTITY GENERATORS

Certain hazardous waste generators are conditionally exempt small quantity generators (SQGs) and as such qualify for a set of special (reduced) requirements for the management of their hazardous wastes. These requirements are found at 40 CFR 261.5.

For all hazardous wastes except those classified as acutely hazardous, a generator is a conditionally exempt small quantity generator as long as no more than 100 kg (220 pounds or roughly one-third of a 55-gallon drum) of hazardous waste is generated in that calendar month. In addition, no more than 1000 kg of hazardous waste can be accumulated onsite at any one time if the generator is to retain the conditionally exempt SQG status.

In effect, the conditionally exempt SQG is given an exemption from the notification requirement and Parts 262–268 and 270 and 124 of the regulations in exchange for keeping his small quantities of hazardous wastes moving offsite to one of the several types of facilities given as management options in the regulations. The small quantity exclusion level applies to the total amount of all hazardous waste streams at the facility, not to individual waste streams.

Conditionally exempt SQGs should have sufficient documentation, such as estimated generation rates for every waste stream, and historical data, such as hauling records, to substantiate their determination. This documentation should include the determinations of the nonhazardous nature of other waste streams, and evidence of the proper disposition and disposal of conditionally exempt hazardous wastes. This information should be readily available, should there be any question on the part of state hazardous waste inspectors, landfill operators, insurance companies, or corporate management.

With regard to wastes classified as acutely hazardous, a different set of exclusion levels holds for generation in a calendar month or accumulation over a period of time. Acutely hazardous wastes are primarily those meeting the 261.33(e) commercial chemical products listings. However, those listings in the 40 CFR 261.31 and 261.32 lists identified by the hazard code ''H'' are also considered acutely hazardous. The 261.31 listings identified by the hazard code ''H'' primarily refer to wastes potentially contaminated with dioxins and furans. Currently there are no 261.32 listings identified by the hazard code ''H.'' For acutely hazardous wastes, if more than 1 kg (2.2 pounds) is generated in a calendar month, or if more than 100 kg of any residue or contaminated soil, water, or debris from the cleanup of a spill of such materials is generated, then all quantities of any acutely hazardous and other hazardous wastes are subject to full regulation, including notification. Full regulation in this instance means those regulations applicable to >1000 kg/month generators ("large quantity generators").

It must be emphasized that the regulation of conditionally exempt small quantities of hazardous waste begins when the exclusion level is exceeded for either hazardous wastes or acutely hazardous wastes. At that time any quantities of both types of wastes that are onsite are subject to regulation, including notification requirements.

Regulatory requirements for conditionally exempt SQGs are discussed in Chapter 7.

100–1000 KG/MONTH SMALL QUANTITY GENERATORS

Certain hazardous waste generators qualify for certain limited requirements by virtue of determining that less than 1000 kg (2200 pounds) of hazardous wastes are generated in a calendar month. These generators have more stringent requirements than conditionally exempt SQGs, but less rigorous requirements than (large quantity) generators accumulating hazardous wastes for less than 90 days.

These 100–1000 kg/month generators, sometimes referred to simply as SQGs, are allowed to accumulate hazardous wastes onsite in containers or tanks for up to 180 days (six months) under certain conditions, without needing a RCRA permit or interim status. No more than 6000 kg can be accumulated in the 180-day time period, which is equivalent to the maximum (1000 kg) generation rate in each of the six months of accumulation.

These conditions also include notifying EPA and obtaining an EPA identification number, as well as shipping wastes offsite under manifest within the 180-day time period to RCRA TSD facilities. There are container handling, emergency preparedness, and emergency planning requirements as well.

It is important to note that some states have different, more inclusive definitions of hazardous waste. This may cause a generator who would otherwise be considered a SQG to fall into a more regulated category. Many states consider used oil to be a hazardous waste for purposes of their regulatory program. Regulatory requirements for 100–1000 kg/month SQGs are discussed in Chapter 8.

GENERATORS WHO ACCUMULATE HAZARDOUS WASTES

Generators who accumulate, sometimes termed "90-day generators," "90-day accumulators," or "large quantity generators," qualify for a special, somewhat reduced set of requirements. Generators who accumulate less than 90 days perform the act of storage of a hazardous waste, and as such are technically owners or operators of a TSD facility, subject to the 40 CFR Part 265 standards and the RCRA permit requirements.

The regulatory definition of "storage," found at 40 CFR 260.10, is "the holding of hazardous waste for a temporary period, at the end of which the hazardous waste is treated, disposed of, or stored elsewhere." The regulations, however, allow owners or operators of otherwise fully regulated storage facilities to comply only with the standards found at 40 CFR 262.34(a) under the following circumstances:

1. The owner or operator of the storage facility is also the person generating the wastes.
2. The waste is stored in containers or tanks in the accumulation area(s) for less than 90 days.

The 90-day accumulation period is only allowed for generators who place the waste in containers (typically 55-gal drums or 20-yd³ roll-off boxes) or tanks. Containers are distinguished from tanks on the basis of portability, not in terms of size or capacity. Containers are portable devices, while tanks are stationary devices. The 90-day accumulation period is not available to generators storing in surface impoundments (pits, ponds, or lagoons) or waste piles.

As may be guessed, the most important aspects of the 40 CFR 262.34 standards are when the 90-day accumulation period begins, and how a generator/accumulator is to conduct HWM activities during this period. With the exception of containers in *satellite* accumulation areas, the accumulation period begins when the first drop of hazardous waste is placed in the container or tank. It is at this time that the date of accumulation (sometimes referred to as the accumulation start date) should be clearly marked on the container. The container may continue to be filled for several days or weeks. However, unless the container is located in what is termed a "satellite accumulation" area, the time when the accumulation period starts is clear, beginning with the initial filling activities.

This provision caused economic hardship on many large quantity generators with one or more individual waste streams generated in relatively small amounts. In order to retain 90-day accumulator status, such generators were required to ship mixed or partially-full drums of such wastes offsite on a quarterly basis. The per-unit transportation and disposal rates were typically much higher than for segregated waste materials shipped in full containers.

In 1985, EPA promulgated a regulatory modification allowing some limited relief in such situations. These regulations, sometimes referred to as the satellite accumulation provisions (40 CFR 262.34(c)), allow the accumulation period to begin the date the 55-gallon container is completely full. As with any exemption or other regulatory relief, there are definite qualifying conditions.

Generators are allowed to accumulate up to 55 gallons of hazardous waste in satellite accumulation areas. These are areas at or near any point of generation where wastes initially accumulate (e.g., paint booths, vapor degreasers) that are under the control of the operator of the process generating the waste.

There are other conditions for satellite accumulation areas, including proper containerization of the wastes in the area, and marking/labeling of the container. Once full, the accumulation start date must be marked on the container, and the container must be moved to a 90-day accumulation area within three days after being filled to capacity. Satellite accumulation provisions are more fully discussed in Chapter 9. As currently interpreted, both 100–1000 kg/month SQG and 90-day generators can take advantage of the satellite accumulation provisions.

A generator can have any number of satellite accumulation areas or 90-day accumulation areas. Satellite accumulation areas are subject to the requirements found at 40 CFR 262.34(c). Ninety-day accumulation areas are subject to the requirements outlined in 40 CFR 262.34(a). Generators with TSD facilities can also have multiple satellite and 90-day accumulation areas. Such areas can be enlarged and relocated without regard to the restrictions on modifications to permitted facilities or facilities under interim status. In addition, accumulation areas are permissible for use with onsite recycling operations. It is important to note that the regulatory conditions associated with satellite and 90-day accumulation areas must be met in order for a generator to qualify for the reduced requirements for such areas.

At the end of the 90-day accumulation period, the containers in 90-day accumulation areas must be moved to an onsite TSD or shipped under manifest to an offsite TSD. There is a provision for a single, 30-day extension of the 90-day deadline, provided that the extension is necessary due to unforeseen, temporary, or uncontrollable circumstances. This extension is discretionary on the part of EPA or the authorized state regulatory agency, and written application should be made for the extension prior to the expiration of the 90-day accumulation period (40 CFR 262.34(b)).

It must be emphasized that the 90-day accumulation option is not as easy as it may seem from a first reading of 40 CFR 262.34. This is because 262.34 references requirements found elsewhere in the regulations, primarily in 40 CFR Part 265. There are definite requirements for generators who choose to use this option, and it is implicit that generators who fail to comply with these requirements are not (or will not be) allowed to take advantage of the accumulation area option. The requirements for generators who accumulate are discussed in Chapter 9.

GENERATORS WHO STORE HAZARDOUS WASTE

Manufacturing facilities unable to take advantage of the 262.34 accumulation provisions for all of their hazardous waste storage activities are subject to the full brunt of the hazardous waste regulations. Many manufacturing firms filed Part A of the RCRA permit application, and qualified for interim status. As owners or operators of hazardous waste storage facilities, such manufacturing firms must comply with all applicable provisions of 40 CFR Part 265, "Interim Status Standards for Owners and Operators of Hazardous Waste Treatment, Storage, and

Disposal Facilities,'' and at the same time pursue a full RCRA permit under the provisions of 40 CFR 270 and 124.

As briefly discussed in Chapter 1, manufacturing firms that own or operate a hazardous waste storage facility had to meet certain requirements to qualify for interim status. Remember that interim status is a legal construct whereby a TSD facility is treated as if it had a permit, until a final administrative disposition is made on its permit application. The three requirements necessary for a manufacturing firm to qualify for interim status, found at 40 CFR 270.70, are:

1. The hazardous waste TSD area(s) must have been in existence on or before November 19, 1980, or the effective date of any applicable statutory or regulatory amendments.
2. The owner or operator must have complied with the notification requirements of Section 3010 of RCRA on or before August 18, 1980, or the specified deadline of any applicable amendments.
3. The owner or operator must have filed Part A of the RCRA permit application (EPA forms 3510-1 and 3510-3) on or before November 19, 1980, or the specified deadline of any applicable amendments.

The interim status period ends, by statutory provision, on November 8, 1992 for storage facilities, unless a complete Part B permit application was submitted by November 8, 1988. Interim status can also be terminated by failure to submit a complete Part B application within six months of call-in by EPA or an authorized state. Interim status also terminates when a Part B permit is issued or denied (40 CFR 270.73).

The possession of interim status is very important, for it prevents the manufacturing firm from having to cease onsite HWM (storage) activities. A manufacturing firm that failed to obtain interim status or that had interim status revoked or terminated and was not yet in possession of a full RCRA permit would have as its only option the accumulation of hazardous wastes onsite in containers or tanks for less than 90 days, pending shipment of the wastes offsite to an interim status or fully permitted TSD facility. In effect, a facility under interim status agrees to comply with the interim status standards (40 CFR Part 265) in exchange for being allowed to continue HWM activities pending submission and final administrative disposition of the Part B permit application.

It is important to note that a storage facility under interim status is not to store hazardous wastes not specified in the Part A application, employ processes not specified in the Part A application, or exceed the design capacities specified in the Part A application (40 CFR 270.71). However, there are provisions under certain circumstances for revising the Part A application to accommodate changes during interim status (40 CFR 270.72).

If interim status is not obtained, or is terminated or revoked, a manufacturing firm must accumulate less than 90 days in containers or tanks while preparing a complete RCRA permit application (both Part A mentioned above and Part B, as specified at 40 CFR 270 Subpart B). Quarterly shipments to offsite TSD facilities must continue until a RCRA permit is obtained by meeting the applicable permit standards (40 CFR Part 264). Facilities granted a full RCRA permit must comply with the 40 CFR Part 264 standards. It is important that the state hazardous waste management agency be contacted regarding the existence of any state permit requirements, imposed in addition to federal RCRA permit requirements. See Appendix B for a list of such agencies nationwide.

Manufacturing facilities that have not yet filed a permit application may still be granted interim status or treated as if they have interim status. EPA has the authority to issue what are called "interim status compliance letters." This document could, under some circumstances, grant interim status to facilities unable to meet the requirements for interim status. An interim status compliance letter would, in effect, allow the TSD facility to be treated by the EPA as if it had interim status for any enforcement purposes. This provision may prove valuable to manufacturing firms that, for some reason or another, failed to accurately determine their regulatory status or that desire to change that status.

EPA has recognized that some generators filed what are termed "protective" or "precautionary" Part A permit applications. In other words, generators applied for a RCRA permit even though they were not required to do so by the regulations. For the most part, such protective filings were made by generators who wanted to have interim status for a storage area in case they had to hold wastes onsite longer than 90 days. Also, some SQGs mistakenly filed in the event they later became large quantity generators. Generators are encouraged to withdraw protective Part A filings unless they intend to obtain, and are capable of obtaining, a full RCRA permit for their TSD facility. Many manufacturing firms with onsite TSD storage facilities under interim status are finding that such facilities are not necessary or economical for efficient onsite storage or offsite treatment or disposal. The regulatory requirements for interim status storage facilities at manufacturing firms are discussed in Chapter 11.

GENERATORS WHO SHIP HAZARDOUS WASTE OFFSITE

Although most hazardous waste-generating manufacturing firms will also fall in this category, it is useful to discuss the transportation of hazardous wastes separately. Shipping hazardous wastes offsite places a generator under the regulatory jurisdiction of both the EPA and the DOT. Transportation of hazardous wastes comes under the purview of the DOT because hazardous wastes are considered a subset of DOT-regulated commodities known as "hazardous materials." DOT regulates the transportation of hazardous materials under the authority of the federal Hazardous Materials Transportation Act (HMTA). The Hazardous Material Transportation Regulations (HMTR) issued to implement the HMTA are found at 49 CFR Parts 171–199.

By making a distinction between onsite and offsite transportation, the regulations regarding the transportation of hazardous wastes are not made to apply to movement of hazardous wastes within the generator's property. This allows for onsite HWM activities to take place without the necessity of a manifest or the other DOT pretransportation requirements.

The definition of onsite found in the regulations (40 CFR 260.10) is unnecessarily complex and lengthy, but basically refers to property owned by the generator that is the same as or contiguous to the property on which the waste was generated. This property may be divided by public or private right(s)-of-way (highways or railroads), and no manifest is necessary for moving the wastes across (as opposed to along) such right(s)-of-way to get the wastes to other portions of the generator's property.

The preamble to the regulations clarified the definition of onsite (and thus the exemption from DOT transportation requirements) when wastes are moved across public or private right(s)-of-way. The entrance and exit of geographically contiguous property—for the movement of hazardous waste onsite without a manifest—must be directly across from each other (45 FR 33069).

The transportation of hazardous wastes offsite—from the site of generation to a TSD facility—must take place in compliance with both EPA and DOT standards. EPA and DOT standards make frequent reference to each other. This joint rulemaking was used to minimize confusion and duplication. However, generators who offer hazardous waste for offsite transportation (known as "shippers" in DOT jargon) must have a working knowledge of both EPA and DOT regulations, and must use transporters ("carriers" in DOT jargon) who likewise know and abide by both EPA and DOT standards.

Hazardous waste transportation is further discussed in Chapter 10.

GENERATORS WHO TREAT HAZARDOUS WASTE BY AN EXEMPTED METHOD

The regulatory definition of treatment, found at 40 CFR 260.10, is very comprehensive and inclusive. Treatment means "any method, technique, or process, including neutralization, designed to change the physical, chemical, or biological character or composition of any hazardous waste so as to neutralize such waste, or so as to recover energy or material resources from the waste, so as to render such waste nonhazardous, or less hazardous; safer to transport, store, or dispose of; or amenable for recovery, amenable for storage, or reduced in volume."

This regulatory definition is significant in that, with few exceptions, persons owning and operating hazardous waste treatment facilities must apply for and obtain a RCRA permit. The broad nature of the definition of treatment and the associated requirements make any exemptions all the more important. Please note that recycling and reclamation operations would require a RCRA treatment permit, unless exempted.

The interim status and permanent status standards found at 40 CFR 265 and 264 and the permit requirements of 40 CFR 270 do not apply to certain hazardous waste treatment operations, processes, and methods. EPA has informally allowed generators to treat hazardous waste onsite in accumulation containers and tanks without a RCRA permit, under certain circumstances. Such treatment must be in conformance with 40 CFR 262.34, which among other things limits the accumulation period and requires compliance with 40 CFR 265 Subpart I ("Use and Management of Containers") and Subpart J ("Tank Systems"), as appropriate.

EPA stated in the preamble to the final small quantity generator regulations (51 FR 10168—March 24, 1986) that treatment could occur in a generator's accumulation tanks and containers without a permit, provided that treatment was performed in accordance with 40 CFR 262.34. EPA has taken the common sense position that treatment should not be unduly discouraged, particularly in view of restrictions on the direct land disposal of untreated hazardous wastes. EPA does, however, give the caveat that generators should not make large investments in onsite treatment facilities based on the assumption that they will be indefinitely exempted from permitting requirements. Also, state regulatory agencies may or may not recognize this informal exemption.

There are formal exemptions as well. These are specifically outlined in the regulations themselves, as opposed to preamble explanations, interpretations, or policy statements.

Of most interest to manufacturing facilities would be the exemptions granted at 40 CFR 265.1(c) to:

1. the owner and operator of a facility managing recyclable materials described at 40 CFR 261.6(a)(2) and (3), except to the extent that such recyclable materials are regulated at 40 CFR Part 266. Recyclable materials so described include:
   (a) hazardous wastes burned for energy recovery (*Note:* applies only when performed in industrial boilers and furnaces meeting certain requirements. See 40 CFR 266 Subpart D.)
   (b) characteristically hazardous used oil burned for energy recovery (*Note:* applies only when performed in industrial boilers and furnaces meeting certain requirements. See 40 CFR 266 Subpart E.)
   (c) recyclable materials from which precious metals are reclaimed (40 CFR 266 Subpart F)
   (d) spent lead-acid batteries that are being reclaimed (40 CFR 266 Subpart G)
   (e) industrial ethyl alcohol that is reclaimed
   (f) used batteries (or used battery cells) returned to a battery manufacturer for regeneration
   (g) characteristically hazardous used oil that is recycled in a manner other than by being burned for energy recovery
   (h) scrap metal
   (i) certain petroleum refining waste
   (j) certain iron and steel industry wastes

(*Note:* Other hazardous wastes that are recycled are subject to the requirements for generators, transporters, and storage facilities. With respect to such wastes, it is the recycling process itself that is exempt from regulation. Transportation and storage activities prior to the recycling process are fully regulated.)

2 . a generator accumulating waste onsite in compliance with 40 CFR 262.34

3. the owner or operator of a totally enclosed treatment facility, as defined in 40 CFR 260.10

4. the owner or operator of an elementary neutralization unit or a wastewater treatment unit as defined in 40 CFR 260.10 (*Note:* These units must meet the regulatory definition of ''tanks,'' with the resulting aqueous discharge subject to Clean Water Act requirements.)

5. the owner or operator of a publicly owned treatment works (POTW, also known as a sewage treatment plant) that treats, stores, or disposes of hazardous wastes

6. persons carrying out treatment or containment activities necessary to immediately respond to releases or threatened releases of hazardous waste

7. a transporter storing manifested shipments of hazardous waste at a transfer station for a period of 10 days or less

8. the addition of absorbent material to waste in a container or the addition of waste to the absorbent material in a container, provided that these actions occur at the time waste is first placed in the containers, and the requirements of 40 CFR 265.17(b), 265.171, and 265.172 are complied with

It is important to note that the owner or operator of an exempted treatment facility can still be a generator of hazardous wastes, due to any spills or releases, or due to the pollution control residuals or sludges resulting from treatment operations. Remember that sludges resulting from the treatment of listed hazardous wastes are listed hazardous wastes, by virtue of the ''derived-from'' rule (40 CFR 261.3(c)(2)). Shipments of hazardous waste from such a facility must comply with the standards for generators found in 40 CFR Part 262. The 90-day accumulation provisions for generators (40 CFR 262.34) may be used by such facilities for the hazardous wastes generated at that facility. Chapter 12 explains the meaning and implications of exclusions 1–8 listed above.

GENERATORS WHO MARKET OR BURN USED OIL FUEL

EPA modified the RCRA regulatory scheme in 1985 with respect to used oil recycled by being burned for energy recovery. Most used industrial oils are managed in this manner. Previously, characteristically hazardous wastes, except sludges, were not subject to regulation when managed by recycling. EPA proposed listing used oil as a hazardous waste and later withdrew its proposal. EPA is now expected to re-propose management standards and make a listing decision for used oil around April of 1991.

Used oil, as defined at 40 CFR 266.40, means any oil that has been refined from crude oil, used, and as a result of such use, is contaminated by physical or chemical impurities. Used oil exhibiting a characteristic of hazardous waste *without* being a listed hazardous waste under the mixture rule or the derived-from rule is regulated under 40 CFR 266 Subpart E, when burned for energy recovery. Such used oil is termed "used oil fuel." Used oil fuel includes any fuel produced from used oil by processing, blending, or other treatment.

Quality specifications have been established for used oil fuels (40 CFR 266.40(e)). If used oil burned for energy recovery cannot meet these specifications, it is termed "off-specification used oil fuel." Marketers and burners of used oil fuel are subject to certain administrative controls found at 40 CFR 266 Subpart E, and off-specification used oil fuel can only be burned in industrial boilers or furnaces, as specified at 40 CFR 266.41(b).

Generators of used oil that is burned for energy recovery are not subject to 40 CFR 266 Subpart E except to the extent that the used oil is found to be off-specification, and the generator markets such oil directly to burners, or burns the off-specification used oil fuel him- or herself.

Generators shipping their used oils to marketers should be aware of the requirements imposed on marketers and burners of used oil fuel, in order to be able to evaluate the compliance status of these parties.

The requirements for generators that market or burn used oil fuel are discussed in Chapter 13.

Hazardous waste, and mixtures of used oil and hazardous waste (except from conditionally exempt SQGs) when burned for energy recovery, are regulated as hazardous waste fuel at 40 CFR 266 Subpart D.

UNDERGROUND PRODUCT STORAGE TANK OWNER/OPERATORS

The Hazardous and Solid Waste Amendments of 1984 (HSWA), which reauthorized RCRA, extended the scope of the RCRA regulatory program to include certain previously unregulated underground storage tanks. Subtitle I of RCRA, as amended, gave EPA the statutory authority to regulate underground storage tanks (USTs) *not* containing hazardous wastes. This action created an entirely new category of potentially regulated parties under RCRA, that of owners and operators of USTs containing petroleum (including used oils) and hazardous substances (other than hazardous wastes). Regulations for these underground storage tanks are codified at 40 CFR Part 280. Please note that underground and other storage tanks containing hazardous wastes had been regulated since the inception of the HWM program in 1980.

The RCRA Subtitle I program focuses on underground storage tanks, as defined, containing regulated substances, as defined. An underground storage tank is defined as any one tank or combination of tanks (including underground pipes connected thereto) used to contain an accumulation of regulated substances. To be considered underground, the tank must have at least 10% of its volume below ground, including the volume of pipes attached to the tank. Excluded from the definition are the following:

1. farm and residential tanks having a capacity of no more than 1100 gallons used for storing motor fuel for noncommercial purposes
2. tanks storing heating oil for consumptive use on the premises where stored
3. septic tanks
4. pipeline facilities
5. surface impoundments, pits, ponds, or lagoons
6. storm water or wastewater collection systems
7. flowthrough process tanks
8. liquid traps or associated gathering lines directly related to oil or gas production and gathering operations
9. storage tanks in an underground area (i.e., basement) if the tank is situated on or above the floor surface
10. any pipes connected to any tank described in items 1–9, above

The regulatory definition of regulated substances includes CERCLA hazardous substances, *except* hazardous waste. Petroleum, including crude oil or any fraction thereof which is liquid at standard conditions of temperature and pressure, is also considered a regulated substance. Regulated substances include such petroleum-based substances as motor and jet fuels, distillate fuel oils, residual fuel oils, lubricants, petroleum solvents, and used oils.

The CERCLA list of hazardous substances is found at 40 CFR Part 302. Hazardous wastes are not considered regulated substances for purposes of the 40 CFR 280 UST program, even though for CERCLA purposes hazardous wastes are a subset of hazardous substances.

Notification Requirements

Owners and operators of nonexempt underground storage tanks are to notify their respective state regulatory agency. Notification is required for USTs that have been used to store regulated substances since January 1, 1974 and that are in the ground as of May 8, 1986. Notification is also required for USTs containing regulated substances that are brought into use after May 8, 1986. Notification is not required for tanks taken out of operation prior to January 1, 1974, or for tanks removed from the ground prior to May 8, 1986. This notification, on EPA form 7530-1, was to have been submitted by May 8, 1986. A copy of EPA form 7530-1 (September 1988 revision) is included as Appendix F of this book. Appendix II of 40 CFR Part 280 provides a list of state agencies (with mailing addresses and telephone numbers) designated to receive the notifications. Non-notifiers are subject to potential civil penalties of $10,000 per tank.

A variety of information items is necessary to complete the notification form. These items are:

1. current status of tank
2. estimated age
3. estimated total capacity
4. material of construction
5. internal protection, if any
6. external protection, if any
7. piping (material of construction)
8. substance currently or last stored
9. additional information for out-of-service tanks
   (a) estimated date last used
   (b) estimated quantity of substance remaining
   (c) whether the tank was filled with an inert material

Certification of compliance for new tanks (those installed after December 22, 1988) is a mandatory element of the UST notification from. Certifications are necessary with respect to installation, release detection and corrosion protection systems, and financial responsibility compliance. Chapter 14 discusses requirements and options for owners/operators of underground product storage tanks. Final EPA regulations promulgated September 23, 1988 (53 FR 37194–37212) established standards for the design, construction, and installation of UST systems. General operating requirements are specified, as are requirements for release detection systems and release reporting. Mandatory actions are to be taken to investigate and respond to releases. Corrective actions are required to ensure adequate cleanup. Requirements for out-of-service tanks are specified, including closure procedures. Financial responsibility is to be demonstrated for the costs of corrective action, and for compensating third parties for damages resulting from releases.

This concludes Section I. Armed with an understanding of the regulatory definitions of solid and hazardous wastes, the reader should now be able to make the following determinations:

1. What are all the wastes at the facility?
2. Which of the wastes are hazardous, using the regulatory definitions?
3. What HWM activities are occurring at the manufacturing facility, and which regulatory categories best fit such activities?

The next section explains the regulatory standards that apply to each of the regulatory categories. Sections III and IV will then provide guidance in setting up a corporate compliance program.

SECTION II

Regulatory Standards and Responsibilities

This section explains the regulatory standards for the categories of hazardous waste management activities most common at manufacturing firms. In many cases, a manufacturer will fall into more than one category, especially when hazardous wastes are shipped to an offsite TSD facility. Knowledge of these regulatory standards is the necessary base on which to build a comprehensive environmental management program, using the guidance given in Section III.

CHAPTER 6

Nonhazardous Industrial Solid Wastes

Although this book is primarily focused on the proper management of wastes defined as "hazardous," some discussion of standards and good practices for the management of nonhazardous waste is necessary. This chapter is included because most manufacturing firms generate both hazardous and nonhazardous wastes, as well as wastes not meeting the regulatory definition of solid waste.

Federal and state statutes and regulations govern the disposition of nonhazardous solid wastes. However, such wastes are controlled primarily at the state level. Many states have special provisions for nonhazardous solid wastes from industrial sources, often termed "special wastes."

At the federal level, RCRA prohibits "open dumping" of solid waste at Section 4005. The term "open dump" is defined as "any facility or site where solid waste is disposed of which is not a sanitary landfill which meets the criteria promulgated under Section 4004 and which is not a facility for the disposal of hazardous waste." The federal guidelines promulgated under RCRA Section 4004 are the "Criteria for Classification of Solid Waste Disposal Facilities and Practices" (40 CFR Part 257). These federal criteria are important because they have been used as guidelines by the various states in amending statutes prohibiting open dumps and in revising state sanitary landfill regulations. EPA has proposed, at 53 FR 33405–33422, to revise 40 CFR Part 257 and to establish a new part (40 CFR Part 258), establishing criteria specific to municipal solid waste landfills.

As mentioned earlier, nonhazardous industrial wastes are subject to regulation primarily at the state level, and should be managed properly. Proper management, for the most part, means disposing of such wastes at a sanitary landfill holding a valid state permit. At a minimum, such landfills must be well operated and should meet all of the requirements of the state's sanitary landfill regulations. A generator would be well advised not to send any nonhazardous industrial wastes to landfills under litigation or to landfills being "phased out" or closed due to poor geology or evidence of groundwater contamination. Such information can be obtained from state solid waste management agencies (Appendix B). Ask to talk to the field inspector for the site in question.

Some sanitary landfills have been or will be placed on the National Priority List (NPL) for EPA enforcement action under CERCLA authority. Placing a sanitary landfill on the Comprehensive Environmental Response, Compensation and Liability Information System (CERCLIS) list is often a prelude to evaluation with respect to the NPL. Use of such landfills for industrial waste disposal should be avoided at (almost) all cost.

It must be noted that some state regulatory agencies require a generator to obtain prior written approval from them for land disposal of hazardous and nonhazardous industrial wastes within state boundaries. These approvals, often called "permits" or "approval letters," are typically either for a specific waste stream (type) for a set period of time or for each shipment of wastes. The approval is usually good only at a specific landfill.

A written request is typically required for such an approval, accompanied by relevant information on the waste and the process generating it. A detailed chemical and physical analysis is usually necessary. The purpose of such evaluations is to prevent the disposal of hazardous wastes at sanitary landfills as well as the land disposal of incompatible wastes and wastes unsuitable for land disposal. Such requirements are intended to prevent or delay adverse environmental impacts from the landfill site. This in turn limits the liability of the waste generator, because wastes unsuitable for land disposal are not placed in a landfill, but are required to be managed by other technologies, typically some form of waste treatment.

Some aqueous wastes can be treated by an onsite NPDES-permitted wastewater treatment plant, or can be discharged into a municipal sewer system for treatment by the publicly owned treatment works (POTW). The latter option should, of course, be exercised only in accordance with general pretreatment requirements (40 CFR Part 401), applicable categorical pretreatment standards (40 CFR Parts 405–471), and any applicable local sewer ordinance.

One final point regarding land disposal needs to be made. Many nonhazardous industrial waste streams generated in large volumes have been managed on the property of the generator without the benefit of any permits. Onsite waste management practices such as using wastes as fill material, accumulating wastes in piles or heaps, and onsite dumping have been used for wastes such as fly ash, bottom ash, foundry sand, and various sludges. Liquid and semisolid wastes have been stored or dewatered in industrial surface impoundments (pits, ponds, and lagoons), or have been spread on land. Such onsite activities should be reevaluated in light of prohibitions on open dumping. The use of nonpermitted offsite property for the disposal of such wastes should also be eliminated.

It is important to note that all such activities may need a permit from the state solid waste management agency and/or the state water pollution control agency.

These requirements are relatively recent in many states.

It is advisable to work with state officials to close such waste management operations, or to upgrade them to meet permit standards. The use of unpermitted property for waste disposal can have serious consequences, so the proper goal should be to monitor and properly close past sites, to monitor, close, or upgrade existing sites, and to use only permitted sites in the future.

The generator of industrial nonhazardous wastes can be held responsible for onsite and offsite environmental impacts under CERCLA, as amended, as well as under state environmental statutes.

CHAPTER 7

Conditionally Exempt Small Quantity Generators

As mentioned in Chapter 5, a conditionally exempt small quantity generator qualifies for a set of special (reduced) requirements, provided the small quantities of hazardous waste are managed properly and not allowed to accumulate on-site above certain levels. The very limited regulatory requirements for hazardous waste generated by conditionally exempt small quantity generators are found at 40 CFR 261.5.

A generator who, in a calendar month, generates less than 100 kg (220 pounds or roughly 25–30 gallons) of hazardous wastes and who never accumulates more than 1000 kg (2200 pounds) onsite is a conditionally exempt small quantity generator. Different exclusion levels apply to generators of waste materials with a 261.33(e) listing or those identified as acutely hazardous wastes by the hazard code "H." These types of wastes and exclusion levels are explained in Chapter 5.

REDUCED REQUIREMENTS

The hazardous wastes of a conditionally exempt SQG are not subject to full regulation, including the notification requirement, as long as the conditionally exempt SQG complies with a set of reduced requirements (found at 40 CFR 261.5(g)):

1. The SQG must determine the hazardous/nonhazardous nature of all solid wastes generated, as specified at 40 CFR 262.11 and as discussed in Chapter 4.
2. The SQG must not accumulate quantities of hazardous or acutely hazardous waste greater than their respective exclusion levels. The exclusion level, or

maximum accumulation, for hazardous waste (other than acutely hazardous) is 1000 kg (2200 pounds) onsite at any one time. There is no maximum time period for accumulation, provided the exclusion level is not exceeded.

3. The SQG must treat or dispose of hazardous wastes in an onsite facility, or ensure delivery to an offsite facility which does any of the following:
   a) beneficially uses or reuses or legitimately recycles or reclaims the waste
   b) has a permit or interim status as a RCRA TSD facility
   c) is permitted, licensed, or registered by a state to manage municipal or industrial solid waste (e.g., sanitary landfills)

MIXING HAZARDOUS AND NONHAZARDOUS WASTES

Paragraph 261.5(h) holds that a mixture of nonhazardous wastes and hazardous wastes subject to the reduced requirements is subject to those reduced requirements only, unless the mixture meets any of the characteristics of a hazardous waste. Remember that one of the requirements for a small quantity generator is to determine if any of the solid wastes (including mixtures of hazardous and nonhazardous waste) are hazardous. The reduced requirements for mixtures (that do not possess any of the characteristics) apply even if the mixture exceeds the applicable quantity limitations. However, paragraph 261.5(i) holds that if a hazardous waste exceeding the exclusion level is mixed with a solid waste, then the entire mixture is subject to full regulation.

The regulatory provision allowing disposal of hazardous waste from conditionally exempt SQGs at sanitary landfills, as well as that allowing the mixing of hazardous and nonhazardous waste, may appear viable and useful from a theoretical perspective. From a practical standpoint, however, solid waste collection firms and sanitary landfill operators typically do not knowingly allow even small quantities of hazardous wastes to be mixed with plant garbage and trash. Many such firms require all commercial and industrial customers to certify that hazardous wastes are not being mixed with the nonhazardous solid wastes being offered for transportation and disposal.

Mixing hazardous and nonhazardous waste ("waste maximization") is usually not a good idea, and typically results in increased difficulty in securing proper and legal disposal, as well as higher per-unit disposal prices than would be the case if the hazardous wastes were kept separate from the nonhazardous waste. Further, many commercial industrial and/or hazardous waste management firms require all customers to notify and obtain an EPA Identification Number, as well as to utilize a Uniform Hazardous Waste Manifest for shipping purposes. These requirements, although not mandated by the regulations, are imposed by commercial TSD facilities as a condition of utilizing their services. Their internal procedures and requirements typically do not allow case-by-case evaluation of and special exception for the hazardous wastes from firms claiming conditionally exempt SQG status.

100–1000 KG/MONTH GENERATORS

Should a conditionally exempt SQG accumulate wastes in quantities exceeding the applicable exclusion levels for nonacutely hazardous wastes (1000 kg), or generate more than 100 kg of hazardous waste in a calendar month, then the more stringent regulatory requirements for the 100–1000 kg/month generator category apply to all of the accumulated wastes. This 100-1000 kg/month generator category allows the accumulation of up to 6000 kg of hazardous waste, with a maximum accumulation period of 180 days (270 days if the TSD facility is over 200 miles away). The accumulation start date for the 180-day onsite accumulation period is the first day the accumulated wastes exceed 1000 kg.

Should a conditionally exempt SQG exceed the applicable exclusion level for acutely hazardous waste (1 kg/100 kg), then the requirements for generators of greater than 1000 kg/month apply. The requirements for generators who accumulate less than 90 days are discussed in Chapter 9. The requirements for generators who store onsite are discussed in Chapter 11. The accumulation start date for acutely hazardous wastes is the date the applicable exclusion level is exceeded.

If total quantities of onsite nonacutely hazardous waste accumulation are foreseen to exceed 1000 kg (i.e., to meet disposal facility minimum quantity requirements), then it might be beneficial to comply with the requirements of the 100–1000 kg/month generator category. The accumulation start date for nonacutely hazardous waste is the day onsite accumulation exceeds 1000 kg. Regulatory agency inspectors would likely assume manufacturing firms to fall into this category or that of 90-day accumulator, unless thorough records regarding generation rates were available.

Many of the requirements applicable to 100–1000 kg/month generators represent common sense and good management practices, and should be followed regardless of their legal applicability. These requirements are discussed in Chapter 8.

CONCLUSION

The reduced requirements for conditionally exempt SQGs found at 40 CFR 261.5 do indeed give qualifying manufacturing firms greater leeway in managing their small quantities of hazardous waste. As mentioned earlier, such small quantities can be accumulated for an indefinite period, as long as the exclusion levels are not exceeded. Of course, such small quantities should be managed by reuse or recycling, if possible. Remember that small quantity generator status does not allow the manufacturing firm to improperly manage its hazardous wastes.

At the present time, many of the wastes of conditionally exempt small quantity generators are managed by sanitary landfill and thus are disposed of in conjunction with municipal solid wastes (i.e., trash, garbage, and refuse). Hazardous waste regulations do not require a conditionally exempt small quantity generator to tell the landfill operator that hazardous wastes are being shipped to the landfill, nor do they require that the generator alert the landfill operator of the hazards posed by the waste, or of any precautions to be taken during disposal. Regardless of the absence of legal language to that effect, good business practices, common sense, and self-interest all point to the propriety of such a notification to the landfill operator. Even small quantities of hazardous wastes can cause serious problems at the landfill when the wastes are compacted by a hot bulldozer and mixed with other wastes. The potentials for fire or explosion are obvious.

In some states, the federal standards for conditionally exempt small quantity generators have been preempted by more stringent state requirements. Exclusion levels in the state regulations may be smaller, additional waste materials may be considered hazardous, and/or additional requirements may be imposed. Prior written approval from the state regulatory agency may be necessary to take even small quantities of hazardous waste to a municipal landfill. Again, it is necessary to check with your state solid waste official.

CHAPTER 8

100–1000 Kilogram/Month Generators

Generators of between 100 and 1000 kg of hazardous waste in a calendar month (SQGs) have a set of limited regulatory requirements to comply with. These requirements are more stringent than those for conditionally exempt SQGs, but offer more flexibility than available to "large quantity" (>1000 kg/month) generators. Many of the substantive requirements are the same for both 100–1000 kg/month SQGs and large quantity generators. However, much more flexibility is provided in terms of the time periods allowed for accumulating wastes and making arrangements for legal and proper offsite treatment and disposal. Many commercial TSD facilities have minimum pickup quantity or minimum invoice requirements. The longer accumulation period for 100–1000 kg/month SQGs allows the accumulation of sufficient quantities of waste to qualify for more economical unit transportation and treatment/disposal rates.

Manufacturing firms classified as 100–1000 kg/month SQGs are allowed to accumulate up to 6000 kg of hazardous waste onsite in containers or tanks for up to 180 days without a RCRA permit or interim status, under certain conditions. The accumulation period can be extended to 270 days, provided that the TSD facility is over 200 miles from the generator and that the 6000-kg accumulation limit is not exceeded.

Generators of 100–1000 kg/month, in order to maintain their regulatory status, are to comply with the following requirements:

1. Determine if any solid wastes generated are hazardous wastes (40 CFR 262.11). This assumes that a waste inventory was conducted.
2. Determine whether any of the hazardous wastes generated are subject to the 40 CFR Part 268 Land Disposal Restrictions (40 CFR 268.7).

3. Notify EPA and obtain an EPA identification number prior to offsite shipment (40 CFR 262.12). This assumes that the appropriate regulatory category was selected.
4. Keep records of any test results, waste analyses, or other determinations made in accordance with 40 CFR 262.11 for at least three years from the date the waste was last shipped offsite (40 CFR 262.40(c)). This assumes a formal, written hazardous waste determination for each waste material identified.
5. The quantity of hazardous waste accumulated onsite cannot exceed 6000 kg (13,200 pounds). The period of accumulation is up to 180 days, unless the designated TSD facility is over 200 miles away. In such instances, the maximum accumulation period is 270 days (40 CFR 262.34(d) and (e)).
6. Each container or tank holding hazardous waste must be clearly marked with the words "Hazardous Waste" and the accumulation start date (40 CFR 262.34(d)(4)).
7. Accumulation in containers must be in compliance with 40 CFR 265 Subpart I, "Use and Management of Containers," except §265.176. Accumulation in tanks must be in compliance with §265.201 of 40 CFR 265 Subpart J, "Tank Systems" (40 CFR 262.34(d)(2) and (3)). (A discussion of container accumulation requirements can be found in Chapters 9 and 23 of this book.)
8. Comply with the preparedness and prevention requirements of 40 CFR 265 Subpart C. This includes obtaining appropriate emergency equipment, maintaining adequate aisle space, and making arrangements with local emergency authorities. (A discussion of 40 CFR 265 Subpart C can be found in Chapter 9 of this book.)
9. Utilize transporters and TSD facilities with EPA identification numbers (40 CFR 262.12(c)).
10. Follow DOT requirements regarding packaging, labeling, marking, and placarding (40 CFR 262.30–33).
11. Utilize the uniform hazardous waste manifest for offsite shipments (40 CFR 262.20). Equivalent state-issued forms may be required (40 CFR 262.21). Follow requirements for the use of the manifest (40 CFR 262.23). There are limited Exception Reporting requirements if a signed return copy is not received within 60 days of shipment. There is also an exemption from manifest requirements under certain circumstances involving contract recycling (40 CFR 262.20(e)). Keep return copies of the manifest (i.e., with the signature of the designated TSD facility owner/operator) for three years from the date the waste was accepted by the initial transporter (262.40(a)). (Manifest requirements are discussed in Chapters 1 and 10.)
12. Comply with requirements for emergency planning and notification of releases:
    a) Designate an emergency coordinator.
    b) Post near the telephone the emergency phone numbers and the location of emergency equipment.
    c) Inform employees of waste handling and emergency procedures.
    d) Respond to emergencies, and notify the National Response Center at (800)424-8802 of any emergencies that could threaten human health outside the facility, or when a spill has reached surface waters. (Note: State and local spill reporting requirements may also apply.)

The emergency planning and notification requirements, found at 40 CFR 262.34(d)(5), are reprinted below in their entirety.

The generator complies with the following requirements:

(i) At all times there must be at least one employee either on the premises or on call (i.e., available to respond to an emergency by reaching the facility within a short period of time) with the responsibility for coordinating all emergency response measures specified in paragraph (d)(5)(iv) of this section. This employee is the emergency coordinator.

(ii) The generator must post the following information next to the telephone:

(A) The name and telephone number of the emergency coordinator;

(B) Location of fire extinguishers and spill control material, and, if present, fire alarm; and

(C) The telephone number of the fire department, unless the facility has a direct alarm.

(iii) The generator must ensure that all employees are thoroughly familiar with proper waste handling and emergency procedures, relevant to their responsibilities during normal facility operations and emergencies;

(iv) The emergency coordinator or his designee must respond to any emergencies that arise. The applicable responses are as follows:

(A) In the event of a fire, call the fire department or attempt to extinguish it using a fire extinguisher;

(B) In the event of a spill, contain the flow of hazardous waste to the extent possible, and as soon as is practicable, clean up the hazardous waste and any contaminated materials or soil;

(C) In the event of a fire, explosion, or other release which could threaten human health outside the facility or when the generator has knowledge that a spill has reached surface water, the generator must immediately notify the National Response Center (using their 24-hour toll free number 800/424-8802). The report must include the following information:

(1) The name, address, and U.S. EPA Identification Number of the generator;

(2) Date, time and type of incident (e.g., spill or fire);

(3) Quantity and type of hazardous waste involved in the incident:

(4) Extent of injuries, if any; and

(5) Estimated quantity and disposition of recovered materials, if any.

Generators of 100–1000 kg/month of hazardous waste that exceed the maximum accumulation limit of 6000 kg or that accumulate hazardous waste for more than 180 days (270 days if the TSD facility is over 200 miles away) become operators of a hazardous waste TSD facility, fully subject to the requirements of 40 CFR Parts 264 and 265, as well as the permit requirements of 40 CFR Part 270.

A single 30-day extension is available if hazardous wastes must remain onsite longer than the applicable accumulation period due to unforeseen, temporary, and uncontrollable circumstances. This extension is discretionary on the part of the EPA or authorized state. A petition for such an extension should be submitted prior to the expiration of the accumulation period.

CHAPTER 9

Generators Who Accumulate Hazardous Wastes Onsite Less Than 90 Days

The regulatory category of "generators who accumulate" describes at least a portion of the hazardous waste management activities at many manufacturing firms. This category encompasses container or tank storage (termed "accumulation" if less than 90 days) before offsite or onsite waste management at TSD facilities.

The regulatory provision allowing storage by generators for less than 90 days without a permit (40 CFR 262.34, "Accumulation Time") was established to prevent undue interference with the manufacturing process itself. Such interference was avoided by not requiring RCRA permits for areas where hazardous wastes are accumulated temporarily. The regulations found at 262.34 specify a set of standards applicable to such temporary accumulation. As mentioned in Chapter 5, the accumulation option is not as easy as merely moving the waste offsite or to an onsite TSD facility within 90 days. Section 262.34 makes extensive references to regulatory standards found in 40 CFR Part 265. Although the regulatory standards for generators who accumulate are not as comprehensive as those for generators who store at an onsite TSD facility, the 262.34 standards still necessitate a compliance program. Generators are allowed to have any number of 90-day accumulation areas onsite, provided each is managed in accordance with the regulatory standards of 262.34.

Generators who accumulate less than 90 days are well advised to have arrangements with or approvals from more than one commercial (offsite) TSD facility for each hazardous waste generated. Such arrangements will minimize the possibility of exceeding the 90-day accumulation period and becoming subject to enforcement action. A single extension of up to 30 days to the accumulation period is possible when necessary due to unforeseen, temporary, and uncontrollable circumstances (262.34(b)). Such circumstances could include inclement weather, work stoppages at either the generator's facility or the TSD facility, delays in obtaining laboratory analyses or commercial TSD approvals, and rejected (off-specification) shipments returned by commercial TSD facilities. Note that the initial arrangements with commercial TSD facilities often take 30–60 days, due to the delays associated with independent laboratory analyses and internal review and contracting requirements.

Extensions to the 90-day accumulation period are discretionary on the part of the regulatory agency, and are evaluated on a case-by-case basis. Requests for such extensions should be in writing, and should be submitted prior to the expiration of the 90-day accumulation period.

GENERAL OBLIGATIONS OF GENERATORS

The regulatory requirements for manufacturing facilities as generators of hazardous wastes are quite simple. However, matters become much more involved when such manufacturing facilities accumulate or store hazardous wastes onsite, or offer hazardous wastes for offsite transportation.

Manufacturing facilities, as generators of hazardous waste, have the following general obligations:

1. Determine if any solid wastes generated are hazardous wastes, according to the regulatory definitions (40 CFR 262.11). This assumes that a waste inventory was conducted (Chapter 3).
2. Determine whether any of the hazardous wastes generated are subject to the 40 CFR Part 268 Land Disposal Restrictions (40 CFR 268.7).
3. Notify the EPA of hazardous waste activity on EPA Form 8700-12, and obtain an EPA identification number before treating, storing, disposing of, transporting, or offering for transportation any hazardous waste. This assumes that the appropriate regulatory category was selected (Chapter 5). Generators are also not to offer their hazardous waste to transporters or TSD facilities not having a valid EPA identification number (40 CFR 262.12).
4. Keep records of any test results, waste analyses, or other determinations made in accordance with 40 CFR 262.11 for at least three years from the date the waste was last sent to onsite or offsite TSD (40 CFR 262.40(c)). This assumes a formal, written hazardous waste determination for each waste material identified (Chapter 4).

Additional requirements come into play when hazardous wastes are offered for transportation offsite, including use of the Uniform Hazardous Waste Manifest. Other additional requirements include submission of biennial reports every even-numbered year covering the types and amounts of hazardous wastes shipped offsite during the previous calendar year. Biennial reports must also include waste minimization efforts and results. There are recordkeeping requirements as well, with respect to manifests, biennial reports, and any exception reports. Such requirements are discussed in more detail in Chapter 10.

A generator is allowed, at 40 CFR 262.34(a), to accumulate hazardous waste onsite for 90 days or less without a permit or interim status (obtained by applying for a permit), provided that:

1. The waste is placed in containers and the generator complies with Subpart I of 40 CFR Part 265, or the waste is placed in tanks and the generator complies with Subpart J of 40 CFR Part 265 (except §265.197(c) and §265.200).
2. The date on which each period of accumulation begins is clearly marked and visible for inspection on each container.
3. While being accumulated onsite, each container and tank is labeled or marked clearly with the words "Hazardous Waste."
4. The generator complies with the requirements for owners or operators of TSD facilities, as specified at Subparts C and D of 40 CFR Part 265 and at §265.16.
5. Upon cessation of container accumulation activities in a particular area, all hazardous wastes are removed and properly disposed of. Any equipment, structures, or soils contaminated by hazardous waste releases must also be removed and properly disposed of, or decontaminated (§265.114).

   Such activities must be sufficient to ensure that there will be no postclosure escape of hazardous waste or hazardous constituents to the groundwater, surface waters, or the atmosphere.

   Control of any postclosure releases is required to the extent necessary to protect human health and the environment (§265.111).

A generator who accumulates hazardous wastes longer than 90 days is considered an operator of a storage facility, fully subject to the requirements of 40 CFR Parts 264 and 265, as well as the permit requirements of 40 CFR Part 270. However, 262.34(b) provides that the generator may petition to be granted an extension to the 90-day period (see discussion above).

The following are the regulatory references found in §262.34 along with an explanation of the requirements found at each section. Although the explanation closely follows the regulatory language, the reader is advised to go to the regulations themselves should there be any question as to meaning or relationships to other regulatory requirements. Please remember that authorized state programs can have more stringent requirements. The requirements of the state hazardous waste management regulations would take precedence over the federal regulations discussed here.

While reading the remainder of the chapter, it should be kept in mind that the Part 265 regulations extensively referenced in §262.34 are oriented primarily toward commercial (offsite) HWM facilities. When the term "facility" is used in the regulations, it is used to refer to HWM facilities. While this would include all or nearly all of the buildings, equipment, and operations of a commercial hazardous waste management facility, the term "facility" has much more limited application with respect to the operations of a manufacturing firm. The regulatory term "facility" would have meaning for the purposes of regulatory compliance only for areas where hazardous wastes are treated, stored, or disposed of onsite. Such areas are often termed "HWM units." For generators who accumulate, the term "facility" would apply only to areas where hazardous wastes are accumulated. For ease in interpreting the regulations, generators could substitute the term "HWM unit" where "facility" is used in its larger context. The regulatory definition of "facility," found at 40 CFR 260.10, is as follows:

> . . . all contiguous land, and structures, other appurtenances, and improvements on the land, used for treating, storing, or disposing of hazardous waste. A facility may consist of several treatment, storage, or disposal operational units (e.g., one or more landfills, surface impoundments, or combinations of them).

USE AND MANAGEMENT OF CONTAINERS (40 CFR 265 Subpart I)

Condition of Containers (§265.171)

Containers used for holding hazardous waste must be in good condition. If the container becomes damaged or deteriorated or begins to leak, the wastes should be transferred to a container that is in good condition.

There are several reasons for using containers that are in good condition for accumulating hazardous waste. First, the container may be onsite for up to three months (90 days) and it is costly and time-consuming to repackage or overpack deteriorated containers prior to shipment. Second, the containers must be in such a condition as to withstand the stresses involved in transportation. The offsite treatment/disposal facility may be hundreds of miles distant. Each container will be bounced around in the back of a semitrailer, even if proper loading, blocking, and bracing procedures are followed. Incidents resulting from container failure during highway transportation have the potential for widespread and costly contamination, not to mention the attendant publicity and regulatory agency scrutiny. Third, transporters of hazardous waste are typically unwilling to accept marginal containers for shipment, for reasons mentioned above. Holding the transporter while such containers are repackaged may result in demurrage charges, while having the driver leave such containers onsite may result in violations of the 90-day accumulation period.

Typically, wastes are accumulated in the containers that are going to be used to ship the wastes offsite. This minimizes the time, labor, and spills or leaks that accompany the transfer of wastes from one container to another. Drums that are damaged, become deteriorated, or begin leaking are to have the contents transferred to another appropriate DOT-specification container, or the entire damaged or deteriorated drum is to be overpacked into a salvage drum (Figure 9.1).

The Part 262 requirements for generators specify that hazardous waste must

Figure 9.1(a) and 9.1(b). Salvage drums.

be packaged (containerized) according to DOT specifications before offsite trans-portation (40 CFR 262.30). Usually, this requirement is met by placing the wastes in 55-gallon drums that meet DOT specifications for the type of material to be transported. For the most part, this means that the standard DOT-specification 17H open-head drum (Figure 9.2) is used for solids and sludges, and the stan-dard DOT-specification 17E closed-head (bung-opening) drum (Figure 9.3) is used for liquids.

Figure 9.2. Standard DOT-specification 17H open-head drum.

Figure 9.3. Standard DOT-specification 17E closed-head (bung-opening) drum.

Compatibility of Waste with Container (§265.172)

Containers used for holding hazardous wastes must not be deteriorated by the waste. The container or liner must be compatible with the wastes to be stored.

Certain waste materials, such as acids and caustics, require plastic drums, steel drums with liners, and/or drums made of stainless steel. If there is a question as to the proper type of container, refer to the DOT Hazardous Materials Table (49 CFR 172.101) for reference to the packaging requirements for specific hazardous materials. Other sources of information include barrel vendors, HWM vendors, or regulatory agencies such as the DOT and the state hazardous waste management agency.

When containers are reused for waste shipments (as allowed at 49 CFR 173.28), the wastes should be compatible with residues (if any) remaining in the container. Chapters 10 and 23 discusses the regulatory requirements for reuse of containers.

Management of Containers (§265.173)

Containers holding hazardous wastes must always be closed during storage. The only time containers can be opened during storage is to add or remove waste.

Containers should be kept closed for at least two reasons. First, open containers of waste materials with volatile constituents will allow such constituents to escape, potentially resulting in additional employee chemical exposures. If the contents are flammable, the vapors may pose a fire hazard. Second, open containers are more subject to releases during movement within the facility, or if the container is bumped during storage.

The requirement for closed containers is typically interpreted to mean that the bungs in a tight-head drum must be in place and tight. For open-head drums, typically the head (and head cover gasket) must be present and the ring-and-bolt closure assembly holding the cover must be secured tightly. These specific examples for closed containers meet the DOT requirements that containers be sealed prior to transportation.

For partially full containers, particularly those at satellite accumulation areas, the requirement for containers to be closed can be interpreted more liberally. Remember, however, the reasons for closed containers. For drums in the process of being filled, it is reasonable to allow some flexibility with respect to container closure. Such flexibility is necessary to make it as convenient as possible for employees filling such containers to ensure that the container is closed after each addition of waste to the container.

Funnels are often used to minimize losses when filling containers with liquid wastes, such as spent solvents or paint-related materials. Open funnels remaining in the bung opening are obviously a violation of the requirement for closed containers. However, several safety equipment suppliers have available funnels that thread into the 2-inch bung opening, and that have a hinged lid for the funnel. The Justrite Model 8-207 and the Protectoseal Model 5510 also have flame arresters in the funnel neck, as well as a fusible link that will automatically close the lid in case of fire. Clearly, these types of funnels should meet the regulatory requirement for closed containers, providing the lid is placed on the funnel after each use.

For open-head drums, lever-ring type closure assemblies can be used in-plant to secure the head. These are much more convenient for employees than the ring-and-bolt type closures. Remember that ring-and-bolt type closures are required for open-head drums prior to transportation. The lever-ring closures should be for in-plant use only. Also, temporary drum lids are available or can be readily fabricated to meet the in-plant requirements for closed containers. However, these typically would be used for waste solids that do not have free liquids, are non-volatile, and are chemically compatible with the lid. Filter press solids from wastewater treatment operations would typically fall into this category. Tarps are used routinely on 20- and 30-yd³ roll-off boxes for closure purposes. Roll-off boxes and similar portable devices are considered "containers" under RCRA.

Containers holding hazardous wastes are to be managed to avoid rupturing or damaging the container or otherwise causing the container to leak.

This requirement is in the form of a performance standard, and as such it is deliberately vague in comparison with prescriptive standards such as those mandating closed containers. It can be interpreted to require several things, including:

- not stacking containers higher than their structural strength allows
- not storing containers in a location or manner which may result in the containers being damaged by vehicular traffic
- utilization of drum handling equipment capable of minimizing the potential for damage
- not opening the drum or otherwise handling the drum in such a manner as to prevent it from being closed again
- not allowing the drum to become damaged from expansion of the contents

Compliance may require not stacking containers more than two high, as well as physically locating the containers away from areas with heavy traffic. Many drums have been punctured or damaged by forklift traffic and improper use of forklift equipment. Methods and equipment used to move containers should not damage or deform the container. Typically, drum clamps or "parrot-beak" adapters are used on forklifts when handling containers, or the containers are placed on good quality pallets. If pallets are used, containers should not overhang the edges.

Another consideration is the opening of containers. Drums are occasionally punctured to provide venting for faster transfer of materials from container to container. Samples are occasionally taken from drums with rusted or frozen bungs by driving a spike through the lid to provide access to sampling equipment. Neither practice is recommended. Needless to say, such actions prevent the container from ever again being closed. Repackaging or overpacking would then be necessary for onsite accumulation or offsite transportation.

Additionally, solvents or other materials with high vapor pressure should not be overfilled or stored in black drums in direct sunlight. Many leaks have been caused by expansion of the contents of a waste container. Occasionally, vapor pressure will permanently deform the drum, making it unsuitable for shipping. Also, some waste materials must be protected from freezing. Containers with frozen contents are typically considered off-specification at commercial TSD facilities, subject to rejection or surcharges. There are several means of avoiding leaks from expansion of contents, including:

1. allowing expansion room when filling drums
2. storing containers under a roof in ventilated areas
3. using storage areas not in direct sunlight when wastes must be stored outside
4. using light-colored drums when wastes are stored outside

Considerations in container handling and storage are discussed in Chapter 23.

Inspections (§265.174)

Weekly inspections must be made of container storage (accumulation) areas, looking for leaks or other evidence of actual or potential releases.
Although not specifically mentioned, there should be some documentation of: (1) what is looked for during an inspection; (2) that the inspections are actually performed; and (3) any remedial action taken as a result of conditions discovered during the inspection. Inspection items should include container and storage area condition, as well as the condition of safety, emergency, spill control, and personal protective equipment used for hazardous waste purposes.

An inspection checklist could be developed to ensure that whomever inspects the facility on a weekly basis knows what to look for. By developing a checklist, the inspections could be assigned to employees not highly trained in waste management, i.e., security guards, safety personnel, foremen, or laborers.

A sample inspection form is provided in Figure 9.4. Some items not required to be inspected by the regulations have been included for possible use. Inspection of these items would constitute good management practice. Merely performing inspections and noting deficiencies is not enough. Remedial actions must be taken to prevent small problems from becoming big problems. For example, the contents of corroded or damaged drums should be transferred to another container or overpacked into a salvage container before the contents leak out of the original container. To promote this, the sample inspection sheet has spaces for the type of remedial action necessary, for the remedial action performed, and for the date it was performed. The person inspecting must sign the form, and ideally, the form should be reviewed by a person responsible for hazardous waste management before the form is filed. The inspection forms become part of the record-keeping system necessary to document compliance, and should be kept for at least three years from the date of inspection.

Figure 9.4. Sample inspection form.

Area _____

Date and Time _____

Inspector's Name _____

Reviewed By _____

Number of Containers _____

Number of Empty Containers _____

Inspection Item	Potential Problems	Status (Acceptable or Unacceptable)	Remarks or Observations	Remedial Actions Necessary	Date Remedial Actions Performed
Adequate Aisle Space	Barrels placed too close to properly inspect or remove.				
Container Stacking	Containers stacked more than two high; containers not on pallets.				
Condition of Containers	Deteriorated, damaged, corroded, rusted, or leaking drums; drums damaged or leaking from expansion of contents.				
Sealing of Containers	Containers not stored closed; containers without bungs or lids; bungs or lids not tight on container.				
Labels	Labels identifying generator, contents, and accumulation date are missing or faded.				

Figure 9.4, cont'd.

Inspection Item	Potential Problems	Status (Acceptable or Unacceptable)	Remarks Remedial or Observations	Actions Necessary	Date Remedial Actions Performed
Accumulation Start Date	Not present or not readable; 90-day period has passed.				
Segregation of Materials	Wastes not segregated by type; wastes not in proper aisle.				
Emergency Equipment	Fire extinguishers, spill absorbent, or other necessary equipment is not functional or available at the accumulation area.				
Drainage	Water or other liquids are ponding around drums.				
Evidence of Leakage	Deteriorated containers are leaking; pumping of wastes or filling of containers has resulted in spilled material not cleaned up.				
Storage Pad	General damage; cracks, uneven settlement, or erosion of storage pad.				
Unacceptable Material	Presence of containers of garbage or refuse in storage area; presence of equipment or materials that are not supposed to be in the area.				

Figure 9.4, cont'd.

Inspection Item	Potential Problems	Status (Acceptable or Unacceptable)	Remarks Remedial or Observations	Actions Necessary	Date Remedial Actions Performed
Surrounding Area	Presence of activities or equipment posing a potential source of spark or flame (examples include welding, intense heat sources, smoking areas).				
"No Smoking" and "Danger—Keep Out" Signs	Absence of such signs; deterioration of or damage to such signs.				
Separation of Incompatible Wastes	Incompatible wastes in same general area; incompatible wastes not separated by a berm, dike, or wall.				

Special Requirements for Ignitable or Reactive Waste (§265.176)

Areas where ignitable or reactive hazardous wastes are stored (accumulated) should be located at least 50 ft from the facility property line. Such wastes should also be separated and protected from sources of ignition or reaction, e.g., open flames, smoking, cutting, welding, hot surfaces, frictional heat, sparks, and radiant heat. "No Smoking" signs should be posted wherever there is a hazard from ignitable or reactive wastes.

It is possible that the plant layout or plant operations are such that ignitable or reactive wastes cannot be separated and protected from ignition sources without violating the rule requiring 50 ft of separation from the property line. If this is the case, physical structures may be necessary, such as storage buildings meeting certain fire-resistive ratings. Contact the state regulatory agency, the state fire marshal, and your company's insurance carrier in such situations. A request for a variance from the application of 40 CFR 265.176 could be filed with EPA and the state regulatory agency. Chapter 23 contains additional considerations for the location of hazardous waste container storage areas within the overall plant property.

Special Requirements for Incompatible Wastes (§265.177)

Incompatible wastes, or incompatible wastes and materials, must not be placed in the same container for storage purposes. Further, hazardous waste cannot be placed in an unwashed container that previously held an incompatible waste or material.

Incompatible hazardous wastes and wastes incompatible with nearby materials must be separated or protected from each other by means of a dike, berm, wall, or other device.

The purpose of the separation or protection is to prevent any adverse consequences to human health or the environment that could result if incompatible wastes are mixed. These adverse consequences include fires, explosions, and container rupture or failure resulting in liquid releases or emissions of toxic gases, fumes, vapors, or mists. Such mixtures could result, for example, if a container leaked or was accidentally punctured due to deterioration of the container.

The generator is to determine whether any two waste streams are incompatible. Guidance in this determination is provided by Appendix V of 40 CFR 265. Appendix V provides examples of potentially incompatible wastes, waste components, and materials. These wastes, components, and materials are organized into groups for purposes of outlining the potential adverse consequences of contact between incompatible groups. Appendix V is reprinted in Appendix G of this book for handy reference. Another source of information regarding potentially incompatible wastes is "A Method for Determining the Compatibility of Hazardous Wastes" (EPA-600/2-80-076, U.S. EPA (1980)).

Examples of incompatible material combinations include acids and cyanides, strong acids and strong alkalies, flammable or combustible materials and oxidizers, and solvents and corrosives. Hazards posed by potential incompatibilities are a strong incentive for segregating different waste materials. Other reasons for waste stream segregation are discussed in Chapters 18 and 23.

Separation of incompatible wastes can be accomplished by segregated storage in different locations. Sufficient distance should be allowed to avoid any possibility of mixture or contact should containers of one (or both) type(s) of waste be damaged or leak. It is important to remember that strong corrosives, when spilled or leaked, can rapidly deteriorate any nearby containers. This is another possible avenue of contact between incompatible wastes that can be avoided by separation. Physical separation of containers of incompatible materials should be at least 20 feet. Greater distances may be necessary, depending upon the circumstances.

Protection of incompatible wastes from contact with each other is the second alternative offered for meeting the regulatory requirements for storing incompatible wastes. Protection refers to physical structures or devices that will prevent the contact of incompatible wastes, even if containers of both types leak simultaneously. Protective devices, including berms, dikes, or walls, allow incompatible wastes to be stored in closer proximity. If waste storage areas are bermed or diked to prevent mixing, any drainage ways and collection sumps must be designed as separate systems to prevent mixing of any collected liquids from the separated areas, which may contain the incompatible wastes.

Tank Systems (40 CFR 265 Subpart J, except 265.197(c) and 265.200)

Regulatory requirements for the accumulation of hazardous wastes in tanks are discussed at the end of this chapter.

Accumulation Start Date (40 CFR 262.34(a)(2))

The accumulation start date for each container is to be marked clearly on each container. The accumulation start date marking must be visible for inspection.

The requirement that the accumulation start date be marked and clearly visible for inspection allows ready determination by regulatory personnel of the compliance of a generator who accumulates with the 90-day limitation. Markings can be applied by indelible marker, spray paint and stencils, or a label with a space for marking the start date. Compliance may require that the accumulation start date be marked on opposite sides of the container. Alternatively, all containers could be oriented so as to have the start date marking easily visible. Either alternative will necessitate some form of orderly arrangement of containers, such as in rows that are two containers wide, stacked no more that two containers high. Rows of containers would be separated by sufficient aisle space (2.5 feet) to allow ready inspection of the accumulation start date.

The accumulation start date is the date hazardous waste is first placed in the container or tank. The sole exception is for any container(s) being filled in areas qualifying as satellite accumulation areas. Satellite accumulation areas are areas at or near any point of generation where wastes initially accumulate, under the control of the operator of the process generating the waste. Satellite accumulation containers are marked with the accumulation start date the day the accumulation limit is exceeded. This is normally when a single 55-gallon container is filled.

Satellite accumulation areas are not subject to the full set of regulatory standards for 90-day accumulation areas. Generators are allowed to accumulate up to 55 gallons of hazardous waste (or 1 quart of acutely hazardous waste) in satellite accumulation areas. If multiple waste streams are generated at a single satellite accumulation area, such wastes can be placed in separate containers. Containers are to be marked with the accumulation start date the day the accumulation limit of 55 gallons (or 1 quart of acutely hazardous waste) is exceeded. Such containers are to be moved to a 90-day accumulation area within three days of accumulation start date marking, or else the generator must ensure the satellite area's conformance with requirements for 90-day accumulation areas.

There are other limited, common-sense requirements for satellite accumulation areas. These requirements, in summary form, are:

1. Wastes must be placed in containers that are in good condition.
2. Wastes must be compatible with the containers.
3. Containers must always be closed, unless wastes are being added or removed.
4. Containers must be marked with the words "Hazardous Waste" or with other words that identify the contents of the containers.

Generators are permitted to have as many satellite accumulation areas as circumstances allow. Typical satellite areas are near vapor degreasers, paint booths, and parts cleaning operations. There is no time limit on the length of the satellite accumulation period, as long as the accumulation quantity limit (55 gallons) is not exceeded. Once satellite containers are moved to a 90-day accumulation area, the full complement of regulatory standards applies.

Labeling (40 CFR 262.34(a)(3))

Each container or tank must be labeled or marked clearly with the words "Hazardous Waste."

As with the accumulation start date, the words "Hazardous Waste" can be marked with indelible marker, painted on the container or tank, or attached by a label. Many generators find it more convenient to affix the DOT markings necessary for shipment at the time waste is first placed in drums. These markings are found on hazardous waste shipping labels available commercially from Labelmaster and others (Figure 9.5). Such labels often have an area to mark the accumulation start date as well.

HAZARDOUS WASTE

FEDERAL LAW PROHIBITS IMPROPER DISPOSAL

IF FOUND, CONTACT THE NEAREST POLICE, OR
PUBLIC SAFETY AUTHORITY, OR THE
U.S. ENVIRONMENTAL PROTECTION AGENCY

PROPER D.O.T.
SHIPPING NAME_____UN OR NA#_____

GENERATOR INFORMATION:

NAME_____

ADDRESS_____

CITY_____STATE_____ZIP_____

EPA
ID NO._____

EPA
WASTE NO._____

ACCUMULATION
START DATE_____

MANIFEST
DOCUMENT NO._____

HANDLE WITH CARE!
CONTAINS HAZARDOUS OR TOXIC WASTES

STYLE WM-6

Printed by LABELMASTER, Div of AMERICAN LABELMARK CO   CHICAGO, IL 60646

Figure 9.5.   Hazardous waste shipping label.

PREPAREDNESS AND PREVENTION (40 CFR 265 Subpart C)

A preparedness and prevention program, required of generators who accumulate as well as of generators who store, is just what the name implies. The prevention aspect of such a program involves the proper operation and maintenance of accumulation areas to minimize the risk of fire, explosion, leaks, spills, or damaged containers. The preparedness aspect involves obtaining appropriate equipment, testing and maintaining such equipment, ensuring access to communications or alarm systems, ensuring appropriate aisle space, and attempting to make arrangements and obtain agreements with local emergency authorities and contractors.

The generator who accumulates or stores is not given much guidance or specific directives in establishing a preparedness and prevention program. The regulations merely specify a performance standard—that the storage/accumulation areas be maintained and operated so as to minimize the possibility of a fire, explosion, or unplanned sudden (accidental) or gradual release of hazardous waste or hazardous waste constituents to the air, soil, or surface water.

The major portion of the prevention aspect of a preparedness and prevention program would be the adherence to the standards set out in Subpart I, "Use and Management of Containers," or, if applicable, Subpart J, "Tank Systems." Further discussion of best management practices as they apply to the containerized storage of hazardous waste is found in Chapter 23.

Subpart C is much more specific with respect to the preparedness aspect of a preparedness and prevention program. There are equipment requirements, requirements to test and maintain equipment, requirements for access to alarm systems and external communications, required aisle space for movement of emergency equipment, and required arrangements with local authorities. What follows is a section-by-section explanation of Subpart C, "Preparedness and Prevention."

Maintenance and Operation of Facility (§265.31)

Accumulation areas must be maintained and operated to minimize the possibility of a fire, explosion, or any unplanned sudden or gradual release of hazardous waste or hazardous waste constituents to air, soil, or surface water that could threaten human health or the environment.

This is another performance standard which can conceivably require a number of things. Strictly speaking, any preventable incidents, including chemical releases, can be considered de facto violations of the performance standard. This particular regulatory requirement is cited in many enforcement actions where there is visible evidence of releases (i.e., discolored soil) or lack of adherence to good management practices for which prescriptive requirements have not been set. Principles of container handling and storage are discussed in Chapter 23.

Required Equipment (§265.32)

Generators who accumulate must evaluate the type and degree of hazard(s) posed by each facility waste stream. A full complement of specified emergency equipment is required to be available at each accumulation area unless none of the hazards posed by the wastes handled could require a particular kind of specified equipment. The specified equipment includes:

1. internal communications or alarm system
2. telephone or two-way radio at the scene of operations capable of summoning external assistance
3. portable fire extinguishers

4. fire control equipment (including special equipment using foam, inert gas, or dry chemicals as the extinguishing agent)
5. spill control equipment
6. decontamination equipment
7. water at adequate volume and pressure to supply water hoses, foam extinguishing equipment, or automatic sprinklers or spray systems

Though not specifically identified, serious consideration should also be given to the types and amounts of personal protective clothing and equipment. Such items are necessary for both routine handling and emergency situations. Requirements regarding employee health and safety in a spill response situation are specified by OSHA at 29 CFR 1910.120.

An informal assessment of the risk(s) and hazard(s) posed by hazardous waste activities at each accumulation area will aid in the determination of the necessary personal protection, emergency, and spill control equipment. The risk assessment will involve an evaluation of the sources of risk (the containerized waste materials) as well as the pathways (storm sewers, surface runoff, air transport) and receptors (employees, neighbors, wildlife, livestock, and surface and groundwaters) that can be affected by releases of hazardous wastes.

The selection of the type(s) and amount(s) of emergency and spill control equipment should be made in consideration of the physical/chemical properties and hazards of the wastes in storage, given the likely release scenarios. These scenarios could include spills and leakage, container/tank rupture, fire, explosion, and/or release of toxic gases or vapors, as well as contamination of soil and groundwater. The total volume and number of containers are also of concern here, as is the presence of incompatible wastes.

The presence of potential pathways, or routes of escape, of hazardous wastes or constituents in an emergency situation should be identified. These pathways include nearby storm and sanitary sewers and other surface water drainage ways. Releases can leave the accumulation area via air movement as well as by surface and groundwater transport mechanisms. Releases to soil can contaminate groundwater, become airborne as dust particles, and/or contribute to direct contact. The presence of impermeable secondary containment in the form of a sealed concrete pad with berms can eliminate the soil, surface water, and groundwater pathways.

The proximity of potential receptors must also be considered. Of most importance are employees and neighbors, broadly defined. Fish and wildlife, as well as livestock, are also of concern, as are environmental media such as surface and groundwaters, soil, and air resources. The use of nearby surface waters or groundwater aquifers for drinking water requires special consideration.

There are other factors involved in decisions regarding the type(s) and amount(s) of emergency and spill control equipment. These include the quality of existing emergency procedures, and the presence and resources of internal spill response teams and external emergency response organizations, such as the fire department and spill response contractors. The level of expertise and degree of training of internal teams and external organizations must be taken into account.

Risk reduction measures such as controlling the quantities of wastes in storage, in conjunction with engineering controls such as secondary containment, should be integrated into the decisions as to the type(s) and amount(s) of emergency and spill control equipment necessary and desirable at each 90-day accumulation area. The regulations should be viewed as a starting point, and should not inhibit further evaluation of the type described earlier.

Given that the generator has made a determination as to which kinds of emergency equipment are required at the accumulation area(s), it might be instructive to give examples of each category.

Internal Communications or Alarm System (§265.32(a))

Presumably, this communication or alarm system would be used by workers handling hazardous wastes to summon internal assistance or to recommend evacuation should there be a fire, explosion, or major release while wastes are being moved or transferred from container to container. It is important to note that such a device does not have to be at the actual accumulation area; visual or voice contact between the workers handling hazardous wastes and another employee who does have immediate access to such a device is sufficient.

Just what constitutes an internal communications or alarm system is unclear. The regulations provide little guidance. It is possible that a variety of devices could meet the standard. Such devices include:

1. an internal alarm (such as a pull station alarm) or communication device that would alert a guard station or managerial office of a problem. The guard or plant manager or emergency coordinator could then alert internal spill response teams, summon commercial emergency response vendors, and/or sound a general plant evacuation alarm, if necessary.
2. a telephone system that gives access to a plant intercom
3. a simple alarm, such as a buzzer or compressed air horn, that would alert nearby workers to summon assistance

Telephone or Hand-Held Two-Way Radio (§265.32(b))

It is obvious that this device is meant to summon assistance from external sources. Telephones and two-way radios are given as examples. Once again, employees handling hazardous wastes at accumulation areas are to have immediate access to such a device, unless they are in visual or voice contact with another employee who has such access. If access to alarms and telephones is to be provided through visual or voice contact with another employee (presumably some distance away from the accumulation area), it would be advisable to have a written waste handling standard operating procedure (SOP) to that effect. Such a procedure would require the use of an "extra" employee to man the phone or alarm system whenever wastes are handled (poured, mixed, or transferred in or out of containers) at the accumulation area(s). This written SOP would be necessary to document compliance by use of visual or voice contact rather than having communication devices in the immediate vicinity of each accumulation area.

Portable Fire Extinguishers (§265.32(c))

Fire extinguishers should be available within reasonable distance of waste handling areas; otherwise, any fire started will very likely be out of control once the employee leaves the scene of operations to obtain a fire extinguisher.

The fire extinguishers should be readily available, but not so close to the containerized waste materials that a fire or chemical release would restrict accessibility or evacuation routes, or endanger those attempting to reach the extinguishers. Fire extinguishers should be mounted at strategic locations, preferably just outside the exit(s) from the accumulation area. It is recommended that extinguishers be located not less than 10 feet and no more than 30 feet from flammable liquid storage areas. Please note that OSHA has independent regulatory authority regarding the placement, use, maintenance, and testing of portable fire extinguishers. The OSHA requirements are found at 29 CFR 1910.157. Further guidance on portable fire extinguishers can be found in the National Fire Protection Association (NFPA) publication NFPA 10: "Portable Fire Extinguishers." NFPA 10 specifies both the minimum extinguisher rating and the maximum travel distance for a fire extinguisher of that rating.

Extinguishers available to the accumulation area must be of the proper type to fight the type of fire that could result from the particular wastes present. Flammable and combustible liquid fires require a Class B extinguisher and combustible material (paper, trash, wood) fires require a Class A extinguisher, while fires involving energized electrical equipment require a Class C extinguisher. A fire extinguisher rated ABC is rated for all three types of fires. Class D extinguishers are rated for combustible metal (sodium, magnesium) fires.

Fire Control Equipment

If the type or amounts of ignitable, flammable, or combustible wastes present require more response than is available from portable fire extinguishers, or if the fire department is distant or does not have adequate equipment or training, thought might be given to making fire control equipment available. Such equipment would include:

1. sprinkler systems
2. fire hose stations
3. large, wheeled extinguishers

4. foam-producing equipment
5. inert gas systems

It would make sense to contact your fire insurance carrier and local fire department to obtain input on the types of fire control equipment desirable. It is likely that a reduction in insurance premiums could result from better fire protection.

Spill Control Equipment

Spill control equipment can be as simple as bags of industrial absorbent (expanded clay, corn cob meal, synthetic absorbents) or as complicated as spill booms or vacuum-type spill cleanup devices. Many of the newer synthetic absorbents have much greater absorption capacities than clay or vermiculite. Additionally, they are packaged in readily usable forms, such as tubular socks, pillows, and pads. Areas where liquid wastes are stored or pumped, poured, mixed, or transferred from container to container should have spill absorbent readily available. Likewise, absorbent should be available when moving containers. Ready access to absorbent will minimize cleanup costs and problems.

The decision as to which type(s) of absorbents to obtain should be made in conjunction with the evaluation of disposal options for spent absorbents. Certain absorbents, such as those that biodegrade or release liquids during compression, are unsuitable for land disposal. Land disposal restrictions (40 CFR Part 268) further limit disposal options. If incineration is to be utilized for disposal of spent absorbents, those absorbents with low ash and high Btu values are usually preferable. Prior contact with those TSD facilities likely to be utilized for disposal of spent absorbent materials is advisable.

Decontamination Equipment

The regulations provide no examples of what constitutes decontamination equipment. Obviously, spilled materials will cause areas of the storage pad to need decontamination. Likewise, any protective equipment worn by workers responding to a spill will need to be decontaminated. Employee skin contact may require immediate access to a deluge shower and emergency eye wash. Spill absorbent will need to be collected and containerized. Hand tools, such as shovels and squeegees, may be necessary. Nonsparking tools are required where flammable liquids may be spilled. Contaminated equipment and floors will need to be washed off, and the washwater treated properly. Spills of acids or bases may require neutralization. Contaminated soils may require removal. Any damaged drums may need to be placed in overpack containers.

The need for decontamination equipment will have to be evaluated on a case-by-case basis, taking into consideration the requirements for responding to spills, leaks, or other accidental or gradual releases of hazardous wastes into the environment.

Water at Adequate Volume and Pressure (§265.32(d))

The use of the term ''adequate'' refers to the volume and pressure necessary to properly supply water hoses, automatic sprinklers, spray systems, or foam-producing equipment, as well as certain types of decontamination equipment.

Testing and Maintenance of Equipment (§265.33)

All of the emergency equipment obtained must be tested and maintained as necessary to assure its proper operation in time of emergency. Such equipment includes communication devices, alarm systems, fire protection equipment, and spill control equipment. Normally, emergency equipment will be used infrequently. Infrequent usage, combined with the potential need for such equipment in emergency situations, dictates that it be inspected routinely and maintained as necessary. Testing and maintenance implies and requires a formal, written inspection and evaluation program for such equipment. This inspection and testing program is normally performed internally. Other parties often serve as quality control with respect to testing and maintenance of emergency equipment. Very often, loss prevention specialists from insurance carriers will review and inspect such equipment. Local fire departments often perform similar inspections. Private firms routinely are contracted to provide periodic (at least annual) inspection, testing, and maintenance services for fire extinguishers and fire control equipment. It is important to keep records of the performance of such inspections, to document compliance with this requirement.

Access to Communications or Alarm System (§265.34)

The regulations require that all personnel involved in the pouring, mixing, or handling of hazardous wastes at the accumulation area(s) must have immediate access to an internal alarm or emergency communication device. As explained above, such access must be provided only if such equipment is required due to the hazards posed by the wastes handled at the facility. Also, such access can be provided through visual or voice contact with another employee, as explained earlier.

However, if such devices are to be provided at or near the accumulation area for direct use by waste-handling employees, the meaning of the term ''immediate'' comes into question. Obviously, if the employees have to travel some distance from the accumulation area to obtain access, then the alarm or communication device is not immediately available. Common sense holds that employees will be better able to respond to and control any fires or releases when alarms, communication devices, fire extinguishers, and emergency and personal protective equipment are all close at hand. Perhaps the alarm(s) could be located next to the fire extinguisher(s) just outside of the exit(s) to the accumulation area.

In this manner, the alarm could be activated prior to any fire or spill control efforts. OSHA specifications on minimum and maximum travel distances to fire extinguishers could be used as a guide in locating the alarms so as to ensure immediate access. Alarms should not be so close to the containerized wastes that a fire or chemical release would endanger employees attempting to activate the alarm, or restrict their evacuation options.

This section also requires that external communication devices (telephone or hand-held two-way radio) be immediately available (if required) at the scene of operation if there is ever just one employee on the premises while the "facility" is operating. The definition of "facility" was discussed earlier in the chapter in regard to the orientation of the regulations to commercial (offsite) hazardous waste management facilities. For manufacturing firms, it is unclear whether this provision applies when only one employee is handling wastes, or when there is only one employee at the facility when wastes are being handled (for example, during a plant shutdown for maintenance or renovation).

Given that manufacturing firms normally have many people present, and that waste-handling operations are carried out by two or more persons during regular shifts, and that accumulation areas are usually in the approximate vicinity of plant operations and other employees, the author contends that the latter interpretation holds. However, if waste accumulation is conducted in remote areas of the plant property and normally only one worker is involved, then such immediate access to external communications should be provided along with internal alarm or communication systems.

Required Aisle Space (§265.35)

Adequate aisle space is required for several reasons. Aisle space is needed to inspect containers for deterioration, leakage, and the accumulation start date without having to move the containers to conduct the inspection. Adequate aisle space is necessary for easy access to the wastes, for example, when a particular type of waste is ready for offsite shipment.

This provision of the regulations requires that sufficient aisle space be maintained for emergency purposes. In other words, sufficient aisle space must be maintained "to allow the unobstructed movement of personnel, fire protection equipment, spill control equipment, and decontamination equipment to any area of 'facility' operation in an emergency, unless aisle space is not needed for any of these purposes."

Again, it is important to remember the regulatory definition of "facility" and to recognize the size differences between an accumulation area (where wastes must be moved offsite or to an onsite TSD area within 90 days from the accumulation start date) and the storage area of a TSD facility, where much larger amounts are likely to be present.

For the purposes of accumulation area conformance with aisle space requirements, an orderly arrangement of containers is necessary. Containers should be arranged in rows that are two containers (or one pallet) wide. Containers, if stacked, should be stacked no more than two containers high. Containers can be placed on pallets, if the pallets are in good condition.

A system of main aisles and minor aisles could be devised to ensure the adequacy of the aisle space between rows. The main aisle should be sufficiently wide (minimum of 8 feet) to allow access to the accumulation area by forklifts or fire control equipment. Minor aisles should be sufficiently wide (minimum of 2.5 feet) to allow access for inspection purposes.

Arrangements with Local Authorities (§265.37)

Arrangements with local authorities are potentially a very important element of a preparedness and prevention program. Basically, the required arrangements are to familiarize local public safety agencies with the plant layout, areas where hazardous wastes are stored, and the particular hazards posed by the wastes. As with equipment requirements, arrangements with local authorities are required as appropriate for the type of waste handled at the facility and the potential need for the services of these outside organizations.

The requirements for arrangements with local authorities are not onerous, and such arrangements could promote good relations with municipal public safety officials. Detailed arrangements will become more prominent as a result of the emergency planning requirement of EPCRA (Title III of SARA). These regulatory requirements present a good opportunity for a closer relationship with local emergency organizations, including, perhaps, reciprocal training. In other words, the fire department would train company employees on the use of hand-held fire extinguishers, in exchange for hazardous waste management training by appropriate personnel from the manufacturing firm.

If such arrangements are appropriate for the type of waste handled, and there is a potential need for the services of outside emergency organizations, the generator who accumulates must attempt to make the following arrangements.

Arrangements with Local Police Departments, Fire Departments, and Emergency Response Teams to Familiarize Them with the Operations of the Manufacturing Firm

These aspects include:

1. the overall layout of the manufacturing facility
2. areas where hazardous wastes are accumulated or stored
3. properties of the hazardous waste handled at the facility and associated hazards

4. places where personnel would normally be working
5. entrances to roads inside the facility
6. possible evacuation routes

A few of the above items need to be expanded upon.

Any efforts to familiarize local officials on the plant layout and location of hazardous waste should include information on how and where both hazardous waste and hazardous input materials (such as bulk solvents or chemicals) are stored. Emergencies can occur involving raw material as well as hazardous waste handling or storage. Locations of electrical switchgear and natural gas cutoffs should also be identified.

When discussing the properties of the hazardous materials and wastes handled and the associated hazards, it will be important to include the types of hazards associated with the conditions to which local police and fire personnel may respond. In other words, emergency responders should be informed of the hazards of spilled or leaked material when exposed to air or water. Potentially incompatible waste combinations should be identified, as well as potential consequences of uncontrolled mixing. The hazards resulting from the wastes being combusted or exposed to heat or flame should also be explained. Many materials form toxic gaseous by-products under such conditions. This information would be in addition to the properties of and hazards posed by the wastes under "normal" conditions.

For the purposes of this requirement, representatives of the facility could organize group tours of the plant, distribute written material on the necessary aspects of plant operations, and distribute copies of the facility's contingency plan (discussed in detail below). It is important to document any contacts made with local authorities, any agreements/arrangements, or the refusal by local authorities to enter into any formal agreements/arrangements. Memoranda to the file or letters confirming verbal agreements or mutual understandings could be used. If tours are conducted, a list of attendees and the areas toured, along with the date(s) of the tour(s), should be prepared and kept on file.

At a minimum, these agencies should receive a copy of the contingency plan, under a cover letter inviting them to enter into formal agreements to provide emergency services. The plan should be mailed certified mail, return receipt requested, so as to provide documentation.

Agreements Designating Primary Emergency Authority to a Specific Police Department and a Specific Fire Department

This requirement provides the opportunity to clearly delineate responsibility when a manufacturing facility is located in overlapping jurisdictions. This could be the case in the suburbs of a large city, where township or volunteer fire departments exist in addition to the municipal fire department. Insurance premiums may be reduced if primary response authority is given to the municipal fire department, with backup support provided by the other fire departments.

Agreements with State Emergency Response Teams, Emergency Response Contractors, and Equipment Suppliers

Again, these agreements are required as appropriate, given the type of waste handled and the potential need for such services. It must be stated, however, that many states have very strict spill reporting laws. It is advisable to submit a copy of the contingency plan to the state spill response authority, including a discussion of the types of wastes and materials handled. It is also important to determine the specific state spill reporting requirements. It is recommended that contact be made with commercial emergency or spill response contractors. These contractors should become familiar with the facility prior to being called in to respond to a massive release. These contractors should also have a copy of the contingency plan.

Arrangements to Familiarize Local Hospitals with the Properties of Hazardous Waste Handled at the Facility and the Types of Illnesses That Could Result from Fires, Explosions, Spills, or Leaks at the Facility

Such arrangements should be made with potential emergency medical providers. These providers include ambulance services, industrial clinics, and hospital emergency rooms. These entities should be provided with information such as MSDSs or chemical literature on the types of materials and wastes handled. If at all possible, appropriate MSDS or emergency care information should be given to emergency care providers, such as ambulance personnel, when they come to transport injured or exposed employees to the hospital.

With this and the other required agreements/arrangements, it is important to document the contacts made, the information transferred, and any meetings, tours, etc. Refusal by local authorities should also be documented. If local authorities are particularly reluctant to enter into written agreements, then the required information could be sent to them via certified mail, return receipt requested. The signed receipt would then be attached to a photocopy of the cover letter or the information mailed. This documentation becomes part of the recordkeeping system used to allow ready verification by regulatory officials of the firm's compliance status.

CONTINGENCY PLAN AND EMERGENCY PROCEDURES (40 CFR 265 Subpart D)

A contingency plan incorporating the procedures necessary to respond to emergencies involving hazardous waste is an important document for generators who store or accumulate. The contingency plan can be thought of as the central document of a preparedness and prevention program. In fact, the requirements of Subpart C, "Preparedness and Prevention," and Subpart D, "Contingency Plan and Emergency Procedures," are intertwined.

A contingency plan should be a concise document, a one-stop source of the basic information necessary to respond effectively to an emergency involving hazardous waste. A contingency plan should be designed and written to be useful and to be used. A good contingency plan can serve as the core of the required personnel training program. The plan itself, or a single-page summary, should be posted at or near storage and accumulation areas. Copies should be given to responsible personnel at the manufacturing facility. The plan should be distributed to local emergency authorities.

The requirements for a contingency plan offer plant management and personnel assigned responsibilities for hazardous waste a unique opportunity to plan for emergency incidents at the manufacturing facility. Of course, possible incidents are not limited to those involving the firm's hazardous waste storage or accumulation areas. Incidents can involve the various hazardous materials used in the manufacturing process. Many firms already have documents describing emergency procedures for other incidents, such as the responses necessary to fires or explosions (not involving hazardous wastes), floods, power outages, tornadoes, earthquakes, broken gas or water mains, railroad derailments, or civil disobedience. The planning for emergencies involving hazardous wastes can build on prior efforts.

For the contingency plan to be useful, the persons involved in its preparation should attempt to anticipate the potential types of incidents involving hazardous wastes. Appropriate responses should then be laid out for each type of incident. The more likely types of incidents include:

1. fire
2. explosion
3. leakage from deteriorated or damaged containers or tanks to waterways or sewers
4. spills during routine handling
5. splashes or other exposures involving worker injury
6. spills or leaks producing toxic vapors or fumes
7. contamination resulting from containers ruptured due to build-up of internal pressure

The specific conditions at the manufacturing operation as well as the types of wastes handled may require other potential incidents to be explored. Some sorting and sifting of the less feasible incidents should be performed by assessing the potential incident in terms of the probability of occurrence and the severity of the consequences. Thorough evaluation of sources, pathways, and receptors must be included in the planning process.

Personnel Actions

A description of the actions facility personnel must take in response to fires, explosions, or any unplanned releases of hazardous waste or hazardous waste constituents to air, soil, or surface water at the facility is the core of the contingency plan. This is where the potential incidents and appropriate responses will be laid out.

The regulations lay out a performance standard for these planned responses. The contingency plan must be designed to minimize hazards to human health or the environment from fires, explosions, or any unplanned releases. Further, the provisions of the plan must be carried out immediately whenever there is a fire, explosion, or release that could threaten human health or the environment.

Arrangements with Authorities

A description should be included of arrangements agreed to by local police departments, fire departments, hospitals, contractors, and state and local emergency response teams to coordinate emergency services.

These are the agreements potentially required as a part of the preparedness and prevention program. The agreements should specify the circumstances and signal(s) necessary to activate the arrangements.

List of Emergency Coordinators

A list should be given of names, addresses, and phone numbers (office and home) of all persons qualified to act as emergency coordinators. Where more than one person is listed, one person must be named as primary emergency coordinator and others must be listed in the order in which they will assume responsibility as alternates.

The emergency coordinator has heavy responsibilities to assume in an emergency. At least one employee capable of acting as an emergency coordinator is required at all times to either be on the facility premises or on call. The emergency coordinator has the responsibility for coordinating all emergency response measures and must have the authority to commit resources (incur debts) to carry out the contingency plan. He or she should be a highly capable and well-trained individual. Persons listed as emergency coordinators should be thoroughly familiar with and knowledgeable about:

1. all aspects of the contingency plan
2. all operations and activities at the facility
3. the location and characteristics of wastes handled
4. the location of all records within the facility
5. the facility layout

The emergency coordinator is required to perform certain activities and follow certain procedures in an emergency. Whenever there is an imminent or actual emergency situation:

1. The emergency coordinator (or his designee when the emergency coordinator is on call) must immediately activate internal facility alarms or communication systems to notify all facility personnel, and notify appropriate state or local agencies with designated response roles if their help is needed (265.56(a)(1) and (2)).

2. If there is a release, fire, or explosion, the emergency coordinator must immediately identify the character, exact source, amount, and areal extent of any released materials. This may be done by observation or review of facility records or manifests and, if necessary, by chemical analysis (265.56(b)).

3. At the same time, the emergency coordinator must assess possible hazards to human health or the environment that may result from the release, fire, or explosion. This assessment must consider both direct and indirect effects of the release, fire, or explosion (e.g., the effects of any toxic, irritating, or asphyxiating gases that are generated, or the effects of any hazardous surface water runoff from water or chemical agents used to control fire or heat-induced explosions) (265.56(c)).

4. If the emergency coordinator determines that the facility has had a release, fire, or explosion that could threaten human health or the environment outside the facility, he must report his findings as follows. If his assessment indicates that evacuation of local areas may be advisable, he must immediately notify appropriate local authorities. (*Note:* As a result of Title III of SARA (EPCRA), the appropriate local authorities would consist of the fire department, the local emergency planning committee, and the State Emergency Response Commission.) The emergency coordinator must be available to help appropriate officials decide whether local areas should be evacuated. He must immediately notify either the regional on-scene coordinator (OSC) or the National Response Center using their 24-hour toll-free number (800/424-8802). The report must include (1) name and telephone number of caller; (2) name and address of facility; (3) time and type of incident (e.g., release, fire); (4) name and quantity of material(s) involved, to the extent known; (5) the extent of injuries, if any; and (6) the possible hazards to human health or the environment outside the facility (265.56(d)(1) and (2)).

5. During an emergency, the emergency coordinator must take all reasonable measures necessary to ensure that fires, explosions and releases do not occur, recur, or spread to other hazardous waste at the facility. These measures must include, where applicable, stopping processes and operations, collecting and containing released waste, and removing or isolating containers (265.56(e)).

6. If the facility stops operations in response to a fire, explosion, or release, the emergency coordinator must monitor for leaks, pressure build-up, gas genera-

tion, or ruptures in valves, pipes, or other equipment, wherever this is appropriate (265.56(f).

7. Immediately after an emergency, the emergency coordinator must provide for treating, storing, or disposing of recovered waste, contaminated soil or surface water, or any other material that results from a release, fire, or explosion at a facility (265.56(g)).

8. The emergency coordinator must ensure, in the affected area(s) of the facility, that (1) no waste that may be incompatible with the released material is treated, stored, or disposed of until cleanup procedures are completed; and (2) all emergency equipment listed in the contingency plan is cleaned and fit for its intended use before operations are resumed (265.56(h)).

9. The owner or operator must notify the EPA regional administrator and appropriate state and local authorities that the facility is in compliance with 265.56(h) before operations are resumed in the affected area(s) of the facility (265.56(i)).

10. The owner or operator must note in the operating record the time, date, and details of any incident that requires implementing the contingency plan. Within 15 days after any incident reported to the National Response Center and/or requiring implementation of the contingency plan, he must submit a written report on the incident to the EPA regional administrator and/or the authorized state environmental agency. The report must include (1) the name, address, and telephone number of the owner or operator; (2) the name, address, and telephone number of the facility; (3) the date, time, and type of incident; (4) the name and quantity of material(s) involved; (5) the extent of injuries, if any; (6) an assessment of actual or potential hazards to human health or the environment, where this is applicable; and (7) estimated quantity and disposition of recovered material that resulted from the incident (265.56(j)).

Emergency Equipment List

A list should be included of all emergency equipment at the facility, including the location and a physical description of each item on the list and a brief outline of its capabilities.

The emergency equipment could include, where required, fire extinguishers, spill control equipment, communication and alarm systems, personal protective clothing and equipment, and decontamination equipment. The subject of required emergency equipment was discussed earlier in regard to preparedness and prevention programs.

A point of interest, once again, is the use of the term "facility" in the regulatory language. The author interprets the term "facility" to refer to the hazardous waste management facility (in this case the accumulation area) within the overall manufacturing facility. It is clearly illogical to require a listing of all emergency equipment (e.g., fire extinguishers) at a manufacturing establishment when only those extinguishers or pieces of equipment readily available could possibly be used in case of fire or emergency at the hazardous waste management facility.

Of course, all emergency equipment that could potentially be used to respond in an emergency should be listed, whether physically located at the accumulation area ("facility") or not. For example, some manufacturing plants have internal fire brigades with mobile foam-producing firefighting equipment. Clearly, these capabilities should be included in the equipment list and in the response planning for emergencies.

Some contingency plans have implemented this requirement as a list and a plant layout map. There is a list of equipment, with a physical description and outline of the capabilities of each item, separate from (but attached to) a plant layout map showing the accumulation area(s). Symbols are used to show the location of the items on the list within or near the accumulation area(s).

Evacuation Plan

An evacuation plan is necessary for facility personnel where there is a possibility that evacuation could be necessary. Again, the term "facility" is a point of concern. Obviously, if ignitable or reactive wastes are accumulated or stored inside of buildings or near where employees work, the evacuation plan should be for the entire manufacturing facility, or at least the building where wastes are accumulated. If accumulation areas are outside of or remote from areas where production employees work, the evacuation plan could focus on the evacuation of waste-handling personnel from the accumulation areas and the immediately adjacent areas. Please note that OSHA has independent regulatory requirements for evacuation plans, as an element of what are termed "Emergency Action Plans." The scope and applicability of the emergency action plan requirements are found at 29 CFR 1910.38.

Where an evacuation plan is necessary, it must describe the signal(s) to be used to begin evacuation, the evacuation routes, and alternate evacuation routes if there is a possibility that the primary routes could be blocked by fires or releases of hazardous wastes.

Additional Considerations

EPA has prepared a guidance manual that covers, among other things, the preparation of contingency plans. Entitled "Plans, Recordkeeping, Variances, and Demonstrations for Hazardous Waste Treatment, Storage and Disposal Facilities" (SW-921), this manual presents a suggested outline of some merit for contingency plans. This suggested outline is found in Table 9.1.

As mentioned earlier, the potential types of incidents should have appropriate responses laid out. This is the essence of the contingency plan. For each incident,

Table 9.1. Suggested Outline of Contingency Plan

Facility Identification and General Information

1. Name of facility
2. Location
3. Name, address, and telephone numbers (office and home) of owner and operator
4. Name, title, address, and telephone numbers (office and home) of primary emergency coordinator and alternates
5. Type of facility
6. Facility site plan (including structures, topography, roadways, sewers, pipelines or utility networks, adjacent land uses, and water bodies)
7. Description of hazardous waste activities (including types of wastes, amounts, and capacities)

Emergency Coordinator(s)

1. Primary coordinator
2. Alternate coordinator(s)
3. Emergency duties and authority to commit resources

Implementation of Contingency Plan

1. Decision criteria for each potential type of incident
2. Decision criteria for calling for outside assistance

Emergency Response Procedures

1. Notification phase
2. Control and containment phase
3. Follow-up and cleanup phase

Emergency Equipment

1. Emergency equipment inventory
2. Location of emergency equipment
3. Equipment capabilities
4. Emergency equipment available from other sources

Coordination Agreements

1. Police
2. Fire
3. Other emergency response units
4. Hospital

Evacuation Plan

1. When to evacuate
2. Signals to begin evacuation
3. Primary evacuation routes
4. Alternate evacuation routes

Required Reports

1. Notification of compliance before resuming operations following incident
2. Report on the incident

Identification of Hazardous Materials

1. Material Safety Data Sheets
2. Other reference materials on chemicals present

Source: Modified from "Plans, Recordkeeping, Variances, and Demonstrations for Hazardous Waste Treatment, Storage and Disposal Facilities," SW-921, U.S. EPA, U.S. Government Printing Office (1981), pp. 45–46.

a series of steps should be devised to adequately respond. Also, the equipment, materials, and personal protection (e.g., respirators and protective clothing) necessary to respond to each incident should be identified. The response strategy should specify when to invoke the agreements/arrangements with state and local authorities, and decision criteria for evacuation. Not to be overlooked are decision criteria for contingency plan implementation. Example contingency plan implementation criteria (from SW-921) are found in Table 9.2.

Table 9.2. Example Contingency Plan Implementation Criteria

The contingency plan must be implemented if an imminent or actual incident could threaten the environment or human health.

Spills

- The spill could result in release of flammable liquids or vapors in amounts capable of creating a fire or vapor explosion hazard.
- The spill could cause the release of toxic liquids or vapors in amounts capable of threatening human health.
- The spill exceeds the Reportable Quantity (RQ) for the hazardous substance or waste.
- The spill can be contained onsite, but the potential exists for groundwater pollution due to aquifer contamination.
- The spill cannot be contained onsite, resulting in offsite soil contamination and/or groundwater or surface water pollution.
- The spill reaches sanitary or storm sewers.

Fires

- The fire could cause the release of toxic vapors.
- If the fire spreads, it could ignite materials at other locations at the site or cause heat-induced explosions.
- The fire could spread to offsite areas.
- Use of water or water-and-chemical fire suppressant could result in contaminated runoff.

Explosions

- An imminent danger exists that an explosion could occur, resulting in a safety hazard due to flying fragments or shock waves.
- An imminent danger exists that an explosion could ignite other hazardous waste at the facility.
- An imminent danger exists that an explosion could result in release of toxic material.
- An explosion has occurred.

Source: Modified from "Plans, Recordkeeping, Variances, and Demonstrations for Hazardous Waste Treatment, Storage and Disposal Facilities," SW-921, U.S. EPA, U.S. Government Printing Office (1981), p. 48.

Once the contingency plan is implemented, the emergency coordinator has little latitude regarding certain actions. Section 265.56 specifies in some detail the emergency procedures the emergency coordinator must follow. The requirements of §265.56, "Emergency Procedures," were given earlier in this chapter. The procedures to be followed can be divided into three phases: (1) discovery and notification phase; (2) containment and control phase; and (3) follow-up and cleanup phase. The requirements of §265.56 should be extensively referenced in any contingency plan developed, because of their specificity and mandatory nature.

The contingency plan is not meant to be a static document. It must be reviewed and amended immediately if necessary, whenever:

1. applicable regulations are revised
2. the plan fails in an emergency
3. the facility changes in a way that materially increases the potential for incidents or changes the responses necessary to emergencies
4. the list of emergency coordinators changes
5. the list of emergency equipment changes (265.54(a)-(e))

Copies of the contingency plan and all revisions must, of course, be maintained at the facility. In addition, the plan and all revisions must be submitted to the following organizations that may be called on to provide emergency services:

1. local police departments
2. local fire departments
3. hospitals
4. state and local emergency response teams

These organizations, with the exception perhaps of the hospital(s), should have received a copy of the contingency plan when emergency agreements/arrangements were being developed.

It is necessary to note that any existing Spill Prevention, Control, and Countermeasures (SPCC) plans prepared to comply with Clean Water Act regulations (40 CFR 112) can be amended to meet the contingency plan requirements. SPCC plans need to be amended to incorporate the five elements of a contingency plan discussed above. However, the author recommends that separate contingency plan and SPCC plan documents be maintained.

It is important to note the existence of parallel and overlapping federal, state, and, often, local requirements for reporting spills and other emergency incidents. The various laws and regulations at the different levels of government have different specifications as to what, where, and how much constitutes a reportable incident. The hazardous waste regulations (under RCRA) require immediate oral reporting to the National Response Center (800/424-8802) if there is a fire, explosion, or release, regardless of amount, that could threaten human health or the environment outside the facility. Written reports must be submitted to the appropriate EPA regional office within 15 days of any hazardous waste releases reported to the National Response Center and/or requiring implementation of the contingency plan.

The CERCLA regulations (40 CFR Part 302) require immediate reporting to the federal government, via the National Response Center (800/424-8802), of any release of a hazardous substance to the environment in an amount equal to or exceeding its reportable quantity (RQ) in any 24-hour period. Notification is to be made by any person in charge as soon as he has knowledge of any reportable release. Release is defined to include "any spilling, leaking, pumping, pouring, emitting, emptying, discharging, injecting, escaping, leaching, dumping, or disposing into the environment . . . ," with limited exclusions involving federally permitted releases, releases within the workplace, continuous releases previously reported, and pesticide application. The term "release" is interpreted to include the abandonment or discarding of barrels, containers, and other closed receptacles containing any hazardous substance or pollutant or contaminant (42 USC 9601(22)).

Hazardous substances are listed by technical name, Chemical Abstracts Service (CAS) registry number, and synonyms (if any) at 40 CFR 302.4, along with their individual RQ. Please note that hazardous wastes are a subset of hazardous substances, for release reporting purposes. Each listed hazardous waste and each characteristic of hazardous waste has been assigned a RQ. In many instances, the RQs are relatively low. For example, the RQ for waste classified as characteristically hazardous by virtue of ignitability, corrosivity, or reactivity is only 100 pounds. Table 9.3 identifies common hazardous wastes/substances and their CERCLA/DOT RQs.

RQ releases of CERCLA hazardous substances, as well as SARA "extremely hazardous substances," are also to be reported immediately to the state emergency response commission (SERC) and the local emergency planning committee (LEPC). If more than one state or locality may be affected, all SERCs and LEPCs involved must be notified. Both the SERC and LEPC were established by act of Congress in EPCRA (Title III of SARA). There are also requirements for written reports to be submitted to the SERC and LEPC for each oral notice. These follow-up reports would update the oral notice, and describe the response action taken. The specific requirements for the SERC/LEPC oral notices and written reports are found at 40 CFR 355.40.

It is important that the contingency plan address the various federal, state, and local spill and emergency incident reporting requirements. Inquiries should be made to the following agencies as to their reporting requirements, including their 24-hour telephone numbers, for emergency incidents and accidental discharges, releases, or spills:

1. state air pollution control agency
2. state water pollution control agency
3. state solid/hazardous waste management agency
4. state fire marshal's office
5. state emergency response commission
6. municipal sewage treatment authority
7. municipal water supply authority
8. local health department
9. local air pollution control agency

Table 9.3.   Common Hazardous Wastes/Substances and their CERCLA/DOT RQs

Hazardous Substance	RQ, pounds
Unlisted (characteristic) HW	
D001 Ignitability (I)	100
D002 Corrosivity (C)	100
D003 Reactivity (R)	100
(Please note that these RQs apply to ICR substances which become wastes upon release.)	
EP Toxicity (E)	
D004 Arsenic	1
D005 Barium	1000
D006 Cadmium	10
D007 Chromium	10
D008 Lead	1
D009 Mercury	1
D010 Selenium	10
D011 Silver	1
(Please note that these RQs apply to the waste itself, not merely to the toxic contaminant.)	
Listed HW	
F001/F002 Spent solvents/mixtures	10
1,1,1-Trichloroethane	1000
Trichloroethylene	100
Tetrachloroethylene (perchloroethylene)	100
Methylene chloride	1000
Chlorinated fluorocarbons	5000
F003 Spent solvents/mixtures	100
Xylene	1000
Acetone and methanol	5000
F004 Spent solvents/mixtures	1000
F005 Spent solvents/mixtures	100
Toluene	1000
Methyl ethyl ketone	5000
F006 Electroplating WWT sludges	10
F007 Cyanide plating bath solutions	10
F008 Plating bath sludges where cyanides are used in the process	10
F009 Stripping/cleaning bath solutions where cyanides are used in the process	10
K062 Spent pickle liquor from iron and steel industry	1

10. local civil defense or emergency management agency
11. local fire department
12. local emergency planning committee
13. local or county hazardous materials (HAZMAT) response team

The emergency procedures developed for the contingency plan should include the reporting requirements of the various governmental agencies. Waste-handling employees should be trained to report all spills to an emergency coordinator who will be responsible for contacting the necessary governmental agencies. Reporting may also be necessary or advisable to certain upper management personnel, commercial emergency response vendors, and insurance carriers.

PERSONNEL TRAINING (§265.16)

The final requirement of §262.34 for generators who accumulate is a personnel training program. Employees directly involved with waste management and handling must successfully complete a program of instruction that teaches them to properly perform their duties. This training program can consist of classroom instruction, on-the-job training, or some combination of the two. However, the training program and the training that each individual receives must meet a very broad performance standard: The training program must teach the employees required to take it to perform their duties in a way that ensures the facility's compliance with the Part 265 standards.

The regulations do give some guidance as to what constitutes the minimally acceptable personnel training program. At a minimum, the training program is to ensure that waste-handling employees are able to respond effectively to emergencies. This is to be done by familiarizing such employees "with emergency procedures, emergency equipment, and emergency systems, including where applicable:

(i)   Procedures for using, inspecting, repairing, and replacing facility emergency and monitoring equipment;
(ii) Key parameters for automatic waste feed cut-off systems;
(iii) Communications or alarm systems;
(iv) Resonse to fires or explosions;
(v)   Response to groundwater contamination incidents;
(vi) Shutdown of operations.'' (40 CFR 265.16(a)(3))

The minimal guidance given in the regulations is not, in the author's opinion, sufficient to meet the regulatory performance standard. Employees with waste management duties need to be trained in more than emergency procedures if they are going to perform their duties in a manner ensuring that the facility remains in compliance with the interim status standards found in Part 265. Employees should be trained to be able to handle hazardous wastes in a manner that will minimize the possibility of fire, explosion, or release. Most firms will also want their training program to ensure that the appropriate employees will be able to perform their duties such that the firm's compliance with the standards found in 40 CFR 261 and 262 is maintained.

A good training program would begin with instruction on safe handling of hazardous wastes under normal conditions. This program would likely place a heavy focus on proper handling of hazardous waste containers. The goal of such a program would be to maintain compliance under both normal operating conditions and during emergency situations. This can only be achieved by training in both proper operating procedures and emergency response measures.

It is very likely that several people at a typical manufacturing firm will, due to their involvement in waste management, need to successfully complete the training program. These people will likely have different duties and levels of responsibility. The training program should be structured for the differing information and training needs of the different types of positions involved in HWM. Waste management employees should receive instruction on HWM procedures (including contingency plan implementation) relevant to the duties and responsibilities of the positions in which they are employed. The training program is to be directed by a person already trained in hazardous waste management procedures.

This training was to have been successfully completed by the relevant employees by May 19, 1981. New employees or employees reassigned to waste management duties are to complete the training program within six months after the date of employment or reassignment. Such employees are not to work in unsupervised positions until they have successfully completed the training program. Finally, all employees falling under the training requirement must take part in an annual review of the initial training.

As with many of the other requirements, the training program must be well documented and dutifully recorded. The following documents and records should be considered an integral part of the training program:

1. the job title for each position at the facility related to HWM and the name of the employee filling each job
2. a written job description for each position at the facility related to HWM, including the requisite skill, education, or other qualifications, and duties of employees assigned to each position
3. a written description of the type and amount of introductory and continuing training that will be given each person filling a position at the facility related to HWM
4. records that document that the required training has been given to and completed by each person filling a position related to HWM

Once a curriculum is devised that would meet the requirements of the regulations, the training could be implemented as a series of topics. Topics could include use of fire extinguishers, release reporting requirements and procedures, control of minor spills and leaks, how to conduct inspections of storage areas, containerization and handling considerations for each of the facility's waste streams, etc. An outline would be prepared for each topic. When the topic is presented to the relevant employees, the documentation would include a sign-up sheet of all attending. This sheet would include the topic of the session and the date, place, instructor, and length of the course. A simple true/false or multiple-choice test may be given to document successful completion of each session. This information would be kept as part of the personnel training records.

Any instruction in emergency procedures should include implementation of the contingency plan, which would serve as an instruction manual. Perhaps a select crew of personnel could be recruited for each shift to be intensively trained in contingency plan implementation. Of course, such crews would be supervised and coordinated by the emergency coordinator during any incident requiring implementation of the contingency plan. Internal spill response teams fall under certain OSHA requirements found at 29 CFR 1910.120.

TANK SYSTEMS (40 CFR 265 Subpart J, except §265.197(c) and §265.200)

A detailed explanation of the requirements found at 40 CFR Part 265, Subpart J, "Tank Systems," is beyond the scope of this chapter. It would be more properly the subject of an entire book. Manufacturing firms using tank systems (aboveground, inground, or underground) to accumulate, store, or treat hazardous wastes are well advised to refer directly to the regulations, as well as to certain EPA guidance manuals and directives. These additional information sources include:

1. "Technical Resource Document for the Storage and Treatment of Hazardous Waste in Tank Systems" (12/86) NTIS PB#87-134391.
2. "Questions and Answers Regarding the July 14, 1986 Hazardous Waste Tank System Regulatory Amendments" (8/87) EPA/530-SW-87-012.

Tank systems include any piping, fittings, flanges, valves, and pumps.

The regulatory requirements for tank systems used to accumulate, store, or treat hazardous wastes are quite detailed and comprehensive. Additional obligations and financial expenditures were imposed by EPA in 1986 on owners and operators of such tank systems, with limited exceptions. These additional obligations include annual assessments of the integrity of existing tank systems, pending retrofitting with complete secondary containment by the time existing tanks reach 15 years of age. New tank systems as well as replacement components of existing tank systems must meet stringent design, corrosion protection, and installation standards, including complete secondary containment with interstitial monitoring.

Owners and operators of tanks holding hazardous wastes should carefully evaluate their tank systems, to determine whether such systems are regulated under Subpart J or whether they fall under one of the exemptions or conditional exclusions. The definition of tank, found at 40 CFR 260.10, emphasizes that tanks are *stationary* devices, constructed primarily of *non-earthen* materials which provide structural support. Tanks are thus distinguished from surface impoundments, which are formed from and supported by primarily earthen materials. More importantly, tanks are distinguished from *containers,* which are portable devices. Containers, for hazardous waste regulatory purposes, are differentiated from tanks in terms of their portability, not in terms of size or capacity. Thus, tote tanks and roll-off boxes are considered containers for regulatory purposes.

Another important distinction involves hazardous waste accumulation, storage, or treatment tanks and tanks that are a manufacturing process unit or are part of one. Manufacturing process units include tanks and other process equipment holding materials that, if and when discarded, would be considered hazardous wastes. Examples include electroplating baths, vapor degreasers, parts-cleaning dip tanks, paint booths, rinsing tanks, filtering devices, paint vats, and many others. Manufacturing process units are not regulated unless residues meeting the definition of hazardous waste remain in the unit more than 90 days after the unit ceases to be operated for manufacturing purposes.

There are other exemptions to 40 CFR Part 265, Subpart J, beginning with tanks or tank systems that are part of activities entirely excluded from regulation under Part 265. These exemptions, found at 265.1(c), are largely for certain treatment and recycling activities, and are discussed in Chapter 12. Conditional exclusions from the secondary containment requirements are given to tanks that contain solids with no free liquids, and secondary containment devices such as sumps. Special (limited) requirements are specified for 100–1000 kg/month generators that accumulate hazardous waste in tanks.

The remainder of the discussion on tank systems consists of a listing, in summary form, of the aspects of Subpart J applying to generators that accumulate in tanks. The Subpart J requirements apply to tank systems containing hazardous wastes regardless of whether the tank is above-ground, inground, or underground, unless the tank system is otherwise excluded from regulation.

Assessment of Existing Tank System's Integrity (§265.191)

Existing tank systems without secondary containment per 265.193 must be assessed with respect to their design, structural strength and integrity, and compatibility with the waste(s) to be stored or treated. Specific aspects of the assessment are listed at 265.19(b). The assessment of the tank system's integrity must be certified by an independent, qualified, registered professional engineer. Tanks determined to be leaking or unfit for use must be removed from service, per 265.196.

Design and Installation of New Tank Systems or Components (§265.192)

New tank systems or new components of existing tank systems must be adequately designed, compatible with the waste(s) to be stored or treated, protected from corrosion, installed properly, and tested for leaks prior to being placed in use. A written assessment must be certified by an independent, qualified, registered professional engineer as to the structural integrity of the new tank system or component. Specific aspects of the assessment are listed at 265.192(a). Certification of the installation and of any repairs is also required.

Containment and Detection of Releases (§265.193)

Complete secondary containment, as specified, must be provided for all new tank systems or replacement components prior to their being put into service. Existing tank systems are required to be retrofitted with complete secondary containment by the time the tank reaches 15 years of age, but not before January 12, 1989.

General Operating Requirements (§265.194)

Hazardous wastes or treatment reagents must not be placed in a tank system if they could cause the tank, its ancillary equipment, or the secondary containment system to rupture, leak, corrode, or otherwise fail.

Appropriate controls and practices must be used to prevent spills and overflows from the tank or any secondary containment systems. Minimum types of controls and practices are specified at 265.194(b).

If a leak or spill occurs in the tank system, the requirements of 265.196 must be adhered to.

Inspections (§265.195)

Daily inspections are required of:

1. overfill/spill control equipment
2. aboveground portions of the tank system
3. data gathered from monitoring and leak-detection equipment
4. construction materials and the area immediately surrounding the externally accessible portion of the tank system including secondary containment structures

Cathodic protection systems, if present, must be inspected for specified items on a specified schedule.

All inspections must be documented.

Response to Leaks or Spills and Disposition of Leaking or Unfit-for-Use Tank Systems (§265.196)

Tank systems or secondary containment systems from which there has been a leak or spill, or which are unfit for use, must be removed from service immediately. The following requirements must also be met:

1. Cease use; prevent additional releases or the addition of wastes.
2. Remove waste from the tank system or secondary containment system.
3. Contain and remove visible releases to the environment.
4. Notify the EPA regional administrator within 24 hours of releases not confined by the secondary containment system, unless a report was made to the National Response Center per 40 CFR Part 302.
5. Submit a release assessment, as specified at 265.196(d), within 30 days of the release.
6. Correct the problem by either closing the tank per 265.197 or repairing or replacing the damaged components. Any replacement component is considered a new component per 265.192. Major repairs must be certified by an independent, qualified, registered professional engineer.

Closure and Postclosure Care (§265.197)

At closure, removal or decontamination is necessary for all waste residues, contaminated containment system components, contaminated soils, and structures and equipment contaminated with wastes. These items are to be managed as hazardous waste, unless they do not meet the definition of hazardous waste.

Postclosure requirements are imposed if all contaminated soils are not removed. Such areas are considered hazardous waste landfills for regulatory purposes.

Special Requirements for Ignitable or Reactive Wastes (§265.198)

Ignitable or reactive wastes must not be placed in a tank system except under any of the following conditions:

1. The waste is treated, rendered, or mixed before or immediately after placement in the tank system so that the resulting waste, mixture, or dissolved material is no longer characteristically ignitable or reactive, and 265.17(b) is complied with.
2. The waste is stored or treated in such a way that it is protected from any material or conditions that may cause the waste to ignite or react.
3. The tank system is used solely for emergencies.

Protective distances must be maintained between the tank system and any public ways, streets, alleys, or adjoining property lines per Tables 2-1 through 2-6 of the National Fire Protection Association's "Flammable and Combustible Liquids Code" (1977 or 1981).

Special Requirements for Incompatible Wastes (§265.199)

Incompatible wastes, or incompatible wastes and materials, must not be placed in the same tank system, unless 265.17(b) is complied with. Hazardous waste must not be placed in a tank system that has not been decontaminated and that previously held an incompatible waste or material, unless 265.17(b) is complied with.

General Requirements for Ignitable, Reactive, or Incompatible Wastes (§265.17)

1. Precautions must be taken to prevent accidental ignition or reaction of ignitable or reactive waste. Such wastes must be separated and protected from sources of ignition or reaction. While ignitable or reactive waste is being handled, smoking and open flames must be confined to specially designated locations. ''No Smoking'' signs must be conspicuously placed wherever there is a hazard from ignitable or reactive waste.

2. The treatment, storage, or disposal of ignitable or reactive waste, and the mixture or commingling of incompatible wastes, and incompatible wastes and materials, must be conducted so that it does not:
   a. generate extreme heat or pressure, fire or explosion, or violent reaction
   b. produce uncontrolled toxic mists, fumes, dusts, or gases in sufficient quantities to threaten human health
   c. produce uncontrolled flammable fumes or gases in sufficient quantities to pose a risk of fire or explosions
   d. damage the structural integrity of the device or facility containing the waste
   e. through other like means threaten human health or the environment

CONCLUSION

The regulatory requirements for manufacturing firms as accumulators of hazardous waste are fairly comprehensive. Compliance with these requirements is much more involved than merely moving wastes offsite every 90 days. Compliance will require the implementation of a formal HWM program. Although the requirements for such a program for accumulators are not as extensive as those for interim status or fully permitted storers (discussed in Chapter 11), many of the elements of such a program are shared. Setting up an environmental management program is discussed in Section III of this book.

CHAPTER 10

Generators Who Ship Hazardous Wastes Offsite

The offsite shipment of hazardous waste requires the environmental manager to deal with the regulatory program of the DOT as it meshes with that of the EPA under RCRA. The EPA regulations for generators (40 CFR Part 262) and transporters (40 CFR Part 263) make frequent reference to the DOT Hazardous Materials Transportation Regulations (49 CFR Parts 171–179). In fact, the EPA standards for offsite shipments of hazardous waste are indecipherable without a copy of the DOT regulations. The purpose of this chapter is to explain the hazardous materials transportation program established by DOT, in the context of hazardous waste shipments.

The essence of the EPA/DOT standards for generators who ship offsite is deceptively simple. The generator has the responsibility for determining where the waste will undergo treatment, storage, or disposal, and that the transporter and designated facility have EPA identification numbers and proper permits to accept the generator's particular type of waste. The generator must properly describe the waste using DOT shipping descriptions and ensure that the containers are marked, labeled, and are otherwise both suitable for the particular material and in proper condition for transportation. The waste must be accompanied by a shipping document called a manifest, the transportation vehicle must be placarded, as appropriate, and the generator must verify that the waste is actually received by the designated TSD facility. The many pages of DOT regulations merely apply this scheme to the variations made possible due to the many types of hazardous materials, types of containers, and modes of transportation.

The DOT has the authority to regulate both the inter- and intra-state transportation of hazardous materials, under the authority of the Hazardous Materials Transportation Act (HMTA). DOT promulgated regulations under the authority of HMTA, at 49 CFR Parts 171–179. Their purpose is to promote the uniform enforcement of law amongst the states in order to minimize barriers to interstate commerce, as well as to minimize the dangers to life and property incident to the transportation of hazardous materials. DOT established requirements in these regulations for shipping papers, proper containers, marking and labeling of containers, placarding of vehicles, and incident reporting. The DOT regulations address all modes of transportation (air, rail, highway, waterway, and pipeline), and apply to the transportation of hazardous materials regardless of their end use or ultimate disposition. DOT considers hazardous wastes to be a subset of hazardous materials, and thus regulated by their program.

It is important to remember that the purposes of the DOT regulations (i.e., to minimize the dangers to life and property during transit) are somewhat different than those of the hazardous waste regulations (i.e., protection of human health and the environment from hazardous waste mismanagement). This factor, along with other factors, results in somewhat different terminology, as well as different hazard class definitions between the two programs. For example, DOT uses the terms "shipper" and "carrier" for parties identified by EPA as "generator" and "transporter." Also, the DOT definition of flammable liquid is *not* the same as the EPA definition of ignitable waste. The DOT Poison A or B hazard class definitions do not correspond well with EPA's EP toxicity characteristic. DOT UN/NA identification numbers are not interchangeable with EPA hazardous waste numbers. The DOT and EPA have different definitions of empty container.

For purposes of offsite shipments, it is important to remember that hazardous wastes are a subset of hazardous materials. Recent changes to the DOT regulations have not altered this relationship, but they have resulted in some changes in the proper shipping descriptions and marking requirements for hazardous wastes. DOT has adopted RQs established under CERCLA, replacing the Clean Water Act RQs previously used.

The structure of the DOT regulations is shown in Table 10.1. The hazardous materials transportation regulations are issued by the Research and Special Programs Administration (RSPA) of the DOT. When reviewing *Federal Register*s for revisions to HMTA regulations, one must look under RSPA instead of DOT. Note that the regulations are organized by chapter, subchapter, part, and subpart. This is important to remember, as there are exceptions and exclusions which refer to the requirements of a particular chapter, subchapter, part, or subpart. For example, an exemption from the requirements of a subpart would not mean that a certain hazardous material is completely excluded from the DOT regulations.

In order to understand the DOT regulations, it is necessary to know the definitions utilized for key terms. As many already realize, the regulatory definition of a particular term often differs from the common-sense definition. Therefore, knowledge of regulatory definitions is necessary to determine the applicability and scope of the regulations with respect to a particular situation. The definitions utilized by DOT for certain key terms are given in Table 10.2. Note that many

Table 10.1. DOT Regulations Regarding Hazardous Materials and Wastes

49 CFR Chapter I—Research and Special Programs Administration
    Subchapter C—Hazardous Materials Regulations
        Part 171—General Information, Regulations, and Definitions
        Part 172—Hazardous Materials Table and Hazardous Materials Communication
            Requirements
            Subpart B—Table of Hazardous Materials
            Subpart C—Shipping Papers
            Subpart D—Marking
            Subpart E—Labeling
            Subpart F—Placarding
        Part 173—Shippers—General Requirements for Shipments and Packagings
            Subpart B—Preparation of Hazardous Materials for Transportation
            Subpart C–O—Definitions of Various Hazard Classes
        Part 174—Carriage by Rail
        Part 175—Carriage by Aircraft
        Part 176—Carriage by Vessel
        Part 177—Carriage by Public Highway
        Part 178—Shipping Container Specifications
        Part 179—Specification for Tank Cars

Table 10.2. Key DOT Definitions

Hazardous Material—for transportation purposes, a substance or material, including a hazardous substance, which has been determined by the Secretary of Transportation to be capable of posing an unreasonable risk to health, safety, and property when transported in commerce, and which has been so designated

Hazardous Substance—for transportation purposes, a material, including its mixtures and solutions, that:

1. is listed in the appendix to the Hazardous Materials Table (HMT) at 49 CFR 172.101
2. is in a quantity, in one package, which equals or exceeds the reportable quantity (RQ) listed in the appendix to the HMT
3. when in a mixture or solution, is in a concentration by weight which equals or exceeds the concentration corresponding to the RQ of the material, as shown in the following table:

RQ: lb (kg)	Concentration by Weight	
	%	PPM
5,000 (2270)	10	100,000
1,000 (454)	2	20,000
100 (45.4)	0.2	2,000
10 (4.54)	0.02	200
1 (0.454)	0.002	20

This definition does not apply to petroleum products that are lubricants or fuels.

Hazardous Waste—for transportation purposes, any material that is subject to the Hazardous Waste Manifest requirements of the U.S. EPA, as specified in 40 CFR Part 262

RQ—for transportation purposes, the quantity of a hazardous substance listed as its Reportable Quantity in the Appendix to the HMT entitled "List of Hazardous Substances and Reportable Quantities"

ORM—Other Regulated Materials

n.o.s.—not otherwise specified

of the terms are preceded by the phrase "for transportation purposes." This distinguishes the DOT usage of a particular term from the way the term is defined or used in other regulatory programs. In the case of the definition of hazardous waste, the phrase allows reference to the EPA definition of hazardous waste without specifically including the lists and characteristics of hazardous waste.

Before delving into the body of the DOT regulations, it is instructive to list how the EPA and DOT regulations interact, in the context of a shipment of hazardous waste. It is assumed that the waste material has been determined to be a hazardous waste and that prior arrangements have been made with a commercial TSD facility and transporter and that all parties have agreed to these arrangements. As will be noted by reference to Table 10.3, a whole series of requirements come into play once a generator/shipper decides to have a third party (transporter/carrier) haul his hazardous waste to an offsite hazardous waste management facility.

Table 10.3. EPA/DOT Regulatory Requirements for Generators Who Ship Hazardous Wastes Offsite

1. Select appropriate shipping name, hazard class, and DOT Identification (UN/NA) Number, and determine if RQ is being shipped in a single container. Determine whether the waste is subject to any land disposal restrictions under 40 CFR Part 268.

2. Comply with DOT requirements for packaging, labeling, and marking.

3. Verify that transporter and TSD facility have valid EPA identification numbers and relevant permits.

4. Prepare a uniform hazardous waste manifest, utilizing appropriate form per acquisition hierarchy. Prepare the appropriate notice for restricted wastes shipped to treatment facilities. Prepare the appropriate notice and certification for restricted wastes shipped to land disposal facilities. Prepare the necessary demonstration and certification for soft-hammer wastes shipped offsite.

5. Offer the transporter the appropriate placards.

6. Sign manifest, certifying shipment meets EPA and DOT pretransportation requirements. Signature also certifies that a waste minimization program is in place, and that the method of treatment, storage, or disposal selected is environmentally appropriate.

7. Obtain signature of transporter and date of acceptance.

8. Retain one copy of the signed manifest and give the remaining copies to the transporter.

9. Monitor the manifest tracking system.

10. Submit exception reports, as appropriate.

11. Prepare and submit biennial reports including a description of waste minimization (volume and toxicity) efforts and achievements.

12. Keep records of manifests, land disposal restriction notices/certifications, biennial reports, any exception reports, and waste analyses/determinations.

SHIPPING DESCRIPTION SELECTION

The first requirement is to determine the appropriate shipping name, hazard class, and DOT identification number. These three elements are referred to as the "shipping description." Selecting an appropriate shipping description for a hazardous waste can be difficult, because such waste materials are often variable mixtures of several chemicals, are spent or have become diluted with water or oil, and/or consist of multiple phases. As a result, technical name/chemical name entries in the DOT Hazardous Materials Table (HMT) are often inappropriate for shipping name purposes.

In order to select the proper shipping description for hazardous wastes, it is necessary to be familiar with the definitions of the various DOT hazard classes, and to know how to utilize the DOT Hazardous Materials Table (found at 49 CFR 172.101). Selection of a proper shipping description requires knowledge of the physical properties and the chemical characteristics of the waste material vis-a-vis the DOT hazard class definitions and DOT Hazardous Materials Table. Summaries of the DOT Hazard Class Definitions are found in Appendix I. Table 10.4 lists the various DOT hazard classes; Table 10.5 provides the hierarchy of hazard classes to be utilized when selecting shipping descriptions for waste materials meeting the definition of more than one DOT hazard class.

Table 10.4.   DOT Hazard Classes

Explosive A	Nonflammable Gas
Explosive B	Poison A
Explosive C	Poison B
Blasting Agent	Irritating Material
Flammable Liquid	Etiologic Agent
Combustible Liquid	Radioactive Material
Flammable Solid	Other Regulated Material (ORM)
	ORM-A
Oxidizer	ORM-B
Organic Peroxide	ORM-C
Corrosive Material	ORM-D
Flammable Gas	ORM-E

The generator will have some knowledge of the waste material to be shipped from several sources:

1. analytical result(s) from laboratory testing for the hazardous waste characteristics
2. analytical result(s) from laboratory testing by the designated facility for treatment/disposal purposes
3. MSDS information for the purchased components of the waste material
4. knowledge of the hazard characteristics of the waste in light of the materials and process(es) involved

Table 10.5.   Hierarchy of Hazard Classes

(1) Radioactive material	(10) Corrosive material (solid)
(2) Poison A	(11) Irritating material
(3) Flammable Gas	(12) Combustible liquid (in con-
(4) Nonflammable gas	tainers more than 110 gal)
(5) Flammable liquid	(13) ORM-B
(6) Oxidizer	(14) ORM-A
(7) Flammable solid	(15) Combustible liquid (in con-
(8) Corrosive material (liquid)	tainers less than 110 gal)
(9) Poison B	(16) ORM-E

This listing will *not* apply to materials/wastes specifically identified in the Hazardous Material Table (49 CFR 172.101) or to:

Explosives (173.86)
Blasting agents (173.114(a))
Etiological agents (173.386)
Organic peroxide (173.151(a))

This information is necessary, but may not be sufficient to determine the appropriate shipping description. It is critical that sufficient information be available to determine into which DOT hazard class(es) the waste material is best classified. Remember the DOT hazard class definitions are not the same as the EPA hazardous waste characteristics. For example, a waste material can be characteristically ignitable per EPA regulations, yet not be a DOT flammable liquid. DOT hazard class definitions are found at 49 CFR Part 173, in association with the packaging requirements for materials falling within the hazard class.

If a hazardous waste does not meet any of the technical (chemical name) descriptions, hazardous waste may be assigned a shipping description from the HMT based on the hazard class definitions, the hierarchy of hazard classes, and the shipper's knowledge of the waste material.

Hazardous Materials Table

Selection of the shipping description is done by reference to the HMT (49 CFR 172.101). The HMT designates those materials that are to be treated as hazardous materials for transportation purposes. The HMT is the key to determining the DOT requirements for packaging, labeling, and marking a waste material fitting a particular shipping description. The use of the Hazardous Materials Table (and its appendix, the DOT List of Hazardous Substances and Reportable Quantities) should be mastered by generators/shippers of hazardous wastes. Note that the Optional Hazardous Materials Table, found at 49 CFR 172.102, is for international shipments. The Optional HMT should *not* be used for any domestic hazardous waste shipment.

§172.101 Hazardous Materials Table

(1) +/ A/W	(2) Hazardous materials descriptions and proper shipping names	(3) Hazard class	(3A) Identification number	(4) Label(s) required (if not excepted)	(5) Packaging		(6) Maximum net quantity in one package		(7) Water shipments		
					(a) Exceptions	(b) Specific requirements	(a) Passenger carrying aircraft or railcar	(b) Cargo aircraft only	(a) Cargo vessel	(b) Passenger vessel	(c) Other requirements
	Thiophenol. See Phenyl mercaptan.										
+	Thiophosgene	Poison B	UN2474	Poison	None	173.356	Forbidden	1 gallon	1	5	Shade from radiant heat
	Thiophosphoryl chloride	Corrosive material	UN1837	Corrosive	None	173.271	Forbidden	1 quart	1	1	Keep dry. Glass carboys not permitted on passenger vessels
A	Thiram	ORM-A	NA2771	None	173.505	173.510	Forbidden	No limit			
	Thorium metal, pyrophoric	Radioactive material	UN2975	Radioactive material and Flammable solid	None	173.418	Forbidden	2-5 pounds	1,2	1,2	
	Thorium nitrate	Radioactive Material	UN2976	Radioactive and Oxidizer	None	173.419	Forbidden	25 pounds	1,2	1,2	Separate longitudinally by a complete hold or compartment from explosives
	Time fuze. See Fuze, time										
	Tin chloride, fuming. *See* Tin tetrachloride, anhydrous										
	Tin perchloride. *See* Tin tetrachloride, anhydrous										
	Tin tetrachloride, anhydrous	Corrosive material	UN1827	Corrosive	173.244	173.247	1 quart	1 quart	1	1	Keep dry. Glass carboys not permitted on passenger vessels
	Titanium metal powder, dry *or wet with less than 20% water*	Flammable solid	UN2546	Flammable solid	None	173.208	Forbidden	75 pounds	1,2	5	
	Titanium metal powder, wet with 20% or more water	Flammable solid	NA1352	Flammable solid	None	173.208	Forbidden	150 pounds	1,2	5	
	Titanium sulfate solution *containing not more than 45% sulfuric acid*	Corrosive material	NA1760	Corrosive	173.244	173.297	1 quart	1 gallon	1	4	Shade from radiant heat. Keep dry
	Titanium tetrachloride	Corrosive material	UN1838	Corrosive	173.244	173.247	1 quart	10 gallons	1	1	Keep dry. Glass carboys not permitted on passenger vessels
	Toluene *(toluol)*	Flammable liquid	UN1294	Flammable liquid	173.118	173.119	1 quart	10 gallons	1,2	1	

Figure 10.1. DOT Hazardous Materials Table.

§172.101 Hazardous Materials Table—Contd.

(1) +/ A/W	(2) Hazardous materials descriptions and proper shipping names	(3) Hazard class	(3A) Identi-fication number	(4) Label(s) required (if not excepted)	(5) Packaging (a) Exceptions	(5) Packaging (b) Specific require-ments	(6) Maximum net quantity in one package (a) Passenger carrying aircraft or railcar	(6) (b) Cargo aircraft only	(7) Water shipments (a) Cargo vessel	(7) (b) Passenger vessel	(7) (c) Other requirements
<	Toluenediamine	ORM-A	NA1709	None	173.505	173.510	No limit	No limit			
+	Toluene diisocyanate	Poison B	UN2078	Poison	173.345	173.346	Forbidden	55 gallons	1,3	1,3	Shade from radiant heat
	Toluene sulfonic acid, liquid	Corrosive material	UN2584	Corrosive	173.244	173.245	1 quart	10 gallons	1,2	1,2	
	Torch. See Fireworks, common										
	Torpedo, railway	Class B explosive		Explosive B	None	173.91	Forbidden	200 pounds	1,2	1,2	Passenger vessels in metal lockers only
<	Toxaphene	ORM-A	NA2761	None	173.505	173.510	25 pounds	100 pounds	1,2	1,2	
	Toy caps	Class C explosive		Explosive C	None	173.100 173.109	50 pounds	150 pounds	1,3	1,3	
	Toy propellant device	Class C explosive		Explosive C	None	173.111	50 pounds	150 pounds	1,3	1,3	
	Toy smoke device	Class C explosive		Explosive C	None	173.111	50 pounds	150 pounds	1,3	1,3	
	Toy torpedo. See Fireworks, special										
	2,4,5-TP. See 2,4,5-Trichlorophenoxypropionic acid										
	Tracer	Class C explosive		Explosive C	None	173.105	50 pounds	150 pounds	1,3	1,3	
	Tracer fuze	Class C explosive		Explosive C	None	173.105	50 pounds	150 pounds	1,3	1,3	
	Tractor. See Motor vehicle										
	Trailer or truck body with refrigeration or heating equipment. See Motor vehicle										
	Triazine pesticide, liquid, n.o.s. *(compounds and preparations)*	Flammable liquid	UN2764	Flammable liquid	173.118	173.119	1 quart	10 gallons	1,2	1	
	Triazine pesticide, liquid, n.o.s. *(compounds and preparations)*	Poison B	UN2763	Poison	173.345	173.346	1 quart	55 gallons	1,2	1,2	
	Triazine pesticide, solid, n.o.s. *(compounds and preparations)*	Poison B	UN2763	Poison	173.364	173.365	50 pounds	200 pounds	1,2	1,2	
	Tri-(β-nitroxyethyl)ammonium nitrate	Forbidden									

Figure 10.1, cont'd.

§172.101 Hazardous Materials Table—Contd.

(1)	(2)	(3)	(3A)	(4)	(5) Packaging		(6) Maximum net quantity in one package			(7) Water shipments		
					(a) Exceptions	(b) Specific require-ments	(a) Passenger carrying aircraft or railcar	(b) Cargo aircraft only	(a) Cargo vessel	(b) Pas-senger vessel	(c) Other requirements
+/ ∧/w	Hazardous materials descriptions and proper shipping names	Hazard class	Identi-fication number	Label(s) required (if not excepted)							
A	Trichlorfon	ORM-A	NA2783	None	173.505	173.510	50 pounds	200 pounds	1,2	1,2	
	Trichloroacetic acid, solid	Corrosive material	UN1839	Corrosive	173.244	173.245b	25 pounds	100 pounds	1,2	1	
	Trichloroacetic acid solution	Corrosive material	UN2564	Corrosive	173.244	173.245	1 quart	1 quart	1,2	1,2	Glass carboys in hampers not permit-ted under deck
A	1,1,1-Trichloroethane	ORM-A	UN2831	None	173.505	173.510	10 gallons	55 gallons			

Figure 10.1, cont'd.

The HMT consists of seven major columns, as shown in Figure 10.1. These columns give specific information and/or references to the following information:

1. mode or mode(s) of transportation which are regulated—Column 1
2. proper shipping name—Column 2
3. hazard class—Column 3
4. DOT identification number—Column 3a
5. required labeling—Column 4
6. packaging requirements and exceptions—Column 5

Columns 6 and 7 deal with requirements and limitations for air, passenger-carrying rail, or water shipments, and will not be discussed further. The List of Hazardous Substances and Reportable Quantities appears as an appendix to the current version of the HMT.

Column 1 contains the three symbols (as appropriate): plus (+) and the letters "A" and "W." The appearance of the plus (+) symbol in association with a shipping name fixes the proper shipping name and hazard class for that entry, regardless of whether the material meets the definition of the hazard class associated with the shipping name. As a general rule, if a material meeting a shipping name entry does not also meet the definition of the hazard class associated with that entry, another shipping name must be selected. This rule is ignored when the plus (+) symbol appears in column 1. For example, a generator shipping a waste material best described as "toluene diisocyanate" must use this shipping name along with the Poison B hazard class, even if the waste material was so diluted and reacted that it no longer met the Poison B hazard class definition. Note that relatively few shipping names are preceded by the plus (+) symbol.

The letter "A" in column 1 of an entry limits the application of the regulations to such materials when transported by aircraft, *unless* the material is a hazardous substance or a hazardous waste. The letter "W" in column 1 of an entry limits the application of the regulations to such materials when transported by vessel, *unless* the material is a hazardous substance or a hazardous waste. To generators shipping hazardous wastes, the appearance of the letters "A" or "W" in column 1 of an entry will not exempt the material from DOT regulation, regardless of the transportation mode selected. So, the letters "A" and "W" in column 1 have no meaning for hazardous waste shipments.

Column 2—Hazardous Materials Descriptions and Proper Shipping Names—lists the proper shipping names of those materials designated by DOT as hazardous materials for shipping purposes. Remember that hazardous wastes are *by definition* a subset of hazardous materials, so an appropriate shipping description must be obtained from the HMT. Entries are in alphabetical order, by technical (chemical) name. No trade names are identified in the HMT. Proper shipping names are limited to those entries in roman type. Italicized entries are not to be used. Italicized entries typically are synonyms for and give reference to a shipping name entry in roman type. When a column 2 shipping name entry references another shipping name by the use of the word "see" or "or," then either shipping name can be used, provided both are in roman type.

Column 3 of the HMT contains the hazard class of the material as it must appear on the shipping papers. When the column 3 entry is "Forbidden," that material may not be offered or accepted for transportation. It is important to note that the HMT is not to be used like a menu in a Chinese restaurant. The shipper cannot select any shipping name from column 2 and any hazard class from column 3. The shipping name and hazard class (and identification number) associated with any entry are to be taken as a unit. In other words, if a material can be described by the shipping name in column 2, yet does not meet the definition of the hazard class found in column 3, then the shipping description is not the proper description for that particular material. As discussed above, this rule does not apply when a plus (+) appears in column 1.

If a plus (+) does not appear in column 1, the physical properties and chemical characteristics of the material described by the entry in column 2 should be compared against the definition of the hazard class in column 3. If the characteristics of the material do not meet the definition of any hazard class, including ORM-E, then the material is not regulated as a hazardous material. Note, however, that all hazardous wastes and hazardous substances are *by definition* ORM-E, if they do not meet the definition of any other hazard class. The hazard class identified in column 3 is to be used with Tables 1 and 2 in 49 CFR 172.504 for determining placarding requirements.

Column 3a assigns each hazardous material a UN/NA (United Nations/North America) identification number. The numbers in this column must be placed on shipping papers and marked on packages (containers). Portable tanks, cargo tanks, and tank cars are to be marked with the identification number by use of specification orange panels or placards. The purpose behind the numbering system is to improve the capability of emergency personnel to quickly identify hazardous materials and to ensure that accurate information is available. The UN/NA identification numbers are referenced by the DOT *Emergency Response Guidebook,* carried by most public sector emergency responders. Remember that UN/NA numbers should *not* be confused with EPA hazardous waste numbers.

Column 4 identifies required hazard warning label(s) for packages. Specifications for the 4″ × 4″ diamond-shaped DOT hazard warning labels are found at 49 CFR Part 172, Subpart E—"Labeling." Examples of DOT hazard warning labels for common hazardous wastes are shown in Figure 10.2. Meeting DOT specifications for hazard warning labels is most easily assured by purchasing labels from any one of several label suppliers. Labelmaster in Chicago (800/621-5808) is perhaps the best known.

Column 5 identifies packaging requirements. Column 5a identifies the regulatory sections specifying allowable exceptions, if any. Column 5b gives the regulatory section containing specification packaging options for that hazardous material.

Columns 6 and 7 deal with requirements and limitations for air or water shipments, and will not be discussed further.

Figure 10.2.   DOT hazard warning labels for common hazardous wastes.

Selection of the most appropriate shipping description for hazardous wastes is easier said than done. The DOT regulations give little guidance in this respect. When selecting a proper shipping name for a particular material, the entry in the HMT that most accurately and specifically identifies the material must be used. For waste materials, especially mixtures, the correct technical (chemical) name of a material is not entirely accurate. For other waste materials consisting of proprietary chemicals, a technical (chemical) name is unavailable. Another situation exists when a waste chemical has been so diluted or has become so spent that the waste material described by a technical (chemical) name no longer meets the definition of the associated hazard class. In the event that the correct technical name is not listed, is not accurate, or is not accurately described by the associated hazard class, the next most specific entry must be determined from the other types of descriptions corresponding to the specific hazard class(es) of the material to be shipped. A hierarchy of shipping descriptions has been derived from evaluation of the entries in the HMT. This hierarchy goes from more specific to less specific types of entries. This hierarchy, shown in Table 10.6, can be utilized in selecting the most appropriate shipping description for a particular hazardous waste. The hierarchy is utilized as follows. If an appropriate listing by technical (chemical) name is not found, listings by chemical family are

Table 10.6.   Hierarchy of Shipping Descriptions from the DOT Hazardous Materials Table (49 CFR 172.101)

1. *Listing by Chemical (Technical) Name*
   e.g. Isopropyl alcohol—Flammable liquid—UN 1219
   Sulfuric acid—corrosive material—UN 1830
   Sodium cyanide solution—Poison B—UN 1689

2. *Listing by Chemical Family*
   e.g. Alcohol, n.o.s.—Flammable liquid—UN 1987
   Acid, liquid, n.o.s.—Corrosive material—NA 1760
   Cyanide solution, n.o.s.—Poison B—UN 1935

3. *Listing by Specific End-Use Description Corresponding to Hazard Class of the Material*
   e.g. Compound, cleaning, liquid—Flammable liquid—NA 1993
   Battery fluid, acid—Corrosive material—UN 2796
   Paint related material—Flammable liquid—NA 1263
   Ink—Combustible liquid—UN 2867

4. *Listing by End-Use Description, n.o.s., Corresponding to Hazard Class of the Material*
   e.g. Driers, paint, liquid, n.o.s.—Combustible liquid—UN 1168
   Etching acid, liquid, n.o.s.—Corrosive material—NA 1790
   Petroleum oil, n.o.s.—Flammable liquid—NA 1270
   Disinfectant, liquid, n.o.s.—Combustible liquid—NA 1993
   Drugs, n.o.s.—Combustible liquid—NA 1993

5. *Listing by Hazard Class Description, n.o.s.*
   e.g. Flammable liquid, n.o.s.—UN 1993
   Flammable liquid, poisonous, n.o.s.—UN 1992
   Corrosive liquid, n.o.s.—UN 1760
   Poison B, liquid, n.o.s.—UN 2810
   ORM-A, n.o.s.—NA 1693

6. *ORM-E Listings for Hazardous Substances Not Meeting Any Other DOT Hazard Class Definitions*
   e.g. Hazardous substance, liquid, n.o.s.—ORM-E—NA 9188
   Hazardous waste, solid, n.o.s.—ORM-E—NA 9189

evaluated. If no such listings are found or are appropriate, the listings by specific end-use description and then end-use description n.o.s. (not otherwise specified), corresponding to the hazard class of the material, are reviewed. At this point, the hazard class n.o.s. descriptions are evaluated.

Many hazardous wastes currently being described under a hazard class n.o.s. description would be more properly described by an end-use description. For example, many spent solvents from painting operations manifested as "Waste Flammable Liquid, n.o.s. UN 1993" would be better described by "Waste Paint Related Material, Flammable Liquid, NA 1263."

When evaluating hazard class n.o.s. descriptions, it is necessary to remember the hierarchy of hazard classes when describing materials meeting the definition of more than one hazard class. The hierarchy of hazard classes was shown in Table 10.5. Except when there are hazard class n.o.s. listings which specifically include all hazard classes posed by a material with multiple hazard classes (e.g., flammable liquid, poisonous, n.o.s, UN 1992), the hazard class n.o.s. entry corresponding with the highest hazard class of the material is to be selected. As an example, for a material which is both a flammable liquid and an irritating material per DOT, the "flammable liquid, n.o.s." entry would be utilized, since flammable liquids are higher on the hierarchy than irritating materials, and there is no "flammable liquid, irritating material, n.o.s." entry.

The last resort with respect to shipping descriptions are the ORM-E entries. Recent regulatory changes have resulted in the elimination of all but two ORM-E entries. The remaining ORM-E entries are:

1. Hazardous Waste, liquid or solid, n.o.s.
2. Hazardous Substance, liquid or solid, n.o.s. or ORM-E, liquid or solid, n.o.s.

These entries are to be utilized only after all other levels of the shipping description hierarchy have been evaluated and found not to contain an appropriate entry. Hazardous wastes not meeting the definition of any other hazard class are *by definition* ORM-E, since hazardous wastes are *by definition* a subset of hazardous materials. DOT established the ORM-E hazard class for those materials regulated by EPA as hazardous wastes that would not otherwise pose the level of hazard in transportation requiring DOT regulation.

A good example of an ORM-E hazardous waste would be a wastewater treatment sludge from an electroplating operation. The treatment operations typically performed on electroplating rinsewaters are designed to destroy any cyanides, reduce hexavalent chromium to the trivalent state, and raise the pH to precipitate metals as alkaline metal hydroxides. The dewatered sludges pose little acute environmental or transportation risks, yet are regulated by EPA either as listed wastes or as characteristic wastes (if containing levels of leachable heavy metals in excess of 100 times drinking water standards).

Once an appropriate shipping name/hazard class/identification number combination has been selected from the HMT, it is necessary to note the considerations which apply to how the description will appear on manifests and on containers. Recent revisions to the DOT regulations have added to these considerations when hazardous wastes are shipped. First, the proper shipping name for a hazardous waste must include the word "waste" preceding the shipping name (e.g. waste acetone) unless the word "waste" is included in the description in the HMT.

Secondly, DOT has adopted the RQs EPA established for hazardous substances under CERCLA. All listed and characteristic hazardous wastes have RQs. Common hazardous wastes/substances and their CERCLA/DOT RQs are given in Table 9.4. The DOT List of Hazardous Substances and Reportable Quantities is found as an appendix to the HMT. For hazardous wastes, the generator/shipper should determine the RQ by referring to the entry in the appendix for the hazardous waste number (if the waste is a listed hazardous waste), or the hazardous waste characteristic(s) of the waste. The entries for listed hazardous wastes are at the end of the appendix, following the alphabetical hazardous substance entries. The four entries for the characteristically hazardous wastes are identified together in the alphabetical entries, all preceded by the phrase "Unlisted Hazardous Wastes Characteristic of . . .".

DOT treats RQs somewhat differently than EPA does under CERCLA. Under CERCLA, hazardous wastes are *by definition* hazardous substances, with an RQ for each listed and characteristic waste. DOT considers hazardous wastes to be hazardous substances *for transportation purposes* only when an amount equal to or exceeding the RQ is shipped *in a single package* (i.e., container, roll-off box, or tank truck). The entire amount shipped may greatly exceed the RQ. However, RQ designations on the manifest and containers only become mandatory when a RQ is placed in a single container. The letters RQ must be entered on the shipping paper either before or after the basic description for each hazardous substance. As you may note from review of Table 9.4, many common hazardous wastes have very low RQs, in the 1-pound to 100-pound range. Shipments of these hazardous wastes in 55-gallon containers would result in the hazardous wastes becoming hazardous substances for DOT purposes, requiring the RQ as part of the shipping description, and as part of the container marking requirements.

RQ shipments of hazardous wastes must also have the EPA hazardous waste number(s) entered in association with the proper shipping description. The hazardous waste number(s) are to be placed in parentheses after the basic description. For characteristic wastes, the letters "EPA" and the word(s) "ignitability," "corrosivity," "reactivity," and/or "EP toxicity," in parentheses, as appropriate, may substitute for the EPA hazardous waste number (D001, D002, D003, D004–D017). The EPA hazardous waste number(s) must appear with the shipping description, even though they may also appear in Item I (Waste No.) of the Uniform Hazardous Waste Manifest.

For mixtures of a hazardous material identified by technical name and a *non*-hazardous material, "mixture" or "solution," as appropriate, can be added to the proper shipping name, provided the mixture or solution is not specifically identified in the HMT, and the hazard class of the mixture or solution is the same as that of the hazardous material's HMT entry. For mixtures of two or more hazardous materials, the use of a technical (chemical) name and "mixture" or "solution" is not appropriate, unless the mixture is specially identified in the HMT. In the instances of mixtures of two or more hazardous materials, an appropriate shipping description must be selected using the hierarchy of shipping descriptions.

Before continuing further with the packaging, marking, and labeling requirements, it may be useful to repeat some basic concepts in deriving appropriate shipping descriptions. Any material meeting the definition of a hazard class other than the one specified with a shipping name entry *must* be described by a shipping name that appropriately corresponds with its actual hazard class. The only exceptions are for those entries preceded by a plus (+) in column 1 of the HMT. Also, when utilizing a hazard class n.o.s. description for a waste with multiple hazards, it is necessary to refer to the hierarchy of hazard classes.

PACKAGING OF HAZARDOUS MATERIALS

Column 5 of the HMT contains references to packaging information for the hazardous material listed in Column 2. Some packages are "excepted" from some parts of the regulations provided they meet certain requirements. Column 5a contains a reference to the applicable regulatory section, if any, dealing with excepted packaging.

If Column 5a contains the entry "NONE" *or* if the excepted packaging options specified are not acceptable or feasible, then specification packaging must be used. Column 5b of the HMT refers to the section containing specification packaging requirements.

Container specifications are found in 49 CFR Part 173. All packages of hazardous materials must meet certain general packaging requirements. These are found at 49 CFR 173.24. Most hazard classes have both general and specific packaging requirements, found in 49 CFR Part 173 immediately after the definition section for the particular hazard class.

Certain other aspects of the DOT regulations have a direct impact on packaging of hazardous wastes. DOT allows, at 49 CFR 171.3(e), the use of equivalent specification open-head drums for hazardous wastes containing solids and semisolids, where the use of the required DOT-specification closed-head drum would be impracticable.

DOT allows the use of overpack or salvage drums for damaged or leaking packages containing hazardous materials, under certain circumstances. This exception from specification packaging is found at 49 CFR 173.3(c).

DOT allows the reuse of DOT-specification single-trip containers (STC) and nonreuseable containers (NRC) for hazardous waste shipments at 49 CFR 173.28(p). Such containers can be reused *once* for hazardous waste shipments, prior to reconditioning or disposal. There are certain conditions to reuse under this provision:

1. Except as authorized, the waste must be packaged and offered for transportation in accordance with the regulations.
2. Transportation is by highway only.
3. Package must stand closed for 24 hours prior to shipment.
4. Each package is to be inspected for leakage immediately prior to being offered for transportation.
5. Each package is loaded by the shipper and unloaded by the consignee, unless the carrier is a private or contract carrier.
6. Inconsistent markings and labels should be removed or painted over.

DOT-specification packaging marked as STC or NRC may be reused for the shipment of any corrosive solid, ORM-A, ORM-B, ORM-C, ORM-E, or any material not required to be shipped in DOT-specification packaging. This exception is found at 49 CFR 173.28(n).

MARKING OF PACKAGES

The requirement to mark packages, including containers of hazardous materials, with certain information is found at 49 CFR Part 172 Subpart D. Packages having a rated capacity of 110 gallons or less are to be marked with the proper shipping name and DOT identification number. Containers of certain poisonous materials are also to be marked "Inhalation Hazard," in association with the required labels.

If an amount of hazardous waste equal to or exceeding its RQ is shipped within a single package of 110-gallon capacity or less, then the letters RQ are to be marked on the package in association with the proper shipping name. In addition, the EPA hazardous waste number for listed wastes is to be marked on the package in parentheses, in association with the proper shipping name. For characteristically hazardous wastes, the EPA hazardous waste number can be so marked, or the letters EPA followed by the specific characteristic, as appropriate. Normally, marking the EPA hazardous waste number, in parentheses, will be much less cumbersome.

EPA has additional marking requirements for containers of 110 gallons or less. These requirements are to be met prior to transportation. Such containers are to be marked with the following words and information:

> **HAZARDOUS WASTE**—Federal Law Prohibits Improper Disposal. If found, contact the nearest police or public safety authority or the U.S. Environmental Protection Agency.
> Generator's Name and Address _____.
> Manifest Document Number _____.

Normally, the DOT and EPA marking requirements are satisfied with a commercially-printed decal, of the kind shown in Figure 10.2. The word "waste" need not be part of the shipping description marking if these EPA marking requirements are met.

For materials or wastes falling into the ORM hazard classes, marking requirements apply to packages with rated capacity of 110 gallons or less. Materials or wastes classified as ORM-A, B, C, D, or E must be marked with the ORM designation immediately following or below the proper shipping name. This marking requirement applies despite the fact that the HMT Column 4 shows no labeling requirements for hazardous materials meeting the ORM hazard classes. Many common hazardous wastes, including chlorinated solvents and wastewater treatment sludges, fall into the ORM hazard classes.

LABELING OF PACKAGES

The requirements for hazard class labeling of hazardous material packages are found at 49 CFR Part 172 Subpart E. Labeling refers to placing DOT-specification decals on packages, indicating the hazard class of the contents. Packages, including containers, of hazardous materials are to be labeled as specified in column 4 of the HMT shipping description for the hazardous material. For some hazardous materials, such as combustible liquids and other regulated materials (ORM hazard classes), no labels are required. For other hazardous materials, multiple hazard class labels are specified in the HMT. Portable tanks are not required to be labeled when they are appropriately placarded.

Hazard class labels are to be placed near the shipping name markings. When multiple hazard class labels are required, they must be placed next to each other. Labels on hazardous material packages must reflect the hazard of the hazardous material in the package. "Precautionary" labeling (i.e., use of the "flammable liquid" label on containers of combustible liquids) is prohibited at 49 CFR 172.401. The DOT hazard class definitions, found at 49 CFR Part 173, are very specific. Using common-sense interpretations or the EPA hazardous waste characteristics as substitutes for determining the appropriate DOT hazard class(es) is ill-advised.

PLACARDING OF TRANSPORTATION VEHICLES

Placarding is a joint shipper/carrier responsibility. The shipper must either placard the transportation vehicle or offer the transporter the required placards for the material being offered for transportation, unless the carrier's vehicle is already properly placarded for the material.

Placarding does not apply to etiologic agents or hazardous materials classed as ORM-A, B, C, D, or E. Placards must only be used for materials defined as hazardous materials, and must represent the hazard(s) of the material being transported. Placards must be placed on each side and on each end of the transport vehicle.

The various DOT hazard classes are split into Tables 1 and 2 (found at 49 CFR 172.504) for placarding purposes. Any quantity of hazardous materials classified as:

Class A Explosive
Class B Explosive
Poison A
Flammable solid ("Dangerous When Wet" label)
Radioactive material

must be placarded as provided in Table 1 of 49 CFR 172.504. Very few hazardous wastes would be subject to Table 1 placarding requirements. Most hazardous waste shipments would be subject to Table 2 placarding requirements, shown in Table 10.7.

Table 10.7. DOT Hazardous Material Shipment Placarding Requirements

If the transport vehicle or freight container contains a material classed (described) as:	The transport vehicle or freight container must be placarded on each side and each end:
Class C explosive	dangerous
Blasting agent	blasting agents
Nonflammable gas	nonflammable gas
Nonflammable gas (chlorine)	chlorine
Nonflammable gas (fluorine)	poison
Nonflammable gas (oxygen, cryogenic liquid)	oxygen
Flammable gas	flammable gas
Combustible liquid	combustible
Flammable liquid	flammable
Flammable solid	flammable solid
Oxidizer	oxidizer
Organic peroxide	organic peroxide
Poison B	poison
Corrosive material	corrosive
Irritating material	dangerous

Source: Table 2 of 49 CFR 172.504.

When the gross weight of all hazardous materials covered by Table 2 is less than 1000 pounds, no placards are required on the transport vehicle. For mixed shipments (requiring different placards) of Table 2 hazard classes, the DANGEROUS placard may be used in place of separate placarding for each class. *However,* when 5000 pounds or more of one class of material are loaded at a single facility, the placards specified in Table 2 for that class must be applied. Do *not* combine Table 1 and Table 2 items when determining placarding.

THE UNIFORM HAZARDOUS WASTE MANIFEST

The transportation of hazardous wastes to offsite TSD facilities is to be accompanied by the Uniform Hazardous Waste Manifest (EPA Form 8700-22). The Uniform Hazardous Waste Manifest is the mandatory format for shipping papers that must accompany hazardous waste shipments. In its simplest form, the manifest is a four-part shipping document which serves to track hazardous wastes to ensure delivery to the designated TSD facility. There is a continuation page available (EPA Form 8700-22A) for shipments of more than four types of hazardous wastes. The manifest serves both shipping document and tracking document purposes.

The Uniform Hazardous Waste Manifest (see Figure 1.1) is the mandatory format for hazardous waste shipping papers. However, it is not a universal form, as state hazardous waste management agencies are allowed to require the use of state-supplied manifest forms. Typically, a requirement to use a state-supplied manifest form is accompanied by additional parts of the manifest, as well as by requirements for the generator and/or the designated facility to mail manifest parts or copies to the state regulatory agency.

There is a hierarchy to be followed in determining the appropriate manifest form (not format) to use for shipments of hazardous waste to out-of-state TSD facilities. If the state in which the designated facility is located (the consignment state) supplies the manifest and requires its use, then that state's manifest form must be used. If the consignment state does not supply the manifest, but the state in which the generator is located (generator state) supplies the manifest and requires its use, then the generator state's manifest must be used. If neither the consignment state nor generator state supplies the manifest, then any manifest with sufficient parts or copies meeting the mandatory format of EPA Form 8700-22 can be used.

The Uniform Hazardous Waste Manifest must consist of sufficient copies (sometimes referred to as parts) to provide the generator, each transporter, and the designated facility with one copy each for their records and another copy to be returned to the generator by the designated facility, acknowledging acceptance of the waste shipment. State-supplied manifests typically have two additional copies: one to be mailed to the state regulatory agency by the generator and another to be mailed to the state regulatory agency by the designated facility. Generators required to use state-supplied forms will typically be informed of this by their state regulatory agency and/or their designated TSD facility. The instructions for state-supplied manifests should be carefully read to ensure compliance with the requirements to provide copies to the state regulatory agency.

Completing the Uniform Hazardous Waste Manifest

The Uniform Hazardous Waste Manifest format can be thought of as consisting of three sections. The form itself, consisting of multiple pages, was designed for use on a 12-pitch (elite) typewriter. The form can be completed by hand, using a ball point pen. However, handwritten information usually is illegible on the last few copies of the form. The DOT regulations on shipping papers (49 CFR 172.201) require all copies to be legible and printed (either manually or mechanically).

The upper section of the manifest consists of Items 1–10. These items serve to identify all the parties to the shipment: the generator, the transporter(s), and the designated facility. There are spaces for each party's name and EPA identification number, as well as the generator's mailing address (where the manifests will be kept) and emergency telephone number. There is also a space for the designated facility's site address. Certain areas of the manifest format are shaded. The EPA regulations do not require that any of the items in the shaded area (Items A–K) be completed. States can require the completion of some or all of items A–K, under state regulatory authority. Many of these items, particularly the transporter's telephone number (Items C and F) and the designated facility's telephone number (Item H) should be completed regardless of whether completion is actually required, as a matter of good practice and common sense.

The last five digits of Item 1, following the generator's 12-digit EPA identification number, are to be a unique five-digit number assigned to the manifest by the generator. The number must be unique, so the five digits "00000" or "00001" are not to be reused. The regulations formerly required that a sequential manifest document number be assigned. As a matter of routine and to assist in record-keeping, many generators use the first two digits of the five-digit document number to identify the year of the shipment, with the last three digits being sequential numbers (e.g., 89001, 89002, 89003). Completed manifests (those with a signed copy returned by the TSD facility) are then filed by year.

The final point with respect to the upper third of the manifest is Item 2. Unless continuation sheets (EPA Form 8700-22A) are used, this should be completed as "Page 1 of 1."

The middle section, consisting of Items 11–15, integrates the DOT hazardous material shipping paper requirements into the manifest form. The DOT shipping description(s) (discussed earlier in this chapter) for the waste(s) to be shipped should be entered in Item 11 a–d, as necessary. The DOT shipping description includes the proper shipping name, the hazard class(es), and the DOT identification number preceded by "UN" or "NA," as appropriate. These three elements of a shipping description must be shown in sequence. Unless specifically authorized or required, the shipping description may not contain any code or abbreviation.

For hazardous waste shipments where the amount shipped in a single container equals or exceeds the applicable reportable quantity, the letters RQ, along with the EPA hazardous waste number in parentheses, must be entered in association with the basic shipping description. It is recommended that the letters RQ be placed at the beginning of the shipping description, although they can be placed at the end of the description as desired. The EPA hazardous waste number goes in parentheses after the DOT identification number. The following shipping description serves as an example:

RQ Hazardous Waste, Solid, n.o.s.
ORM-E NA 9189 (F006)

It is not required to place the actual reportable quantity number (e.g., RQ 100 pounds) in the shipping description. If it is felt that the actual reportable quantity for each waste should be shown on the manifest, such information could be placed in Item 15, perhaps in the following format:

"11a. RQ = _____ lb."

Item 12 requires the entry of the number of containers of each waste, as well as the type of container(s) used. Mandatory abbreviations are given for the various types of containers. These abbreviations are:

DM = Metal drums, barrels, kegs
DW = Wooden drums, barrels, kegs
DF = Fiberboard or plastic drums, barrels, kegs
TP = Portable tanks
TT = Cargo tanks (tank trucks)
TC = Tank cars
DT = Dump trucks
CY = Cylinders
CM = Metal boxes, cartons, cases (including roll-offs)
CW = Wooden boxes, cartons, cases
CF = Fiber or plastic boxes, cartons, cases
BA = Burlap, cloth, paper, or plastic bags

Item 13 requires that the total quantity of waste be described, in conjunction with Item 14—the unit of weight or volume used to describe the total quantity. Mandatory abbreviations are also specified for the units of measure entered on the manifest:

G = Gallons (liquids only)
P = Pounds
T = Tons (2000 pounds)
Y = Cubic yards
L = Liters (liquids only)
K = Kilograms
M = Metric tons (1000 kg)
N = Cubic meters

Items I (Waste No.), J (Additional Descriptions for Materials Listed Above), and K (Handling Codes for Wastes Listed Above) are often required to be completed under state regulatory authority. These areas can be used by the generator along with Item 15 (Special Handling Instructions and Additional Information) to make the manifest transmit desirable types of additional information and special instructions. Such information can include:

1. the name and title of the person to whose attention manifest return copies are to be mailed
2. the name and title of the generator's contact person or contractor, as well as appropriate 24-hour telephone numbers, for emergency purposes or for further information. The appropriate Guide Number from the DOT Emergency Response Guidebook could also be specified.
3. instructions to return the shipment to the generator if it or any part is rejected by the designated facility
4. an alternate designated facility, with site address and EPA identification number, in the event an emergency prevents delivery of the waste to the primary designated facility

5. the common names for the waste materials being shipped, along with any TSD facility-assigned waste product codes, contract numbers, or scheduling numbers, sometimes called work order numbers

6. appropriate purchase orders for the treatment/disposal of the waste shipment or other special billing instructions

7. verification that the transport vehicle was properly placarded after loading, or that the appropriate placards were offered to the transporter

8. specification of the method of treatment or disposal to be used by the designated facility on the waste being shipped, as well as any special precautions or other instructions

9. Verification that the appropriate land disposal restriction information, per 40 CFR Part 268, accompanies the manifest. Land disposal restrictions are discussed in more detail later in this chapter.

10. the Reportable Quantity associated with each entry in Item 11, as discussed earlier.

The lower third of the manifest, Items 16–20, is where the parties to the shipment sign the manifest by hand, acknowledging or certifying specific conditions. All parties are to print or type their names, sign their names, and date their signatures. In Item 16, the generator, or typically an employee of the generator in a responsible position, is certifying two statements. First, a certification is made that the shipment is properly described, packaged, marked, and labeled, and is otherwise ready for transportation in accordance with applicable regulations. Second, a certification is made that the generator is complying with waste minimization requirements, and that the method of treatment or disposal selected minimizes present and future threats to public health and environmental quality. Employees of the generator that sign the manifest are allowed to write the words "On behalf of" in the signature block prior to signing their names. This is to address the concerns of employees nervous about personal liabilities for any environmental or public health problems resulting from the waste shipment.

Transporters must acknowledge receipt in Item 17 of the wastes described by the manifest. This is done by entering the name of the person accepting the waste, typically the driver. The transporter's signature is then placed on the manifest, as well as the date of acceptance. EPA Manifest Form 8700-22 allows for the use of two transporter companies. If a second transporter is used, then the first transporter must obtain the name, signature, and date of acceptance from the second transporter. Of course, if two transporters are used, Item 7 (Transporter 2 Company Name and US EPA ID Number) must be completed as well.

Items 19 and 20 are for use by the designated facility owner or operator. If there are any significant discrepancies between the shipment as received and as described on the manifest, the authorized representative of the designated facility's owner or operator is to note the details in Item 19. Significant discrepancies include any differences in the container count, as well as variations of more than 10% in weight for bulk shipments. Significant discrepancies can also exist with respect to the type of waste received, when it differs from the type of waste described on the manifest. Significant discrepancies can result in rejected shipments, which are then returned to the generator, or sent to an alternate designated facility selected by the generator.

If the waste shipment is accepted, or is accepted except as noted in Item 19, the name of the person accepting the waste on behalf of the designated facility must be entered in Item 20, along with his or her signature and date of acceptance. The certification in Item 20 only acknowledges receipt of the manifested shipment, except as noted in the discrepancy indication space (Item 19). It is not certification that the wastes have been or will be properly managed or treated or disposed of by any particular technology.

The manifest is not a "fail-safe" document. Even if state copies are mandated, the system can be circumvented. Receipt of the signed manifest at best signifies that the waste reached the designated facility. Criminals have developed rather clever ways of short-circuiting manifest protections. Any time the transporter conspires with a broker, generator, or TSD facility to illegally dispose of hazardous wastes, the manifest system can be beaten. Further discussion of manifest shortcomings is found in Chapter 23, as part of the evaluation of hazardous waste management vendors.

Using the Manifest

The generator is to complete the appropriate Uniform Hazardous Waste Manifest form, using state-supplied forms when appropriate. The generator is well advised to read the instructions typically printed on the back of state-supplied manifest forms, or the instructions found in the appendix to 40 CFR Part 262. The generator must ensure that the required information is legible on all pages or copies of the manifest. Carbon pages or carbonless forms are typically used. Some state-supplied manifest forms have up to seven identical pages, so using a typewriter becomes necessary to ensure legibility on the bottom copies of the manifest. Any last-minute changes should be initialed by the person signing the manifest.

Upon obtaining the appropriate preshipment acceptance by and making any other necessary arrangements with a commercial TSD facility regarding the particular waste to be shipped, the generator then typically schedules the shipment with both the transporter and designated TSD facility.

The transporter generally inspects the shipment prior to loading, verifying that the containers are in good condition, closed, marked, and labeled. An accurate container count is established and the containers are loaded, properly blocked, and braced to avoid problems during transportation.

The generator then signs and dates the manifest certification. The printed name, signature, and date of acceptance is obtained from the driver. The driver will leave one copy of the manifest for the generator's records, and may need to leave an additional copy for the generator to mail to a state regulatory agency. State-supplied forms often specify which copy or copies are to be left with the generator, either by color or by copy number.

The transporter must take the remainder of the manifest with the shipment, as the manifest serves as the hazardous materials shipping paper for DOT purposes. The first transporter then must deliver the entire shipment to the second transporter, if specified in Item 7, or to the designated facility identified in Item 9. The transporter cannot (and should not be allowed to) select the designated facility for the waste. Remember that the manifest, with the exception of the required signatures, is to be completed by the generator/shipper.

The transporter must obtain the signature of the second transporter, if one is used, and date of delivery or the signature of the designated facility and the date of delivery. The transporter then retains one copy of the manifest with the appropriate signatures, and gives the remaining copies to either the second transporter or the designated facility. The facility retains one copy for its records, and sends to the generator a copy of the manifest with the handwritten signature of the owner or operator of the designated facility. When a state-supplied manifest form is used, then a copy typically is mailed to the state regulatory agency as well.

If the generator does not receive a copy of the manifest with the handwritten signature of the designated facility's owner, operator, or designated representative within 35 days of the date of shipment, the generator is to contact the transporter and/or the designated facility to determine the status and disposition of the waste shipment. If a manifest copy with the handwritten signature of the designated facility is not received within 45 days of the date of shipment, the generator must file an Exception Report with the Regional Administrator of the EPA or the authorized state regulatory agency. The Exception Report must include a copy of the manifest for which no confirmation of delivery was received (termed an ''open'' manifest), as well as a letter signed by the generator explaining the efforts taken to locate the waste shipment, and the results of those efforts.

Having an open manifest beyond the 45-day period allowed is serious business. In effect, it means that the shipment of hazardous waste is lost, despite efforts to determine its status or ultimate disposition. If regulatory agencies discover an open manifest after the 45-day period allowed for obtaining the signed return copy has expired, an enforcement action will be filed. EPA considers an open manifest to be a Class I violation, of serious consequence. Generators are well advised to have a tickler file for manifest copies left by the transporter, pending receipt of the manifest copy signed by the designated TSD facility.

It is recommended that generators investigate open manifests beginning two weeks after shipment. As more time passes from the date of shipment, the likelihood of recovering the waste decreases. Often, the waste was delivered to the designated facility and the signed return copy was mailed, only to be lost in the mail. The generator must then obtain a photocopy of the designated TSD facility's file copy in order to close that particular manifest. Less likely, but not uncommon, is the transporter's holding the manifested shipment pending receipt of additional wastes from other generators, in order to have a full load for delivery to the designated facility.

Transporters are allowed to hold manifested shipments of containerized hazardous wastes up to 10 days. Transporters holding loads longer than 10 days should be asked to return the waste shipment. This is a very delicate situation. Ordering the transporter to deliver late shipments could result in the transporter placing a fake signature on the manifest and sending the appropriate copy back to the generator. Be very careful in these situations. Informing the state regulatory agency of such problems may be in the generator's best interest. Selecting a different transportation firm is probably in order.

Land Disposal Restriction Notices

Generators are required to provide notification to any offsite TSD facility with respect to wastes subject to the Land Disposal Restrictions found at 40 CFR Part 268. This notification becomes, in effect, part of the manifest requirements, because such notices are to accompany each offsite shipment of restricted waste.

These notices are required even if the waste is being sent to a treatment or reclamation facility. Shipments of restricted wastes to land disposal facilities must be accompanied by a notice and certification that the waste meets the applicable treatment standards. Alternatively, the generator must send a notice with the shipment to the land disposal facility that the restricted waste is subject to a case-by-case extension, an exemption, or nationwide variance. Certifications and demonstrations are necessary for so-called "soft-hammer" restricted wastes, pending establishment by EPA of treatment standards, or until May 8, 1990, whichever is sooner.

EPA has established a schedule for evaluating each listed hazardous waste and characteristically hazardous wastes. The purpose of this evaluation is to determine whether land disposal of these wastes is protective of human health and environmental quality. EPA intends to establish concentration levels or methods of treatment which substantially reduce the toxicity of the wastes and/or their potential for migration into the environment. Treatment standards are to be based on Best Demonstrated Available Technology (BDAT). Wastes that do not meet the established standard are prohibited from direct land disposal. Eventually it is likely that only treatment residues will be acceptable for land disposal. The schedule for the EPA evaluations is found at 40 CFR Part 268 Subpart B. All listed and characteristic wastes are to be evaluated by May 8, 1990. Hazardous wastes for which EPA does not establish treatment standards by May 8, 1990 will be prohibited from direct land disposal.

Generators are likely to first encounter the implications of the Land Disposal Restrictions during regulatory agency inspections or, more likely, through interaction with commercial hazardous waste management facilities. Hazardous waste TSD facilities are required to modify their Waste Analysis Plans to reflect the

Land Disposal Restrictions. Commercial TSD facilities should be informing their customers of the existing and upcoming requirements affecting the wastes generated at the customer's facility. Many commercial TSD facilities have placed the information elements of the mandatory restricted waste notices on their own forms, and require customers to use the appropriate form developed by the TSD facility.

The regulatory requirement for determining whether a particular hazardous waste is restricted from land disposal at 40 CFR Part 268 falls squarely on the generator of the waste. Generators are required at 40 CFR 268.7 to test the waste or an extract using the appropriate test method(s), or to use knowledge of the waste to determine if land disposal of the waste is restricted. Testing is normally performed if the generator believes that the waste will meet the appropriate treatment standard, and is thus eligible for land disposal. Using knowledge of the waste in determining the applicability of the land disposal restrictions is appropriate when it is believed that the waste would contain concentrations of hazardous constituents exceeding the appropriate treatment standard. It could be considered a waste of money to perform expensive testing on wastes that would be restricted from land disposal in any event. Using knowledge of the waste may be appropriate, in some instances, in determining whether the waste is subject to certain extensions, exemptions, or any nationwide variances.

The most common notices for shipments of restricted wastes are fairly simple. Requirements for the various types of notices are found at 40 CFR 268.7 ("Waste Analysis and Recordkeeping"). For restricted wastes that do not meet the applicable treatment standard, the notice with each shipment of hazardous waste to the treatment facility must include the following information:

1. the EPA hazardous waste number of the restricted waste
2. the corresponding treatment standard
3. the manifest document number associated with the shipment of restricted waste
4. waste analysis data, where available

Wastes that meet the treatment standard and are being shipped directly to a land disposal facility must be accompanied by a notice and a certification that the waste meets the applicable treatment standards.

The notice to land disposal facilities must include the same information listed above, along with the following certification, signed by an authorized representative of the generator:

"I certify under penalty of law that I personally have examined and am familiar with the waste through analysis and testing or through knowledge of the waste to support this certification that the waste complies with the treatment standards specified in 40 CFR Part 268 Subpart D. I believe that the information I submitted is true, accurate, and complete. I am aware that there are significant penalties for submitting a false certification, including the possibility of a fine and imprisonment."

If a generator's waste is subject to a case-by-case extension under 268.5, an exemption under 268.6, or a nationwide variance under Subpart C of Part 268, with each shipment the generator must forward a notice to the facility receiving the waste, stating that the waste is not prohibited from land disposal. The mandatory elements of such a notice are found at 40CFR 268.7(a)(3).

Generators of restricted wastes for which EPA has not established treatment standards (soft-hammer wastes) are subject to requirements to use the best treatment practically available. Demonstrations and certifications are required prior to offsite shipments of soft-hammer wastes. By statute, there will be no soft-hammer wastes after May 9, 1990.

Generators of soft-hammer wastes who have determined that treatment is impractical or unavailable must submit a demonstration and certification of this finding to the EPA Regional Administrator prior to shipment of such wastes to a landfill or surface impoundment. Such disposal units must also meet the RCRA minimum technology requirements. The mandatory elements of notice of shipment of soft-hammer wastes to offsite facilities are found at 40 CFR 268.7(a)(4). Generator requirements regarding the land disposal restrictions are summarized below. Generators must:

1. Determine, at the point of generation, whether any of the wastes generated are subject to land disposal restrictions.
2. Determine which treatment standard(s) or prohibition levels apply to each restricted waste. Treatment standards typically distinguish between wastewaters and non-wastewaters for each listed waste.
3. Determine, through specified analytical techniques or knowledge of the waste, whether BDAT concentration levels are met or exceeded. Analysis is either on a total constituent basis or of TCLP extract, depending on the type of treatment standard.
4. Comply with storage time limitations (one year presumptive maximum for interim status/permitted storage facilities *or* 90 days for generator/accumulators).
5. Comply with prohibitions on evaporation and/or dilution of restricted wastes as a substitute for adequate treatment.
6. Prepare notifications, demonstrations, and certifications required for onsite or offsite waste management:
   a. When sending offsite for treatment, notify treatment facility in writing of appropriate BDAT concentration levels for waste.
   b. When sending offsite any wastes that meet BDAT concentration levels, provide a notice and certification that the waste meets the applicable BDAT concentration levels.
   c. For soft-hammer wastes, a demonstration and certification must be submitted to the EPA Regional Administrator and be attached to the initial shipment. Notice to the receiving facility is required on all shipments.
   d. When sending offsite any wastes subject to extensions or variances, provide a notice stating that the waste is exempt from the land disposal restrictions.
7. Maintain copies of all notices, certifications, demonstrations, waste analysis data,

and other documentation (for knowledge of waste-based determinations) for at least five years from the date the waste was last sent to onsite or offsite treatment, storage, or disposal.

Recordkeeping

The land disposal restriction notices are, in effect, part of the manifest requirements and would otherwise be subject to the same recordkeeping requirements. Generators are required at 40 CFR 262.40 to keep a copy of each manifest signed by the transporter for three years, or until a copy is received with the signature of the authorized representative of the designated TSD facility which accepted the waste. This latter manifest must be retained at least three years from the date the waste was received by the initial transporter.

If a generator makes the determination whether a waste is subject to the Land Disposal Restrictions solely on knowledge of the waste, all supporting data (MSDS, chemical inventories, etc.) must be maintained onsite. This information would otherwise be subject to the same records retention requirements found at 40 CFR 262.40 for any test results, waste analyses, or other information used in performing the hazardous waste determination per 262.11. Such records must be maintained at least three years from the date the waste was last sent to onsite or offsite treatment, storage, or disposal. However, EPA has determined that a five-year record retention period is appropriate for documents, and supporting information relating to the 40 CFR Part 268 Land Disposal Restriction requirements.

The records retention period for manifests, waste analyses, or Land Disposal Restriction documents is automatically extended during any unresolved enforcement action or as requested by the EPA or authorized state regulatory agency. As a practical matter, manifests, waste analyses, Land Disposal Restriction documents and supporting information should be kept indefinitely, storage space permitting. Such records are often the only means of determining whether EPA claims of liability are justified with respect to the generator's involvement at any facility subject to EPA enforcement under CERCLA.

CONCLUSION

The offsite shipment of hazardous waste requires careful consideration of both DOT and EPA regulatory requirements. Mastering the use of the Hazardous Materials Table in selecting the most appropriate shipping description is critical to determining the appropriate DOT requirements. This information, in conjunction with knowledge of the use of the Uniform Hazardous Waste Manifest and the Land Disposal Restriction notices, allows the preparation of the proper shipping papers. Selection of the transporter and treatment/disposal facility is then necessary to minimize statutory and regulatory liabilities. This topic is covered in Chapter 23.

CHAPTER 11

Generators Who Store Hazardous Wastes Under Interim Status

This chapter is devoted to manufacturing firms with an interim status storage facility on their premises. It will cover the hazardous waste storage practices of manufacturing firms utilizing containers and/or tanks for time periods longer than 90 days. The use of surface impoundments or waste piles for storage of hazardous wastes is not addressed. This chapter is intended to serve as an introduction to and reiteration of the interim status standards (40 CFR Part 265) applicable to storage in containers and tanks. It is not intended, however, to substitute for a close reading of the actual regulatory sections by persons responsible for waste management at manufacturing facilities. Some of the standards applicable to interim status storage facilities have been covered in Chapter 9 as they apply to accumulation areas. The reader is advised to refer to Chapter 9 for further discussion of some topics.

The presence of onsite hazardous waste storage facilities (onsite TSD facilities) at manufacturing firms is becoming less prevalent for the following reasons:

1. increased enforcement activity on the part of EPA and authorized states, coupled with a degree of regulatory control and required procedural activities designed for and more suitable to commercial hazardous waste management operations
2. availability of arrangements with commercial TSD facilities ensuring offsite shipment of hazardous wastes on at least a quarterly (90-day) basis
3. costs of upgrading existing HWM units to meet technical permit standards
4. costs of preparing the Part B permit application

5. costs and difficulty of ensuring ongoing compliance with permit conditions and the 40 CFR Part 264 standards for permitted facilities
6. Part B permit application call-ins and statutory timetables for termination of interim status

The 1984 reauthorization of RCRA set deadlines for the termination of interim status. The purpose of these deadlines included (1) speeding the process of evaluating permit applications and issuing (or denying) final permits, and (2) hastening the closure of substandard HWM facilities and those facilities not seriously interested in pursuing a full RCRA permit. Many manufacturing firms with interim status TSD storage facilities fall in this latter category. As increasingly stringent and costly requirements are imposed on TSD facilities, many manufacturing firms are finding that a TSD storage facility is not necessary for efficient onsite or offsite treatment or disposal.

Interim status for TSD storage facilities terminates November 8, 1992, unless a full RCRA permit has been issued by that date. The interim status period can be extended if a Part B permit application was submitted by November 8, 1988. No additional or newly generated wastes can be placed in interim-status HWM units after November 8, 1992, except as outlined above. Termination of interim status for interim-status TSD storage facilities will require submission of a closure plan for regulatory agency review and approval at least 45 days prior to the date closure is expected to begin. The closure process begins within 30 days after receipt of the final volume of waste. Once the closure plan receives regulatory agency approval, the plan must be implemented and closure certified by the facility owner/operator and an independent registered professional engineer.

The EPA and the state regulatory agency have the option of requesting ("calling in") Part B permit applications in advance of any statutory deadlines. The facility owner/operator then has six months to either submit a complete Part B application or pursue the closure process.

The regulatory definition of storage, found at 40 CFR 260.10, is "the holding of hazardous waste for a temporary period, at the end of which the hazardous waste is treated, disposed of, or stored elsewhere."

As discussed in Chapter 9, certain storage practices do not require a permit. These practices are primarily the accumulation of hazardous wastes in containers or tanks for periods of time under 90 days, provided such wastes are subsequently managed at an onsite recycling facility, or at an onsite or offsite interim-status or fully permitted TSD facility. It must be remembered that even though the permit requirements apply only to storage facilities where wastes are held longer than 90 days, certain interim status standards apply to both accumulation areas and storage facilities.

The interim status standards found at 40 CFR Part 265 are divided into subparts and sections. Overall, however, the Part 265 standards can be categorized into those applicable to all facilities and those applicable only to a particular type of facility. The subparts generally and specifically applicable to storage facilities are:

A "General"
B "General Facility Standards"
C "Preparedness and Prevention"
D "Contingency Plan and Emergency Procedures"
E "Manifest System, Recordkeeping, and Reporting"
G "Closure and Postclosure"
H "Financial Requirements"
I "Use and Management of Containers"
J "Tank Systems"

Subparts A–E, G, and H are generally applicable to the hazardous waste storage facilities covered in this chapter, i.e., container storage facilities and tank storage facilities. Subpart I, "Use and Management of Containers," is specifically applicable to facilities using drums or barrels or other portable storage devices to store or transport hazardous wastes. Subpart J, "Tank Systems," is specifically applicable to facilities using stationary devices constructed of nonearthen materials to contain and store hazardous wastes.

SUBPART A: GENERAL

Subpart A primarily states to whom the requirements of Part 265 do and do not apply. The standards apply to owners and operators of TSD facilities that have met the requirements for interim status. The concept of and requirements for interim status were discussed in Chapter 5. The Part 265 standards apply until final administrative disposition of Parts A and B of the RCRA permit application is made. If a permit is granted, the Part 265 standards no longer apply, as they are replaced by the 40 CFR Part 264 standards for permitted facilities. If a permit is denied, the hazardous waste management facility is to undergo the regulatory closure process. For manufacturing firms with storage facilities, this would require hazardous wastes to be moved offsite every 90 days to a TSD facility under interim status or one possessing a RCRA permit. The interesting exclusions given from the requirements of Part 265 are discussed in Chapter 12.

Subpart A has one other provision worth noting. Section 265.4, "Imminent Hazard Action," states that "Notwithstanding any other provisions of these regulations, enforcement actions may be brought pursuant to Section 7003 of RCRA." The implication is that compliance with all applicable Part 265 standards is not sufficient to prevent enforcement actions brought under the imminent hazard provisions of the RCRA statute.

SUBPART B: GENERAL FACILITY STANDARDS

Subpart B standards encompass a wide range of subjects not readily categorized elsewhere. The owner or operator of every hazardous waste management facility subject to Part 265 must apply for an EPA identification number (265.11). Also, before ownership or operation of a hazardous waste facility can be transferred during its operating life, the new owner or operator must be notified in writing of the requirements governing the facility as set forth in 40 CFR Parts 265 and 270. This notification is to be made by the existing owner or operator (265.12).

These requirements, as well as the other Part 265 standards, apply to the owner of the facility, the operator of the facility, or both. The regulations specify their applicability to an owner/operator for enforcement purposes. This specification of applicability is not necessary for explanatory purposes, so the remainder of the chapter will treat the requirements as if they apply to the facility itself.

General Waste Analysis (265.13)

Before any hazardous waste is placed in storage, a detailed chemical and physical analysis of a representative sample of the waste must be obtained. There should be an analysis of each waste in storage, giving all the information that must be known to store the waste in accordance with the Part 265 requirements. This analysis may consist of a laboratory analysis, and can include published or documented data concerning the waste in question. The waste analysis must be repeated (1) when the process or operation that generates the waste is changed or when different input materials or chemicals are used, if such changes would modify the physical or chemical characteristics of the waste; or (2) as necessary to ensure that it is accurate and up to date. Offsite (commercial) hazardous waste management facilities may require additional analyses, or may themselves perform analyses for information necessary to properly treat or dispose of the waste.

A written waste analysis plan must be developed to serve as a guide to the detailed chemical and physical analyses to be performed. There are four elements of a waste analysis plan that apply to the container or tank storage facilities of manufacturing firms. The plan must specify the sampling procedure(s) to be used to obtain a representative sample of each waste. The sampling procedure chosen depends on the physical state of the waste and the place from which it is to be sampled. The sampling procedure can be one of the methods described in Appendix I of 40 CFR 261, or an equivalent sampling method. The plan must designate the parameters for which each waste will be analyzed. The rationale for selection of these parameters (and presumably why others were not selected) must be explained for each waste.

An explanation of how an analysis for the chosen parameters will adequately supply the information necessary for safe and proper storage of each hazardous waste should be provided. The waste analysis plan must also specify the test methods that will be used to test for the parameters selected. The test methods for the characteristics of a hazardous waste (ignitability, corrosivity, or EP toxicity) should be those mandated by the regulations. Appendix III of 40 CFR 261 specifies appropriate analytical procedures and approved measurement techniques for various organic chemicals and inorganic species. Further guidance on acceptable and approved methods and procedures can be found in "Test Methods for Evaluating Solid Waste: Physical/Chemical Methods" (SW-846, U.S. EPA). If the services of a commercial laboratory are used, the laboratory personnel may be of some help in identifying appropriate test methods for each parameter for each waste. Finally, the waste analysis plan must specify the frequency with which the initial analysis of each waste will be reviewed or repeated to ensure that the analysis is accurate and up to date. Waste analysis plans are to be revised as necessary to conform with Land Disposal Restrictions, codified at 40 CFR Part 268.

Waste Management Facility Security (265.14)

Security provisions are necessary to prevent unknowing entry, as well as to minimize any unauthorized entry to the facility. A storage facility must have either a 24-hour surveillance system (e.g., television monitoring or surveillance by guards) that monitors and controls entry, or a natural or artificial barrier that surrounds the facility (e.g., a fence). If a fence is used, there must be a means of controlling entry through the gate at all times, such as a lock, an attendant, or television monitors.

Manufacturing firms generally meet this requirement by a fence surrounding the entire plant property. Alternatively, the fence can surround only the hazardous waste management unit, i.e., the storage facility. Any gates or entryways must be locked or monitored. At each entrance to the facility and at other locations near the actual storage area, signs must be posted displaying the legend "Danger—Unauthorized Personnel Keep Out." These signs should be visible from any avenue of approach, and should be legible at a distance of at least 25 feet.

General Inspection Requirements (265.15)

The storage facility must be inspected periodically for malfunctions, deterioration, operator errors, and discharges that may cause or lead to the release of hazardous waste constituents to the environment, or a threat to human health. These inspections must be conducted often enough to identify problems and correct them before any releases to the environment or damage to human health.

As with the waste analysis requirements, the inspection requirements begin with a written procedure. In this case, the procedure is called an inspection schedule, rather than a plan. The inspection schedule applies to all monitoring equipment, safety and emergency equipment, security devices, and operating and structural equipment (e.g., dikes and sump pumps) that are required by regulation or that are important to preventing, detecting, or responding to environmental or human health hazards.

The inspection schedule must identify the types of problems to be looked for during the inspection. A specification of the frequency of inspection for each item on the schedule is required. The frequency of inspection is a decision made by the operator. However, this discretion is limited. Loading and unloading areas and other areas subject to spills must be inspected daily when in use. Further, the inspection schedule must include the items and frequencies specified in Subpart I, "Use and Management of Containers," and Subpart J, "Tank Systems." Container storage areas must be inspected at least weekly. For all other items, the inspection frequency should be based on a consideration of both the rate of possible deterioration and the probability of an environmental or human health incident if the deterioration, malfunction, or operator error goes undetected.

Merely performing inspections according to the schedule and noting deficiencies is not sufficient. Deterioration must be remedied, malfunctions corrected, and errors or bad practices stopped before an environmental or human health hazard results. Where a hazard is imminent or has already occurred, remedial action must be taken immediately. Otherwise, corrections must take place in a timely manner as stated above. Some notation of the time frame available to correct the problem should be indicated on the inspection form.

As with many other requirements, the performance of inspections must be documented. The inspections must be recorded in an inspection log or summary. These records must include:

1. the date and time of the inspection
2. the name of the inspector
3. a notation of the observations made
4. the date and nature of any repairs or other remedial actions

These records must be kept for at least three years from the date of inspection. The inspection log requirement can be met by keeping the completed inspection forms in a three-ring binder. Otherwise, the results of the inspections can be summarized and transferred to a master log book. Many facilities have implemented the inspection requirements by developing a set of inspection forms, one form for each type of equipment for which inspection is mandated.

A good inspection schedule and regular inspections are important management tools at a hazardous waste facility. Minor problems associated with the operation of the site are found before they develop into major problems, allowing corrective actions to be initiated quickly.

Personnel Training (265.16)

Personnel training is one of the most critical aspects of the proper management of a hazardous waste facility. Employees must be trained so that they understand the hazardous properties of the wastes they are handling, know how to perform their routine waste management tasks, and know what to do in responding to an emergency situation. Personnel training is covered in Chapter 9, as it is also a requirement for generators who accumulate.

General Requirements for Ignitable, Reactive, or Incompatible Wastes (265.17)

Precautions are mandated at facilities handling ignitable or reactive wastes to prevent accidental ignition or reaction. Mandated precautions include separation and protection of such wastes from sources of ignition or reaction, including (but not limited to):

1. open flames
2. smoking
3. cutting and welding
4. hot surfaces
5. frictional heat
6. sparks (static, electrical, or mechanical)
7. spontaneous ignition
8. radiant heat

Ignitable and reactive wastes should be stored in areas where smoking is prohibited. "No Smoking" signs must be posted wherever there is a hazard from ignitable or reactive waste. Finally, smoking and open flames must be confined to designated locations while ignitable or reactive wastes are being handled.

Section 265.17 also gives a performance standard for the treatment, storage, or disposal of ignitable or reactive wastes, and for the mixture or commingling of incompatible wastes, or incompatible wastes and materials. Such operations must be carried out so that they do not:

1. generate extreme heat or pressure, fire or explosion, or violent reaction
2. produce uncontrolled toxic mists, fumes, dusts, or gases in sufficient quantities to threaten human health
3. produce uncontrolled flammable fumes or gases in sufficient quantities to pose a risk of fire or explosions
4. damage the structural integrity of the device or facility containing the waste
5. through other like means threaten human health or the environment

To adequately address these requirements for ignitable, reactive, and incompatible wastes, the generator must first know the properties of the wastes in storage, and must determine the probable results of subjecting the wastes to various conditions likely to be encountered in storage (extreme heat, freezing, precipitation, and rough handling). The generator should also determine whether any two waste types are incompatible. Only with such information can the regulatory requirements be addressed and the wastes protected adequately.

Subparts I and J of 40 CFR Part 265 give more specific instruction with respect to the management of ignitable, reactive, and incompatible wastes. Discussion of such requirements as they apply to container storage areas is found in Chapter 9.

SUBPART C: PREPAREDNESS AND PREVENTION

A preparedness and prevention program is an important requirement of storage facilities as well as of accumulation areas. The regulations specify a performance standard in this regard: that the storage facility be maintained and operated to minimize the possibility of fire, explosion, or unplanned sudden or gradual release of hazardous waste or hazardous waste constituents to the air, soil, or surface water. Preparedness and prevention programs are given extensive coverage in Chapter 9.

SUBPART D: CONTINGENCY PLAN AND EMERGENCY PROCEDURES

The contingency plan for the storage facility is one of the more important documents to be developed. The contingency plan should address potential emergency situations involving hazardous wastes. For each situation there should be a complete, organized, and considered response that will adequately manage the situation. The requirements of Subpart D are thoroughly discussed in Chapter 9.

SUBPART E: MANIFEST SYSTEM, RECORDKEEPING, AND REPORTING

Subpart E applies to onsite storage facilities at manufacturing firms only with regard to recordkeeping and reporting requirements. The manifest system requirements found in Subpart E apply only to commercial (offsite) TSD facilities, and do not apply to onsite facilities that do not receive any hazardous wastes from offsite sources. The manifest system requirements applicable to manufacturing firms generating and storing hazardous waste are found at 40 CFR Part 262, and are discussed in Chapter 10.

The major requirement of Subpart E with respect to storage facilities is that of an operating record. The operating record, as the name implies, consists of a set of documents associated with the operation of the hazardous waste management facility. The operating record need not be in a single file or binder. It can be devised as a series of three-ring binders, which can be kept at different locations onsite, if desired. The following information must be recorded and maintained in the operating record until closure of the storage facility:

1. a description and the quantity of each hazardous waste stored, and the method and date of its storage at the facility
2. the location of each hazardous waste within the facility and the quantity at each location
3. records and results of waste analyses and trial tests performed
4. summary reports and details of all incidents (if any) that required implementation of the contingency plan
5. records and results of inspections (these need only be kept for three years)
6. closure cost estimates
7. documentation of any refusal by local authorities to enter into emergency agreements

Elements 3–7 of an operating record are fairly simple. However, some further explanation of elements 1 and 2 is necessary, since the regulations do not specify a format for the required information. The requirements of elements 1 and 2 can be met in a variety of ways. It is important to note that these are not static requirements, but must be revised as wastes are placed in storage or shipped off-site for treatment or disposal.

One means by which these two requirements for operating records can be implemented is by use of an operating log of the format shown in Figure 11.1. Element 1 consists of four different information needs. They are (1) description of the waste, (2) quantity of the waste, (3) method(s) of handling, and (4) date(s) of handling. Appendix I of 40 CFR Part 265 provides additional guidance on meeting these information needs.

The description of the waste must include its common name, the EPA hazardous waste number(s) that apply to the waste, and the physical state of the waste (solid, liquid, or sludge). For characteristic hazardous wastes, the description must also include the process generating the waste. Where the waste contains more than one listed hazardous waste, or where more than one hazardous waste characteristic applies, the waste description must include all applicable EPA hazardous waste numbers.

The quantity of each hazardous waste placed in storage must also be known. The estimated or measured weight (or volume and density, where applicable) can be used. The units of measure must be shown, using the units and symbols specified in Table 1 of Appendix I of 40 CFR Part 265, which are:

Mass
pounds: P
short tons (2000 pounds): T
kilograms: K
metric tons (1000 kg): M

CONTAINER STORAGE AREA OPERATING LOG

Description of Waste (must include common name, physical form, & HW number)	Estimated Quantity (in pounds (P) or gallons (G)	Date Full Container Placed In Storage Area	Method of Storage	Date Removed From Storage	Manifest Number	Treatment/ Disposal Site

Figure 11.1. Sample operating log.

Volume
gallons (U.S.): G
liters: L
cubic yards: Y
cubic meters: C

Density
pounds/gallon: P/G
tons/cubic yard: T/Y
kilograms/liter: K/L
metric tons/cubic meter: M/C

Finally, element 1 requires the method(s) and date(s) of storage to be recorded. Table 2 of Appendix I of 40 CFR Part 265 specifies the handling codes to be used. For storage facilities, the only codes of interest are:

SO1: container (barrel, drum, etc.) storage
SO2: tank storage
SO3: waste pile
SO4: surface impoundment
SO5: other (specify)

The second element of an operating record is not quite as complex. The location of each waste within the facility and the quantity at each location must be known. When there is only one container storage area within a manufacturing facility, the confirmation that the container(s) of interest are in the regulated unit will suffice. This would also apply where a single storage tank is involved. Where multiple hazardous waste areas or tanks are involved, the specific area or tank into which the waste was placed should be identified.

Availability, Retention, and Disposition of Records (265.74)

All records (including plans) required under Part 265 must be made available at all reasonable times for inspection at the facility by duly designated federal or state inspectors. Such records and plans must be furnished on request. The retention period for records required under Part 265 is automatically extended by any unresolved enforcement actions or as requested by the federal or state regulatory agencies.

As might be expected, keeping of records is inevitably accompanied by submission of reports. Storage facilities at manufacturing firms potentially have four different types of reports to submit. They are:

1. biennial reports
2. reports of incidents requiring implementation of the contingency plan
3. certification of closures
4. exception reports

Exception reports deal with manifest copies not returned by the designated off-site TSD, and are discussed in Chapter 10. Closure procedures are discussed later in this chapter. The contingency plan is discussed in Chapter 9.

The biennial report is a requirement of generators as well as of TSD facilities. A manufacturing firm generating hazardous wastes and storing such wastes at an onsite interim-status TSD facility prior to offsite shipment would have to fill out biennial report forms with respect to both activities. Biennial reports are due March 1 of every even-numbered year, covering HWM activities taking place the previous (odd-numbered) year. It is important to note that states may require reports on HWM activities that are more comprehensive and/or more frequent than the biennial report.

SUBPART F: GROUNDWATER MONITORING

This subpart does not apply to facilities that store in containers or tanks. The requirements of this subpart apply to surface impoundments, landfills, waste piles, and land treatment facilities.

SUBPART G: CLOSURE AND POSTCLOSURE

Subpart G, as it applies to the hazardous waste TSD storage facilities at manufacturing firms, requires the preparation of closure plans and mandates certain administrative procedures and certain substantive measures to be taken upon closure. These requirements may seem unnecessary, as most manufacturing facilities have no intention of moving or going out of business. However, in the real world, storage needs change, permits are denied, storage areas are relocated within the overall plant property, plants are sold or closed, and companies go out of business. A closure plan is necessary for all these reasons.

Closure, for regulatory purposes, is the period between the last date that wastes are periodically placed in a storage facility and the submittal of a certification of completion of closure in accordance with an approved closure plan. The regulations specify a performance standard for closure—that the storage facility be closed so as to:

1. Minimize the need for further maintenance.
2. Control, minimize, or eliminate, to the extent necessary to protect human health and the environment, postclosure escape of hazardous waste, hazardous constituents, leachate, contaminated runoff, or hazardous waste decomposition products to the ground- or surface waters or to the atmosphere (265.111).

Additionally, when closure is completed, all contaminated equipment, structures, and soil must either be properly disposed of or be decontaminated by removing all hazardous waste and residues (265.114).

The central document of the Subpart G requirements is the closure plan. The closure plan must identify the steps necessary to completely or partially close the facility at any point during its active life. As with other required documents, records, and plans, the closure plan consists of a set of elements:

1. a description of how each hazardous waste management unit at the facility will be closed in accordance with 265.111
2. a description of how final closure of the facility will be conducted in accordance with 265.111. The description must identify the maximum extent of the operation to be unclosed during the active life of the facility.
3. an estimate of the maximum inventory of hazardous wastes ever onsite over the active life of the facility and a detailed description of the methods to be used during partial and final closure, including (but not limited to) methods for removing, transporting, treating, storing, or disposing of all hazardous waste, along with identification of the type(s) of offsite hazardous waste management unit(s) to be used, if applicable
4. a detailed description of the steps needed to remove or decontaminate all hazardous waste residues and contaminated containment system components, equipment, structures, and soils during partial and final closure, including (but not limited to) procedures for cleaning equipment and removing contaminated soils, methods for sampling and testing surrounding soils, and criteria for determining the extent of decontamination necessary to satisfy the closure performance standard
5. a detailed description of other activities necessary during the partial and final closure period to ensure that all partial closures and final closure satisfy the closure performance standards, including (but not limited to) groundwater monitoring, leachate collection, and runon and runoff control
6. a schedule for closure of each hazardous waste management unit and for final closure of the facility. The schedule must include, at a minimum, the total time required to close each hazardous waste management unit and the time required for intervening closure activities to allow tracking of the progress of partial and final closure.
7. an estimate of the expected year of final closure for facilities that use trust funds to demonstrate financial assurance under 265.143 or 265.145 and whose remaining operating life is less than 20 years, and for facilities without approved closure plans (265.112(b))

The regulatory language with respect to closure plans makes the preparation of closure plans and the necessary regulatory agency review quite difficult, complex, and time-consuming. For container and tank storage facilities, planning for closure is conceptually quite simple. Basically, closure of container storage facilities involves removal of all containers for offsite treatment/disposal. Paved areas subject to spills are cleaned. If the container storage area is on dirt or gravel, any contaminated soil is excavated and removed. Any sumps or catch basins used to collect spills, leaks, or contaminated runoff are emptied and cleaned. Storage facilities using tanks have the tanks emptied and the contents managed offsite.

The tanks, pumps, piping, valves, sumps, etc., are then cleaned, and the rinsate is managed properly.

The difficult aspect of closure involves the verification, through sampling and analysis, that the storage area has been decontaminated to the extent mandated by the closure performance standard (40 CFR 265.111). This is the so-called "clean closure" requirement. HWM units that are not closed in accordance with the closure performance standard are to be closed as land disposal units, subject to postclosure permitting and other requirements.

Differences of opinion regarding the nature and extent of sampling and analyses necessary to verify decontamination, as well as the maximum concentrations of hazardous constituents that satisfy the closure performance standard, are associated with most closure plans. Regulatory agency personnel sometimes fail to realize the impossibility of proving a negative (i.e., that a storage area is *not* contaminated). Analytical and other costs can easily become quite high in the absence of common sense, reasonableness, and a good grasp of applied science.

Regulatory agencies typically interpret the closure performance standard to require removal of all hazardous constituents of concern to background levels. For constituents not naturally occurring, this means that concentrations must be below analytical detection limits. The number of samples and the number and types of analyses in attempting to prove a negative become significant.

A schedule for final closure is also required under element 4. However, Section 265.113 of Subpart G gives certain time limitations on closure activities. Within 90 days after placing the final volume of hazardous waste in storage, or 90 days after approval of the closure plan (whichever is later), all hazardous wastes must be removed from the storage facility, in accordance with the approved closure plan. Completion of closure activities (i.e., decontamination) in accordance with the approved closure plan must take place within 180 days after receiving the final volume of wastes, or 180 days after approval of the closure plan, whichever is later.

The schedule for final closure must, within these time limitations, specify the total time required to close the facility and the time required for intervening closure activities, which allow tracking of the progress of closure. Intervening activities include removal of all wastes, decontamination of equipment, removal of contaminated soil, and analytical verification of the extent of decontamination.

The closure plan may be amended at any time during the active life of the facility (i.e., the period during which wastes are periodically placed in storage). The closure plan must be amended whenever changes in operations or facility design affect closure, or when there is a change in the expected year of closure. Closure plans must also be modified if clean closure is no longer feasible, or if other unexpected events require a modification of the closure plan during the closure process.

The Subpart G closure requirements also include certain administrative procedures for closure. The closure plan for container or tank storage facilities must be submitted to the EPA regional administrator and/or the state environmental agency at least 45 days before the start of final closure activities. Enforcement actions mandating closure require submission of the plan within 15 days of any closure decrees. Closure plans must be submitted within 15 days of termination of interim status, brought about, for example, by failure to submit complete Part B permit applications by administrative deadlines or failure to receive a full permit before statutory deadlines.

The federal or state regulatory agency will approve, modify, or disapprove the plan within 90 days of receipt. During this period, a newspaper notice will provide the public with an opportunity to submit written comments on and request modifications of the plan. A public hearing may be held on the closure plan. If the plan is disapproved, it must be modified and resubmitted within 30 days. The regulatory agency must approve or modify this second plan within 60 days. If this plan is modified by the regulatory agency, it then becomes the approved closure plan. A copy of the modified closure plan is mailed to the owner/operator of the facility.

Once the closure plan is approved by the regulatory authorities, the closure time clock begins. As stated earlier, all waste must be removed from the facility in accordance with the closure plan within 90 days. Within 180 days, closure activities are to be completed. The facility is not officially closed until the owner/operator and an independent registered professional engineer submit certification that the facility has been closed in accordance with the approved closure plan. These certifications are required within 60 days of completion of final closure.

Manufacturing firms with interim-status HWM units, even those closed or undergoing closure, can be subject to corrective action requirements. Corrective action authority extends to all solid waste management units (SWMUs) at the facility. SWMUs are typically pre-RCRA spill or disposal areas. Owners/operators can be ordered to investigate for and remediate any releases from SWMUs.

SUBPART H: FINANCIAL REQUIREMENTS

Manufacturing firms with onsite container or tank storage facilities are required to be financially responsible for the closure costs and potential liabilities associated with such facilities. There are two elements to financial responsibility for such facilities: (1) assurance of sufficient funding to properly close the storage facility in accordance with the closure plan; and (2) maintenance of liability insurance to demonstrate financial responsibility for claims to compensate any injuries to people or property that may result from operation of the facility. Liability insurance for container and tank storage facilities is only required for any sudden and accidental occurrences (fires, explosions, or spills) that result from operation of the facility, although the insurance industry typically will not offer policies without coverage for nonsudden (gradual) accidental occurrences as well.

Disposal facilities and other facilities where wastes are to remain after closure must also have a postclosure plan and provide funds for postclosure monitoring and maintenance. Surface impoundments, landfills, and land treatment facilities must demonstrate financial responsibility for claims for nonsudden occurrences (e.g., groundwater contamination) in addition to responsibility for sudden and accidental occurrences (e.g., fires or explosions).

Closure Cost Estimate (265.142)

Financial responsibility under Subpart H for the closure of storage facilities is integrated with the closure requirements of Subpart G. Section 265.142 of Subpart H requires the preparation of an estimate, in current dollars, of the cost of closing the facility in accordance with the closure plan. This cost estimate must be prepared assuming the circumstances that would make closure of the facility the most expensive. For storage facilities, this point would be when the amount of wastes in storage at the facility is at the facility design capacity. Cost estimates for storage areas without secondary containment (i.e., on bare soil) should include estimates of the expenses associated with contaminated soil removal, as well as analytical efforts necessary to verify the extent of decontamination. Closure cost estimates are to be based on having an independent contractor perform closure activities. The closure cost estimate becomes part of the operating record.

The closure cost estimate must be adjusted annually for inflation. The existing closure cost estimate can be recalculated in current dollars or multiplied by an inflation factor to get an adjusted cost estimate. The inflation factor is derived by dividing the latest annual implicit price deflator for gross national product by the implicit price deflator for the previous year. These figures are published by the U.S. Department of Commerce in its *Survey of Current Business*. The closure cost estimate must also be revised whenever a change in the closure plan increases the cost of closure.

The closure cost estimate provides the basis for the amount of financial assurance required for closure. Federal regulations provide six options for providing such assurance. State hazardous waste regulations may not allow certain of the federal options. The federal options are:

1. closure trust fund
2. surety bond guaranteeing payment into a closure trust fund
3. closure letter of credit
4. closure insurance
5. financial test and corporate guarantee for closure
6. use of multiple financial mechanisms

Each of these mechanisms has very specific requirements as to when it may be used, the wording of the mechanism, cancellation provisions, etc. A detailed explanation of each is beyond the scope of this chapter. Waste management personnel should have the financial personnel at the manufacturing firm read the regulatory language and guidance manuals regarding financial assurance for closure. An examination of the state requirements and alternatives and their similarities to and differences from the federal mechanisms would then be in order. This information would form the basis of an intelligent decision regarding which of the available mechanisms best suits the firm's circumstances.

Liability Requirements (265.147)

As mentioned earlier, financial responsibility includes liability insurance coverage for any bodily injury and property damage to third parties resulting from facility operations. Owners/operators of container and tank storage facilities need to demonstrate liability coverage for sudden accidental occurrences in minimum amounts of $1 million per occurrence and $2 million annual aggregate, exclusive of legal defense costs. Mandatory language for the insurance policy or certificate of insurance is specified at 40 CFR 264.151.

The financial assurance requirements allow owners or operators to demonstrate the capability to self-insure or to obtain a corporate guarantee from a parent corporation. The regulations provide a financial test that, if met, will satisfy the regulatory requirements for liability coverage. Liability coverage may also be provided through the use of a combination of the financial test and liability insurance, or a combination of the corporate guarantee and liability insurance.

SUBPART I: USE AND MANAGEMENT OF CONTAINERS

Subpart I is discussed thoroughly in Chapter 9.

SUBPART J: TANK SYSTEMS

The regulatory requirements outlined in Subpart J are summarized in Chapter 9, with the exception of 40 CFR 265.197(c) and 265.200. Regulatory language found at 265.197(c) specifies certain closure requirements, while 265.200 deals with waste analyses and trial tests.

PERMIT REGULATIONS

Any discussion of the regulatory standards for interim status storage facilities is incomplete without mention of 40 CFR Parts 270, 271, and 124. Part 270 deals with the Hazardous Waste Permit Program, Part 271 specifies the requirements for state program authorization, and Part 124 specifies the procedures for decisionmaking with respect to permit applications.

EPA regulations at 40 CFR Part 270, entitled "EPA Administered Permit Programs: The Hazardous Waste Permit Program," specify:

1. permit application requirements
2. contents of the permit application
3. mandatory conditions to the permit
4. changes to the permit, including major and minor modifications and revocation and termination of permits
5. interim status provisions

"Requirements for Authorization of State Hazardous Waste Programs" is the title and subject matter of 40 CFR Part 271. Here are specified the minimum requirements the various states must meet to obtain authorization to administer the RCRA regulatory program. Authorization from EPA will enable the state to administer the regulatory program in lieu of, and with minimal interference from, EPA. Part 271 also contains the procedures for approval, revision, and withdrawal of authorization for state-run regulatory programs.

Finally, 40 CFR Part 124, "Procedures for Decisionmaking," specifies the procedures to be followed in making permit decisions. These procedures are to be used by EPA as well as authorized states. Extensive public participation will be a part of every permit decision under the RCRA program.

SUMMARY

The regulatory requirements are fairly comprehensive for manufacturing firms as owners/operators of hazardous waste storage facilities. Compliance will require the implementation of a formal and sophisticated environmental management program. An important part of such a program will be the development of the documentation necessary for compliance purposes. Other elements include waste minimization and periodic internal evaluation of program effectiveness. Section III deals with the development of a formal environmental management program.

CHAPTER 12

Generators Who Treat Hazardous Wastes by an Exempted Method

As might be expected, generators treating hazardous wastes by an exempted method have few requirements to abide by, with respect to the exempted portion of their operations. However, blanket exemptions from the regulatory requirements are rare. Most exemptions or exclusions have definite limitations in terms of their nature, scope, and application. Many are conditional in nature, with the exclusion available only if all conditions are met. It is recommended that generators utilizing any of the exempted treatment methods take the effort to document the factual circumstances and rationale behind the decision that the exemption was applicable and legally available to the firm.

The regulatory definition of treatment, found at 260.10, is very comprehensive and inclusive. Treatment means "any method, technique, or process, including neutralization, designed to change the physical, chemical, or biological character or composition of any hazardous waste so as to neutralize such waste, or so as to recover energy or material resources from the waste, or so as to render such waste non-hazardous, or less hazardous; safer to transport, store, or dispose of; or amenable for recovery, amenable for storage, or reduced in volume."

This regulatory definition is significant in that, with few exceptions, persons owning and operating hazardous waste treatment facilities must apply for and obtain a RCRA permit. The broad nature of the definition of treatment and the associated requirements make any exemptions all the more important. Please note that recycling and reclamation operations, unless exempted, would require a RCRA treatment permit. The use of exempted methods for treatment or reclamation will make efforts at reducing the volume and toxicity of wastes destined for land disposal much more economical and viable. The long lead times associated with obtaining a RCRA permit can be eliminated.

The interim status and permanent status standards found at 40 CFR 265 and 264 and the permit requirements of 40 CFR 270 do not apply to certain hazardous waste treatment operations, processes, and methods. EPA has informally allowed generators to treat hazardous waste onsite in accumulation containers and tanks without a RCRA permit, under certain circumstances. Such treatment must be in conformance with 40 CFR 262.34, which among other things limits the accumulation period, and requires compliance with 40 CFR 265 Subpart I, "Use and Management of Containers," and Subpart J, "Tank Systems," as appropriate. The regulatory requirements of 40 CFR Part 265, Subparts I and J, are discussed in Chapter 9.

EPA stated in the preamble to the final small quantity generator regulations (51 FR 10168—March 24, 1986) that treatment could occur in a large or small quantity generator's accumulation tanks and containers without a permit, provided that treatment was performed in accordance with 40 CFR 262.34. EPA has taken the common-sense position that treatment should not be unduly discouraged, particularly in view of restrictions on the direct land disposal of untreated hazardous wastes. EPA does, however, give the caveat that generators should not make large investments in onsite treatment facilities based on the assumption that they will be indefinitely exempted from permitting requirements.

There are formal exemptions as well. These are specifically outlined in the regulations themselves (as opposed to in preamble explanations, interpretations, or policy statements).

Manufacturing firms that conduct onsite recycling operations in closed loop systems are exempted from regulation at 261.4(a)(8). Management of secondary materials in such a manner is analogous to ordinary production operations, and materials so managed are excluded from the definition of solid waste. Such materials cannot, therefore, be considered hazardous wastes. The hazardous waste regulations are not applicable to such operations.

Closed loop recycling systems are those where secondary materials are reclaimed and returned to the original process(es) from which they were generated, where they are reused in the production process. However, only tank storage can be involved, and the entire process is closed by being entirely connected with pipes.

Also, the reclamation method cannot involve combustion; the secondary materials cannot be accumulated longer than 12 months prior to reclamation; and the reclaimed material cannot be used to make a fuel or be used in a manner constituting disposal. This exemption would be available to relatively few reclamation systems. Closed loop recycling systems are properly exempted, as they resemble or could be considered manufacturing process units.

Of most interest to manufacturing facilities would be the following exemptions. These exemptions, found at 265.1(c) and/or 270.1(c)(2), are with respect to the interim status standards (40 CFR Part 265) and/or the permit requirements (40

CFR Part 270). Both exemptions apply, unless noted. These exemptions are granted to:

1. the owner and operator of a facility managing recyclable materials described at 261.6(a)(2) and (3), except to the extent that such recyclable materials are regulated at 40 CFR Part 266. Recyclable materials so described include:
   a) hazardous wastes burned for energy recovery (*Note:* This applies only when performed in industrial boilers and furnaces meeting certain requirements. Limited regulatory requirements are imposed at 40 CFR 266 Subpart D. Storage prior to energy recovery is fully regulated at Part 265 and Part 270.)
   b) characteristically hazardous used oil burned for energy recovery (*Note:* This applies only when performed in industrial boilers and furnaces meeting certain requirements. Limited regulatory requirements are imposed at 40 CFR 266 Subpart E.)
   c) recyclable materials from which precious metals are reclaimed (*Note:* Limited regulatory requirements are imposed at 40 CFR 266 Subpart F.)
   d) spent lead-acid batteries that are being reclaimed (*Note:* Limited regulatory requirements are imposed at 40 CFR 266 Subpart G.)
   e) industrial ethyl alcohol that is reclaimed
   f) used batteries (or used battery cells) returned to a battery manufacturer for regeneration
   g) characteristically hazardous used oil that is recycled in a manner other than by being burned for energy recovery
   h) scrap metal
   i) certain petroleum refining waste
   j) certain iron and steel industry wastes

   Note: Other hazardous wastes that are recycled are subject to the requirements for generators, transporters, and storage facilities. With respect to such wastes, it is the recycling process itself that is exempt from regulation. Transportation and storage activities prior to the recycling process are fully regulated.

2. a generator accumulating waste onsite in compliance with 262.34, except to the extent that 262.34 references certain interim status standards found at 40 CFR Part 265
3. the owner or operator of a totally enclosed treatment facility, as defined in 260.10
4. the owner or operator of an elementary neutralization unit or a wastewater treatment unit as defined in 260.10 (*Note:* These units must meet the regulatory definition of "tanks," with the resulting aqueous discharge subject to regulation under Section 402 or 307(b) of the Clean Water Act. Section 402 deals with NPDES permits for direct discharges, while Section 307(b) deals with the pretreatment of industrial discharges to publicly owned treatment works.)
5. the owner or operator of a POTW (also known as a sewage treatment plant) that treats, stores, or disposes of hazardous wastes (*Note:* Such facilities are deemed to have a permit by rule, provided they comply with 270.60(c).)
6. persons carrying out treatment or containment activities necessary to immediately respond to releases or threatened releases of hazardous waste

7. a transporter storing manifested shipments of containerized hazardous waste at a transfer station for a period of 10 days or less

8. persons adding absorbent material to waste in a container or adding waste to the absorbent material in a container, provided that these actions occur at the time waste is first placed in the containers, and the requirements of 265.17(b), 265.171, and 265.172 are complied with

9. the owner or operator of facilities used solely for the treatment, storage, or disposal of hazardous waste excluded from regulation at 261.4, or for hazardous wastes from conditionally exempt small quantity generators. Owners and operators of facilities permitted, licensed, or registered by a state to manage municipal or industrial solid waste are also exempted, if the only hazardous wastes accepted are from conditionally exempt small quantity generators.

It is important to note that the owner or operator of an exempted treatment facility can still be a generator of hazardous wastes, due to any spills or releases, or to the pollution control residuals or sludges resulting from treatment operations. These residuals may be subject to the Land Disposal Restrictions, found at 40 CFR Part 268. Remember that sludges resulting from the treatment of listed hazardous wastes are themselves listed hazardous wastes, by virtue of the so-called "derived-from" rule (261.3(c)(2)). Shipments of hazardous waste from such a facility must comply with the standards for generators found in 40 CFR 262. The 90-day accumulation provisions for generators (262.34) may be used by such facilities for the hazardous wastes generated at that facility.

CONCLUSION

The use of exempted treatment methods is a legal method of avoiding some of the regulatory burdens associated with the present HWM system. The exemptions are designed to afford the required degree of environmental protection without unduly inhibiting the exempted treatment or recycling activities. In many cases, such as the exemptions for wastewater treatment, there are still regulatory standards to abide by. It is assumed that generators will use good judgement and best management practices in treating their hazardous wastes under one of the exemptions.

CHAPTER 13

Generators Who Market or Burn Used Oil Fuels

The purpose of this chapter is to provide a brief discussion of the used oil fuel regulatory scheme, promulgated by EPA in 1985. A large percentage of used industrial oils are recycled by being burned for energy recovery. EPA has regulated the use of used oil fuels and hazardous waste fuels as effective substitutes for virgin fuels at 40 CFR Part 266, "Standards for the Management of Specific Hazardous Wastes and Specific Types of Hazardous Waste Management Facilities." Used oils and hazardous waste fuels are considered effective substitutes for virgin fuels when they have a heat value greater than approximately 5000 Btu per pound.

EPA's intention in establishing Part 266 was to allow appropriate types and levels of regulatory controls to be placed on certain recycling scenarios. Theoretically, such "custom-tailoring" of regulatory controls on specific recycling activities will adequately protect public health and environmental quality, without placing undue disincentives on the reclamation of specific types of wastes. Currently, Part 266 addresses precious metals recovery (Subpart F) and reclamation of lead-acid batteries (Subpart G), as well as used oil and hazardous wastes burned for energy recovery.

"Used Oil Burned for Energy Recovery" is the title and subject matter of Part 266 Subpart E. Used oil, as defined at 266.40(b), means any oil that has been refined from crude oil, used, and as a result of such use is contaminated by physical or chemical impurities. Used oil that exhibits a characteristic of hazardous waste, and that is *not* a "mixture rule" or "derived-from rule" listed hazardous waste, is regulated under 40 CFR 266 Subpart E, when burned for energy recovery. Such used oil is termed "used oil fuel." Used oil fuel includes any fuel produced from used oil by processing, blending, or other treatment.

Quality specifications have been established for used oil fuels (40 CFR 266.40(e)). If used oil burned for energy recovery cannot meet these specifications, it is termed "off-specification used oil fuel." Marketers and burners of used oil fuel are subject to certain administrative controls found at 40 CFR 266 Subpart E. Off-specification used oil fuel can only be burned for energy recovery in industrial boilers or furnaces, as specified at 266.41(b).

Used oil knowingly mixed with listed hazardous wastes from regulated generators and burned for energy recovery is considered hazardous waste fuel, and is regulated under 40 CFR Part 266 Subpart D, "Hazardous Waste Burned for Energy Recovery."

Generators of used oil that is burned for energy recovery are not subject to 40 CFR 266 Subpart E, except to the extent that the used oil is found to be off-specification, and the generator markets such oil directly to burners, or burns the off-specification used oil fuel himself (266.42). Generators must, however, determine whether the used oils are considered hazardous wastes by virtue of the mixture rule and derived-from rule. Generators of hazardous wastes burned for energy recovery (hazardous waste fuels) are fully subject to the Part 262 standards for generators of hazardous waste. Generators that market or burn hazardous waste fuel are subject to 40 CFR Part 266 Subpart D.

The requirement to determine whether a particular used oil destined for energy recovery is an "off-specification used oil fuel," subject to Part 266 Subpart E, is placed on the marketers and the burners of used oils. In many instances, generators of used oils neither market nor burn their used oils. Used oils are typically collected onsite pending removal to a used oil blending facility. The used oil blending facility acts as the marketer to the industrial boilers and furnaces legally capable of accepting and burning off-specification used oil fuels.

Generators shipping their used oils to marketers should be aware of the requirements imposed on marketers and burners of used oil fuel in order to be able to evaluate the regulatory status of their used oils, as well as the compliance status of the blender and burner.

The initial determination involves whether a particular used oil destined for energy recovery is an "off-specification used oil fuel," per 266.44(e). Used oil fuels are assumed to be off-specification, and thus subject to Part 266, unless analyses and other information substantiates that the used oil is *not* considered off-specification (i.e., it meets the specifications found at 266.40(e)).

The quality specifications for used oil fuels are identified in Table 13.1. Used oil exceeding any specification level is subject to Part 266 Subpart E, when burned for energy recovery.

Please note that the used oil fuel specification does not apply to used oil fuel mixed with hazardous waste, other than hazardous wastes from conditionally exempt small quantity generators subject to 40 CFR 261.5. Also, used oil that contains more than 1000 ppm of total halogens is presumed to be a hazardous waste by virtue of mixture with or derivation from listed halogenated (chlorinated)

Table 13.1. Used Oil Fuel Specification

Constituent or Property	Allowable Level
Arsenic	5 ppm maximum
Cadmium	2 ppm maximum
Chromium	10 ppm maximum
Lead	100 ppm maximum
Flash point	100°F minimum
Total halogens	4000 ppm maximum

solvents. This is a rebuttable presumption. Generators can rebut this presumption by demonstrating that the used oil does not contain hazardous waste. This can be done by review of the MSDSs for purchased components of the used oils, in conjunction with a thorough review of waste solvent handling and segregation practices. If the MSDS lists halogenated constituents, such as chlorinated paraffins, and review of solvent management practices indicates no possible mixtures, inadvertent or deliberate, of listed halogenated wastes and the used oil, then the presumption of hazardousness can be rebutted. Otherwise, the used oil is to be managed as a hazardous waste fuel. The alternative method of rebutting the presumption of hazardous waste status for the used oil involves the review of MSDSs and solvent management practices in conjunction with laboratory analyses, such as gas chromatography/mass spectrometry (GC/MS), for halogenated hazardous constituents listed at 40 CFR 261, Appendix VIII. Used oil containing more than 100 ppm of any individual listed halogenated solvent (F001–F002) is to be managed as a hazardous waste fuel. Even if the presumption of mixing is successfully rebutted, used oil fuels with more than 4000 ppm total halogens are considered "off-specification used oil fuels," regardless of the absence of listed halogenated hazardous constituents.

Parties that market or burn off-specification used oil fuels are subject to a rather limited set of requirements, largely administrative. These requirements, in summary form, are:

1. Notify EPA of used oil management activities and receive an EPA identification number.
2. Off-specification used oil fuel can only be burned in an industrial furnace or boiler as specified at 266.41(b).
3. Determine whether the used oil is subject to regulation under Part 266 Subpart E. The used oil fuel could be a hazardous waste fuel, subject to Part 266 Subpart D, by virtue of the mixture or derived-from rules. Alternatively, the used oil fuel could meet or exceed the quality specifications, and be subject only to recordkeeping requirements. Used oil is assumed to be off-specification, and thus subject to regulation, in the absence of analytical data and other information showing otherwise.
4. Utilize an invoice system for shipments of off-specification used oils, containing the following information:
   a) an invoice number
   b) the EPA identification numbers of the shipper and the receiving facility

c) the names and addresses of the shipping and receiving facilities
d) the quantity of off-specification used oil to be delivered
e) the date(s) of shipment or delivery
f) the following statement: "This used oil is subject to EPA regulation under 40 CFR Part 266."

   Note that used oils meeting the definition of flammable liquid (flash point <100°F) or combustible liquid (flash point >100°F and <200°F) are considered "hazardous materials" for shipping purposes, subject to the DOT Hazardous Materials Regulations found at 49 CFR Parts 100-177.

5. Obtain or provide the required one-time certification notice for the first shipment of off-specification used oil to each marketer or burner. The certification notice states that the party has notified EPA, and if the recipient is a burner, that only industrial furnaces or boilers meeting the requirements of 266.41(b) will be utilized for energy recovery.

6. Keep copies of used oil invoices for at least three years from the date the invoice was prepared or received. Keep a copy of each certification notice that is received or sent for at least three years from the date of the last transaction with the person sending or receiving the certification notice.

CONCLUSION

As mentioned earlier, the requirements imposed on marketers and burners of used oil fuels do not apply to generators performing neither activity. However, generators of used oils burned offsite for energy recovery should be aware of the regulatory requirements, in order to minimize liabilities under CERCLA. Generators should determine whether their used oil fuel is off-specification, and utilize the services of used oil vendors that are competently managed, in compliance with the Part 266 regulations, and that are adequately insured. The subject of evaluating offsite treatment or disposal vendors is discussed in Chapter 23.

CHAPTER 14

Owners/Operators of Underground Product Storage Tanks

The Hazardous and Solid Waste Amendments of 1984 (HSWA), which reauthorized RCRA, expanded the scope of the RCRA regulatory program to include certain previously unregulated underground storage tanks. Subtitle I of RCRA, as amended, gave EPA the statutory authority to regulate underground storage tanks (USTs) containing materials other than hazardous wastes. This action created an entirely new category of potentially regulated parties under RCRA, that of owners and operators of USTs containing petroleum (including used oils) and hazardous substances (other than hazardous wastes). Regulations for these underground storage tanks are codified at 40 CFR Part 280. Please note that underground and other storage tanks containing hazardous wastes had been regulated since the inception of the HWM program in 1980.

The RCRA Subtitle I program focuses on underground storage tanks, as defined, containing regulated substances, as defined. An underground storage tank is any one tank or combination of tanks (including underground pipes connected thereto) which is used to contain an accumulation of regulated substances. To be considered underground, the tank must have at least 10% of its volume below ground, including the volume of pipes attached to the tanks. Excluded from the definition are the following:

1. farm and residential tanks having a capacity of no more than 1100 gallons used for storing motor fuel for noncommercial purposes
2. tanks storing heating oil for consumptive use on the premises where stored
3. septic tanks
4. pipeline facilities

5. surface impoundments, pits, ponds, or lagoons
6. storm water or wastewater collection systems
7. flow-through process tanks
8. liquid traps or associated gathering lines directly related to oil or gas production and gathering operations
9. storage tanks in an underground area (i.e., basement) if the tank is situated on or above the floor surface
10. any pipes connected to any tank described in items 1–9, above

The regulatory definition of regulated substances includes CERCLA hazardous substances, *except* hazardous waste. Petroleum, including crude oil or any faction thereof which is liquid at standard conditions of temperature and pressure, is also considered a regulated substance. Regulated substances include petroleum-based substances such as motor and jet fuels, distillate and residual fuel oils, lubricants, petroleum solvents, and used oils.

The CERCLA list of hazardous substances is found at 40 CFR Part 302. Hazardous wastes are not considered regulated substances for purposes of the 40 CFR 280 UST program, even though for CERCLA purposes hazardous wastes are a subset of hazardous substances.

NOTIFICATION REQUIREMENTS

Owners and operators of nonexempt underground storage tanks are to notify their respective state regulatory agency. Notification is required for USTs that have been used to store regulated substances since January 1, 1974 and that are in the ground as of May 8, 1986. Notification is also required for USTs containing regulated substances that are brought into use after May 8, 1986. Notification is not required for tanks taken out of operation prior to January 1, 1974, or for tanks removed from the ground prior to May 8, 1986. This notification, on EPA form 7530-1, was to have been submitted by May 8, 1986. A copy of EPA form 7530-1 (September 1988 revision) is included as Appendix F of this book. Appendix II of 40 CFR Part 280 provides a list of state agencies (with mailing addresses and telephone numbers) designated to receive the notifications. Nonnotifiers are subject to potential civil penalties of $10,000 per tank.

EPA finalized proposed rulemaking regarding technical requirements for USTs in the September 23, 1988 *Federal Register* (53 FR 37082-37212). These rules are codified at 40 CFR Part 280 ("Technical Standards and Corrective Action Requirements for Owners and Operators of Underground Storage Tanks (UST)").

This rulemaking fulfilled statutory mandates under Section 9003 (Subtitle I) of RCRA to establish requirements for leak detection, leak prevention, and corrective action for all USTs containing regulated substances, as necessary to protect human health and the environment. Final rules regarding financial responsibility for releases and corrective action were addressed in the October 26, 1988 *Federal Register* (53 FR 43321–43383).

The regulatory definitions of "underground storage tank" and "regulated substances," as used in the 40 CFR Part 280 UST program, were discussed earlier. In the final rules, EPA excluded and/or deferred from regulation certain UST systems that would otherwise be fully subject to 40 CFR Part 280 requirements.

The following UST systems are excluded from the requirements of 40 CFR Part 280:

1. any UST system holding hazardous wastes or mixtures of hazardous wastes and other regulated substances
2. any wastewater treatment tank system that is part of a wastewater treatment facility regulated under Section 402 or 307(b) of the Clean Water Act
3. equipment or machinery containing regulated substances for operation puposes, such as hydraulic lift tanks and electrical equipment tanks
4. any UST system whose capacity is 110 gallons or less
5. any UST system containing *de minimis* concentrations of regulated substances
6. any emergency spill or overflow containment UST system that is expeditiously emptied after use

Other tank systems have had certain elements of the regulations deferred from application. Only the so-called "interim prohibition" (§280.11) and release response and corrective action requirements (40 CFR 280 Subpart F) apply to the following types of UST systems:

1. wastewater treatment tank systems
2. any systems containing radioactive materials that are regulated under the Atomic Energy Act of 1954, as amended
3. any UST system that is part of an emergency generator system at a nuclear power generation facility regulated by the Nuclear Regulatory Commission
4. airport hydrant fuel distribution systems
5. UST systems with field-constructed tanks

A separate deferral is identified from the requirements for release detection (40 CFR 280 Subpart D) for UST systems storing fuel solely for use by emergency power generators.

INTERIM PROHIBITION FOR DEFERRED UST SYSTEMS

Pending final EPA regulations on the design, construction, and corrosion protection of new UST systems listed at 40 CFR 280.10(c) as deferred from regulation, EPA has imposed the so-called "interim prohibition," at §280.11. The interim prohibition is a ban on the installation of new UST systems deferred from regulation at 40 CFR 280.10(c) unless:

1. The system will prevent releases due to corrosion or structural failure for the operational life of the UST system.
2. The system is cathodically protected against corrosion, constructed of noncorrodible material or steel clad with a noncorrodible material, or designed so as to prevent the release or threatened release of any stored substance.
3. The material used in the construction or lining of the UST system is compatible with the substance to be stored.

TECHNICAL STANDARDS

The final EPA regulations regarding USTs containing regulated substances will have a major impact on existing underground storage tanks or any replacement USTs. These final regulations were published in the September 23, 1988 *Federal Register*. The effective date for these rules is December 22, 1988.

The technical standards are quite comprehensive and have requirements for design, construction, and installation. Corrosion protection, spill prevention, and release detection are required as well. There are very detailed and stringent release reporting, investigation, and corrective action requirements. Financial responsibility for third-party damages and the costs of corrective action is required. Existing tanks must be upgraded to new tank standards, or removed from service and closed. Finally, standards for removing tanks from service (''closure'') are provided.

Following is a brief summary of the technical standards for UST systems containing regulated substances. The reader is encouraged to carefully review the regulations themselves, the preamble language, and EPA guidance manuals that will be prepared in response to questions regarding interpretation and implementation of this major rulemaking effort.

The major elements of this rule, as identified by EPA in the preamble language at 53 FR 37098, are:

- New UST systems must be designed and constructed to retain their structural integrity for their operating life, in accordance with national consensus codes of practice. All tanks and attached piping used to deliver the stored product must be protected from external corrosion. Cathodic protection must be monitored and maintained to ensure that UST systems remain free of corrosion.

- Nationally recognized industry installation standards must be followed in placing new UST systems in service. Owners and operators of new USTs must certify that proper installation procedures were followed and identify how the installation was accomplished.

- Owners and operators of both new and existing UST systems must follow proper tank filling practices to prevent releases due to spills and overfills. In addition, owners and operators of either new or upgraded UST systems must use devices that prevent overfills and control or contain spills.

- Tanks must be repaired in accordance with nationally recognized industry codes. These national codes include several tests that must be conducted to ensure quality repairs.

- To close UST systems, industry recommended practices must be followed. The UST system can be removed from the ground or left in place after removing all regulated substances and cleaning the tank, filling it with an inert substance, and closing it to all future outside access. In addition, owners and operators must perform an assessment at the time of UST closure to ensure that a release has not occurred at the site. If a release has occurred, then corrective action must be taken.

- Release detection must be instituted at all UST systems. For petroleum UST systems, several methods will be allowed, although tank owners and operators must adhere to requirements concerning their use. In addition, owners and operators must follow special requirements for pressurized delivery lines. Petroleum UST systems are not required to have secondary containment with interstitial monitoring. All new or upgraded UST systems storing hazardous substances, however, are required to have secondary containment with interstitial monitoring, unless an alternate release detection method is approved by the implementing agency.

- Generally, release detection at existing UST systems must be phased in over a five-year period based on the age of the tank. The oldest UST systems (usually unprotected from corrosion) are required to phase in release detection within one year, and the newest tank systems (usually protected from corrosion) by the end of the five-year period. Release detection for all pressurized delivery lines must be retrofitted within two years.

- Periodic tank tightness testing (every five years) combined with the monthly inventory control is allowed at new or upgraded UST systems for 10 years after new tank installation or existing tank upgrade. After 10 years, monthly release detection is required.

- Either monthly release detection or a combination of annual tank tightness testing with monthly inventory control is required of substandard existing USTs until they are upgraded. Existing UST systems must be upgraded or closed within 10 years of the effective date of the final rule, or within one to five years if a release detection method is not available that can be applied during the required phase-in period for release detection. Upgrading of petroleum UST systems includes retrofitting of corrosion protection and both spill and overfill controls at all tanks. Upgrading of hazardous substance UST systems also includes secondary containment and interstitial monitoring or an alternate release detection method approved by the implementing agency.

- Tank owners and operators must report suspected releases. Indications of a release must be reported to the implementing agency, including positive results from release detection methods, unless the initial cause of the alarm has been immediately investigated and the alarm found to be false. After reporting suspected releases, owners and operators must perform release investigation and confirmation tests and, where a release is confirmed, must begin corrective action.

- Owners and operators of leaking UST systems must follow measures for corrective action. Immediate corrective action measures include mitigation of safety and fire hazards; removal of saturated soils and floating free product; and an assessment of the extent of further corrective action needed. A corrective action plan would be required for long-term cleanups addressing groundwater contamination, although these cleanups could begin upon notification of the implementing agency by the owner and operator. Cleanup levels would be established on a site-by-site basis as approved by the implementing agency.

RISK MANAGEMENT

The presence of underground storage tanks poses potential environmental problems. Spills and overfills are the most common causes of releases. Many releases result from piping system failures. Releases from underground piping occur about twice as often as releases from the tank portion of the UST system. Corrosion, poor installation and workmanship, accidents, and natural events (e.g., freeze-thaw of soil) are the four major causes of piping failure. EPA estimates that approximately 25% of existing UST systems would fail tests using current leak testing methods, and that loose tank fittings or faulty piping causes 84% of these tightness test failures. Releases from UST systems will require response, cleanup, and remediation of natural resources. Therefore, proper management of underground tanks is strongly recommended to assure the integrity of such tanks. Proper management is also necessary to reduce potential financial liabilities and minimize costs associated with product loss.

It is suggested that existing and underground piping systems tanks be leak tested, utilizing a test method capable of detecting leaks greater than 0.1 gallons per hour (the current EPA performance standard). This testing should be performed annually, pending retrofitting to new tank standards, replacement of the tanks, or their removal from service. In addition, detailed inventory reconciliation records should be kept on a monthly basis as a rough indication of the integrity of the tank. The regulations allow other release detection alternatives, including automatic tank gauging, soil vapor monitoring, groundwater monitoring, and interstitial monitoring between the tank and a secondary barrier, provided certain performance requirements are met.

Consideration should be given to removal of existing USTs from service, according to 40 CFR 280 Subpart G, as well as existing consensus codes (API [American Petroleum Institute], NFPA, PEI [Petroleum Equipment Institute], etc.) and best engineering practices. At a minimum, all out-of-service tanks should be emptied and cleaned to the extent possible. Once the tanks are emptied and cleaned, a decision should be made whether to remove them from the ground or fill them with an inert material. This decision should be based on both risk and cost considerations. An assessment of the UST site for the presence of contamination is a requirement at closure. If contamination is discovered, corrective action must be taken in accordance with 40 CFR 280 Subpart F.

The economics of purchasing petroleum and hazardous substances in bulk should be carefully evaluated in view of the increased costs of owning and operating any existing underground storage tanks. The availability and cost of required financial responsibility mechanisms (e.g., environmental impairment liability insurance) is likely to be the most critical factor in these evaluations for small and medium-sized businesses. If a decision is made to continue operation of any UST system, planning and budgeting for costs involved with regulatory compliance should begin prior to the effective date(s) of the regulations. Given the age of many existing underground tanks, retrofitting with corrosion protection and leak detection is unlikely to be cost-effective. Serious consideration should be given to replacement tanks.

If bulk storage remains necessary, the advantages and disadvantages of above- and below-ground storage tanks should be reviewed. Existing, proposed, and foreseeable regulatory requirements should be examined. Secondary containment is recommended for any new above- or below-ground bulk storage system, even in the absence of regulatory requirements.

SUMMARY

Ownership and operation of existing underground storage tanks currently presents definite environmental risks. Recent regulatory initiatives will impose additional costs on owners and operators of such tanks. Certain decisions have to be made regarding existing underground storage tank systems. Proper planning and effective decisionmaking can minimize the risks, costs, and liabilities associated with underground product storage.

SECTION III

Developing the Corporate Environmental Management Program

Many manufacturing firms have yet to implement a comprehensive compliance program to meet the hazardous waste management requirements. Even fewer firms have taken steps necessary to develop and implement an environmental management program. There are many reasons for this situation, not the least of which are the complexity of the regulations and the lack of guidance. Environmental management involves more than mere regulatory compliance. Risk reduction and waste minimization measures are important elements in any environmental management program. Periodic evaluation of the success of the program in meeting stated goals is also a necessary element.

The chapters in Section II specified the requirements applicable to the various regulatory categories. Guidance and interpretation of the regulations were provided when necessary. Section III builds on the information given previously, enabling waste managers to integrate the various requirements into a comprehensive environmental management program.

CHAPTER 15

Key Factors Regarding the Management of Chemical Risks

The purpose of this chapter is to outline 10 key factors that need to be addressed by corporate and environmental managers. Each of these factors involves, broadly speaking, the management of chemical risks, thus affecting firms using hazardous chemicals and generating hazardous wastes. Failure to adequately consider and respond to the challenges posed by these factors can have potentially devastating effects on the profitability and long-term viability of the firm.

This chapter assumes the reader has a general understanding of CERCLA (Superfund). The time frame for consideration of the 10 factors has been shortened by the reauthorization of RCRA (the Hazardous and Solid Waste Amendments of 1984) and the 1986 reauthorization of CERCLA, known as the Superfund Amendments and Reauthorization Act (SARA). In these amendments, Congress has reduced the alternatives available with respect to the factors and has set some definitive goals, policies, timetables, and deadlines.

I. THE USE OF LANDFILLING AS A DISPOSAL METHOD

The first factor in chemical risk management regards the use of landfills as a disposal method for hazardous wastes generated by the firm. The key question is not whether to minimize their use, but how. It is now recognized that hazardous waste landfills, even so-called "secure chemical landfills," involve long-term uncertainties regarding potential environmental impacts. The primary long-term concern is the potential for migration of hazardous waste constituents to the groundwater. Secure chemical landfills, i.e., those engineered sites with dual synthetic liners, leachate collection and leak detection systems, and impermeable cover systems upon closure, should minimize the possibility of environmental impact. The groundwater monitoring systems required should detect contamination before offsite environmental and human health damage has occurred. Required financial responsibility, usually in the form of insurance, for harm to people and property offsite will ensure that compensation is available. Closure and postclosure financial responsibility instruments will help to ensure that the disposal firm will have adequate funds available when the landfill is finally closed. All this should minimize the possibility that generators sending hazardous wastes to a secure chemical landfill will be held responsible for costs associated with closure, cleanup, and offsite damages to people and property, as could be possible under CERCLA and the common law.

The safeguards established by RCRA standards and regulations for hazardous waste landfills do not eliminate the long-term uncertainties and liabilities associated with land disposal. There are other considerations as well. The body of federal and state environmental regulations, requirements, and restrictions is ever-expanding and becoming increasingly stringent. The consensus of what constitutes an "acceptable" environmental impact is moving toward lower and lower levels of discharge. Also, the capabilities of analytical techniques and equipment to measure extremely low levels of environmental contaminants now exceed our scientific understanding of the impacts and risks involved at such low levels. At the same time, the difficulty and costs involved in designing, constructing, operating, and managing disposal sites have continually increased, along with the costs of controlling or cleaning up past or present discharges to the environment, especially to the groundwater.

Price increases and the decreasing availability of land disposal are going to be facts of life for those firms disposing of their hazardous wastes by land. Prices are going to increase not only as a result of the additional costs imposed on disposal firms resulting from governmental requirements, but also due to the action of market forces. There will shortly be a crisis in the availability of land disposal capacity. Stated simply, existing hazardous waste landfills are being closed, while few new sites are being established. This is due to a variety of factors, not the least of which is local opposition to hazardous waste sites. The economic result of scarcity in a good or service with a relatively inelastic demand is higher prices. Our legislative bodies at the federal, state, and local level are doing their part to use the precepts of economics to move generators away from land disposal. Their method is to tax the land disposal of hazardous waste.

The final reasons for reducing land disposal come from HSWA. Congress has directed EPA to review and, if necessary, ban the land disposal of specific hazardous wastes, while at the same time requiring generators to establish internal programs to reduce the volume and toxicity of the wastes they produce. In addition, generators must certify on each manifest that the management method chosen for each waste is the most environmentally acceptable. EPA has already acted to restrict the land disposal of untreated F-listed solvent wastes and the so-called California List and First Third listed wastes. EPA has promulgated a schedule for the establishment of land disposal restrictions on all listed and characteristic hazardous wastes (40 CFR 268 Subpart B). EPA has until May 8, 1990 to establish standards for all listed and characteristic hazardous wastes; after that date, land disposal of hazardous wastes without treatment standards will be prohibited by statutory provisions in RCRA. The handwriting is on the wall: *"Minimize the use of hazardous waste landfills as a disposal method."*

II. USE OF THE WASTE MANAGEMENT HIERARCHY

The second factor involves the management of hazardous wastes overall, assuming that land disposal is of limited availability and used only as a last resort. Obviously, the best thing to do would be to implement source control measures to minimize or even eliminate the generation of hazardous wastes by the firm. Elimination of wastes is not always feasible, from the standpoints of either technology or economics. However, efforts toward this noble end are required by HSWA, as discussed above. Certifications on manifests and biennial reports on the progress of such efforts are mandated. From the point of view of internal cost control, it is important to realize the effects that increased treatment/disposal costs and increased disposal taxes are going to have on the firm. The short-term and long-term liabilities, the increased complexity of regulatory compliance, increased regulatory scrutiny, and insurance considerations all point to the desirability of minimizing a firm's exposure to risks and liability by reducing the volume and toxicity of wastes generated. After all, for manufacturing firms, the principal concern is production of a product, not management of the resulting residuals.

There are many actions that can be taken to minimize the generation of hazardous wastes, including modifying the production process and/or utilizing different inputs to the process. An example of the former would be the use of electrostatic painting or powder coating as a replacement for the traditional air-pressure spray painting in a waterfall paint booth. The latter is illustrated by the substitution of alkaline cleaning solutions for chlorinated organic solvents for the degreasing of metal parts. Even such mundane and obvious measures as good housekeeping and prudent inventory control can have significant impacts on the volume of waste generated.

What is needed and what is available is a systematic means of thinking about the waste management "problem." This systematic method of evaluating waste management alternatives is called the Waste Management Hierarchy. It is a series of seven decisionmaking steps to be used sequentially, in decreasing order of desirability. The first step, the most favorable and best option, is to prevent the generation of waste. The last step, the least favorable or last resort option, is land disposal. The Waste Management Hierarchy is presented below.

1. Prevent the generation of waste (both hazardous and nonhazardous).
2. Minimize the proportion of unavoidable waste that is hazardous.
3. Separate and concentrate waste streams so as to make further management activities more feasible.
4. Reuse internally or offsite through waste transfers.
5. Recycle or reclaim material values or utilize energy values in the waste.
6. Maintain unavoidable and nonreclaimable wastes in a form amenable to treatment. Perform such treatment operations as incineration, neutralization, oxidation, reduction, precipitation, filtration, drying, etc., to render the effluent and residue less hazardous and reduced in volume.
7. Manage remaining hazardous wastes or treatment residues by secure land disposal, and manage remaining nonhazardous wastes by sanitary landfill.

The application of the Waste Management Hierarchy is discussed in Chapter 18.

III. ONSITE/OFFSITE HAZARDOUS WASTE MANAGEMENT ACTIVITIES

The diligent application of the Waste Management Hierarchy addresses what is to be done with the wastes generated by the firm. The next factor deals with the question of where such waste-related activities are to take place. The third factor is the determination of whether regulated hazardous waste treatment, storage, or disposal activities should be conducted onsite (on the generator's property) or offsite (on the property of a commercial hazardous waste management firm). For manufacturing firms, this is best divided into two separate considerations:

- Should a firm continue regulated (permitted) storage and disposal operations onsite?
- What types of treatment activities should be conducted onsite, and what types offsite?

Regarding the continuation of onsite storage and disposal operations requiring a permit at a manufacturing facility, the choice is clear. Either do all that is necessary to obtain a full RCRA permit for storage or disposal, or close the existing interim-status onsite storage or disposal facilities. The vast majority of manufacturing firms regulated under the interim status standards for TSD facilities are simply storing hazardous wastes in containers onsite longer than 90 days. This decision requires the firm to comply with interim status standards (40 CFR Part 265) and to upgrade the facility to the technical standards (40 CFR Part 264) once a Part B permit application is requested by the EPA or an authorized state regulatory agency. The technical and administrative (substantive and procedural) standards found in 40 CFR Parts 264 and 265 are quite involved, and to a large extent would not be applicable to manufacturing firms that chose to *accumulate* their wastes onsite less than 90 days rather than *store* their wastes onsite longer than 90 days. The manufacturing firm must weigh the benefits of being able to hold wastes onsite indefinitely against the costs associated with obtaining and keeping a RCRA permit. These costs are not inconsequential and are in part associated with:

1. preparation of the permit
2. upgrading the facility
3. insurance
4. financial responsibility for closure/postclosure requirements
5. administrative overhead—technical, legal, accounting, and managerial involvement
6. corrective action for environmental impacts caused by past activities

When their Part B permit applications are called in, many manufacturing firms with interim-status storage facilities are finding that such a facility is not necessary for efficient onsite or offsite treatment or disposal. In these instances, certified closure of the interim-status storage facility is the most appropriate alternative. HSWA sets deadlines for the termination of interim status, in order to speed the process of issuing or denying RCRA permits. Interim status for storage facilities terminates November 8, 1992, unless a RCRA permit was issued or a Part B application was submitted by November 8, 1988.

There are obviously other considerations involved besides costs, including the complexities of regulatory compliance, the limited availability of insurance, and the geological suitability of onsite property for certain hazardous waste management activities. These considerations are usually site- and waste stream-specific and warrant thorough evaluation. Very close scrutiny is necessary for onsite surface impoundments as a storage method. A secondary aspect of the onsite/offsite factor has to do with treatment activities. Certain considerations are worthy of note:

1. Some onsite treatment activities do not require RCRA permits (i.e., reclamation, elementary neutralization, and wastewater treatment).
2. Onsite treatment (''pretreatment'') in some instances is mandatory under Clean Water Act regulations for any discharges to sewers or waterways. Inadequate consideration of onsite wastewater treatment activities can have adverse impacts on regulatory compliance and liabilities, especially when surface impoundments are a part of the treatment process.
3. Treatment activities can reduce the volume and toxicity of waste streams, which reduces land disposal costs and potential liabilities.
4. Restrictions on land disposal promulgated by the EPA under the provisions of HSWA have resulted in mandatory treatment prior to land disposal.

By now, it should be realized that treatment is a necessary and desirable management method. An example may help to illustrate some of the considerations in the onsite/offsite decision. Solvent reclamation can be performed either onsite or offsite. The prices charged for offsite reclamation may make the purchase of onsite equipment economically attractive to large-volume solvent waste generators. Onsite reclamation will certainly reduce the risks of transporting hazardous wastes and the liabilities associated with a commercial TSD facility. The total cost per reclaimed gallon of solvent may be lower if reclamation is performed onsite. However, segregation of waste streams is necessary if the reclaimed solvent is to be reused in the original process. In addition, there will always be some residues (still bottoms) from the distillation process requiring disposal. Other considerations, such as the quality and purity of the reclaimed solvent, training of the equipment operators, and scheduling of wastes to be reclaimed against requirements for reclaimed solvent, should be taken into account in assessing an onsite solvent reclamation program.

IV. EVALUATION OF OFFSITE FACILITIES

The fourth factor concerns the evaluation of actual offsite facilities prior to shipping wastes for treatment or disposal, and reevaluation periodically thereafter. Earlier factors discussed the what and where (offsite/onsite) of managing wastes, while this factor deals with the question, "By whom?" In the evaluation of offsite (commercial) hazardous waste management facilities, the following nine elements should be considered.

Permits

This element includes appropriate federal and state hazardous waste permits (or interim status) as well as other environmental permits for discharges to waterways, sewers, or air. Permits are a necessary condition of doing business with a commercial facility, but they are not in themselves sufficient.

Facilities and Physical Plant

During a facility tour, it is necessary to verify the existence and working condition of processing and ancillary equipment necessary to facility operations. Determine the types and amounts of security, emergency, and spill control equipment. Are storage tanks and drum staging areas paved and diked? Do the landfill cells have dual liners and leachate collection systems? Is there adequate groundwater monitoring? Is there a working laboratory onsite?

Management and Personnel

It is necessary to evaluate the qualifications of facility personnel and their ability to properly operate a business in a complex and highly regulated field. An intimate knowledge of regulatory requirements is a must. Evaluate the training program for laborers, especially in emergency response procedures. Is there a formal safety program? Do employees follow safety procedures and wear appropriate personal protective clothing and equipment?

Housekeeping and Operations

This category is relatively easy to evaluate. Observe the general level of housekeeping, particularly around off-loading and waste transfer areas. Observe the employees' waste handling practices. Inspect drum storage areas. Is there an excessive inventory for facility operations? Are small spills promptly cleaned up? Housekeeping is a good indicator of managerial quality, compliance status, and to a lesser extent, financial stability.

Insurance/Financial Responsibility/Financial Stability

Determine whether the facility has obtained the minimum insurance coverages required by regulation. Determine whether the required financial instruments have been established to cover the costs of closure/postclosure activities. Are the amounts of these instruments reasonable given the nature of operations? Evaluate whether the firm is sufficiently stable financially to continue operations over the long term. A Dun & Bradstreet report may be of some value in this matter, as can annual reports for publicly held companies.

Regulatory Compliance Status

Review the facility's compliance programs and documentation. Ask to see past inspection reports from the EPA or state environmental agencies, as well as the facility's written response to these reports. Ask about past enforcement actions or compliance problems. Determine whether the facility is meeting any improvement deadlines specified as permit conditions or in consent decrees. Verify your findings with the state environmental agency.

Environmental Impact

Evaluate the facility's impact on air, groundwater, and surface water resources. Review the possibility of direct contact with wastes, as well as fire and explosion. Current operations should be evaluated, as well as any pre-RCRA disposal activities that may cause adverse environmental impacts. Both the probability of releases and the severity and magnitude of the effects should be taken into account. It is important to remember that CERCLA and common law liabilities still apply to facilities which have achieved compliance with RCRA requirements.

Community Relations

This element is often difficult to assess, unless the facility has an adversarial relationship with its neighbors and the community. Responding to the fears, concerns, and complaints of "neighbors," however far away, is an important aspect of maintaining good community relations. Once relations with the community deteriorate, it is often impossible to restore trust and confidence. Disgruntled neighbors, effectively organized, can disrupt operations and attract media and regulatory scrutiny to the facility. Frequently, organizations are established with the sole purpose of closing an existing facility or preventing a new facility from being built. Local politicians, once summoned to the cause, can create many legal roadblocks to continued operations.

Relative Contribution or "Other Deep Pockets"

Make a conscious attempt to find out who the customers of the facility are, their financial strength, and their relative contribution to the amounts of waste being processed. Potentially Responsible Parties (PRPs) always find it better to be a small fish in a big pond when it comes to CERCLA liabilities.

A detailed discussion of these nine elements can be found in Chapter 22. Suffice it to say that these elements, although interrelated, are discrete. For example, a facility can possess the proper permits, yet not be in compliance with all pertinent regulatory requirements. Further, a facility can possess all necessary permits, and be in compliance with all applicable laws and regulations, and at the same time fail to prevent adverse environmental impacts on the groundwater beneath the facility.

V. EMPLOYEE CHEMICAL EXPOSURE

The fifth factor involves two considerations associated with employee chemical exposure. A decision has to be made, either explicitly or by default, as to what chemicals employees are going to be exposed to, and at what levels. This is more involved than the requirements of the OSHA Hazard Communication Standard. This Standard, often referred to as "Employee Right-to-Know," requires, among other things, that employers inform employees of the presence of and hazards involved with workplace chemical exposure. The OSHA Hazard Communication Standard will ultimately have large and far-reaching effects on all chemical users. The compliance deadline for chemical users in manufacturing operations was May 1986. OSHA has expanded the scope of the Standard to include chemical users in nonmanufacturing operations. At present, there are no requirements for employers to review or choose alternative processes, operations, or inputs that would allow less hazardous chemicals to be utilized. Although the

OSHA Hazard Communication Standard is a much-needed improvement, there are a variety of reasons for manufacturing employers to go beyond the regulatory requirements.

In making decisions concerning which chemicals employees will be exposed to, and at what levels, consideration needs to be given to both the acute (short-term) and chronic (long-term) effects on human health. Special efforts must be made to reduce or eliminate workplace exposure to carcinogens, mutagens, or teratogens. These chemicals, unless used in processes where there is absolutely no employee exposure, place the firm in a situation of unknown risk. For each chemical used, is there an adequate amount of scientific information available to allow an understanding of the potential acute and chronic health effects resulting from exposure at low levels over the long term? Just as efforts to reduce waste generation are desirable, so are efforts to minimize employee exposure to chemicals that place the firm in a situation of unknown risk. Some of the same techniques are available, including process modification and input substitutions. Are alternatives available for chemicals known or suspected to possess carcinogenic, teratogenic, or mutagenic properties? Can production processes be modified to reduce potential employee exposure or to allow the use of "safer" alternative input chemicals? It is important to understand the short- and long-term health effects of workplace levels of exposure to the alternative chemical. Are these effects fully researched, known, or understood?

Another very important consideration regarding employee chemical exposure involves adequately informing employees of the risks of the chemicals to which they are exposed. This transfer of information should be handled with care. Informing employees of the effects of exposure to a particular chemical in an uncontrolled situation where no protective clothing or equipment are used may be useful as a means to teach "respect" for that particular chemical. However, this approach does nothing to enable the employee to make an informed judgment as to the relative risks of exposure under controlled conditions where appropriate process enclosure or ventilation is used, and where work rules requiring protective clothing and equipment are followed. A more balanced approach to effective transfer of chemical risk information is needed, but is clearly beyond the scope of this chapter. Perhaps the concept of workplace health programs, as announced by OSHA (48 FR 54546) could be the vehicle for such a transfer, as well as for employees to make informed choices. These choices include the types and amounts of chemical exposure that are acceptable, and what circumstances regarding chemical exposure may cause employees to pursue grievance procedures or to seek other employment.

Corporate executives should anticipate that some disgruntled employees (e.g., employees that have been dismissed, and employees out of work as a result of layoffs or plant closures), as well as employees with real or imagined health problems, will be using their new-found knowledge of chemicals to attempt to obtain monetary compensation for damages that are allegedly the result of chemical exposure. Various pieces of proposed legislation regarding "toxic torts" will make lawsuits for damages much more viable than at present.

VI. CONTROL OF ONSITE ENVIRONMENTAL IMPACTS

The sixth factor involves several considerations regarding the control of onsite environmental impacts. These impacts can result from past or present operations. Future impacts can be prevented or at least minimized with proper controls and good decisions made prior to the onset of serious problems including enforcement activities. These efforts are necessary to avoid "fouling the nest." Key considerations include:

- Closure of existing onsite *surface impoundments* (also known as pits, ponds, and lagoons) used for waste treatment, storage, or disposal. Surface impoundments will come under increasing regulatory scrutiny and will become increasingly expensive to operate as a result of the minimum technological requirements mandated by Congress in the HSWA. Even when properly engineered with state-of-the-art environmental control technology, surface impoundments are a prime potential source of groundwater contamination.

- *Underground chemical and waste storage tanks* require close attention. Not only can the tanks themselves corrode or otherwise be damaged and leak, but the ancillary pumps and piping can also become a source of leakage and groundwater contamination. The problem of leaking underground storage tanks attracted the attention of Congress in HSWA. A notification program for owners of existing and new tanks became effective May 8, 1986. New tanks must be corrosion-resistant and compatible with their contents.

  EPA has promulgated regulations regarding underground product (petroleum and hazardous substance) storage tanks (53 FR 37194–37212). These regulations are quite comprehensive and have requirements for design, construction, and installation. Corrosion protection, spill prevention, and release detection are required as well. There are very detailed and stringent release reporting, investigation, and corrective action requirements. Financial responsibility for third-party damages and the costs of corrective action will be mandated. Existing tanks must be upgraded to new tank standards, or removed from service and closed. Finally, standards for removing tanks from service ("closure") are provided.

  Existing EPA regulations require secondary containment for all hazardous waste storage and accumulation tanks. It very well may be more economical and more protective of the environment to remove underground storage tanks and install aboveground storage tanks within secondary containment. In the meantime, a program of regular and precise inventory reconciliation and leak testing can serve as a rough measure of the soundness of existing

underground chemical storage tanks. Underground product storage requirements are further discussed in Chapter 14.

- *Pavement of onsite chemical storage and transfer areas and impermeable
  diking of aboveground chemical storage tanks* minimize environmental impact. Pavement and diking are necessary and desirable regardless of whether
  the chemicals involved are unused products or waste materials. Pavement
  and diking prevent surface water and soil contamination in case of spills
  or leaks. Also, any spills that do occur are easier to contain and clean up.
  Careful attention must be given in order to ensure the complete cleanup of
  any spills or leaks that do occur. This is necessary even for minor incidents,
  because of the possible long-term cumulative effects and impacts of small
  spills or leaks. Formal employee training in spill response and cleanup measures is important, particularly in the reporting procedures to alert appropriate management personnel and regulatory agencies. Emergency planning
  and appropriate arrangements with local emergency authorities are also
  necessary and desirable.

- *Preventative measures* to minimize spills and leaks are certainly in order,
  and include engineering measures, regular inspections, and maintenance of
  chemical storage tanks and areas as well as locations where chemicals are
  used. Other preventative measures include training of employees in spot-
  ting potential problems and taking corrective measures. Management concern must be relayed to all employees on this subject.

  It is important to remember that employees can cause serious problems
  by their actions (or lack thereof) regarding hazardous chemicals. Proper
  and adequate supervision, along with mechanical and structural "fail-safe"
  systems, are often the only means available to prevent some employees from
  taking shortcuts, ignoring signs of impending spills or leaks, or even causing releases. Improper attitudes and misinformation on the part of employees
  can have serious consequences on the environmental impact and regulatory
  compliance status of the firm.

VII. RESPONDING TO CONTAMINATION
FROM PAST DISPOSAL PRACTICES

The seventh factor has to do with the firm's strategy for handling situations
involving serious environmental contamination from past disposal practices. These
contaminated areas can be either onsite or offsite. Places where potential problems
can occur include industrial and municipal landfills, dumps, surface impoundments (pits, ponds, lagoons), tank farms, underground storage tanks, reclaimers,
scrap yards, and drum reconditioners. Certain of these sites have already been
identified by CERCLA 103 notifications, or have already been placed on the

National Priorities List or the CERCLIS list. Other sites are often identified as a result of government investigations, environmental audits, or internal reviews of past disposal practices. Some sites are uncovered during inspections prior to industrial property acquisitions.

Regardless of how a particular site is identified and brought to management's attention, a decision has to be made on whether to take a proactive or reactive approach to the situation. At one extreme would be sites on the National Priorities List where the responsible party allows the EPA to manage the cleanup and sue the company later under a cost-recovery action. At the other extreme would be a site identified internally that is voluntarily remediated at the responsible party's expense.

When dealing with the EPA and state environmental agencies, many potentially responsible parties have found it to their advantage to take a proactive approach. This initially involves performing or having contractors perform certain activities, including the determination of the extent of the problem and appropriate remedies. Companies may then undertake cost-effective remedial measures with respect to problem situations. The "generator committees" formed in response to some Superfund sites have found that they can perform studies and cleanups much less expensively than can the EPA. Certain performance guarantees, liability limitations, and cost ceilings can be obtained from contractors. In any event, cooperative proactive measures can limit legal expenses, which can often exceed the cost of physical cleanup of the site.

VIII. ORGANIZATION FOR REGULATORY COMPLIANCE

The eighth factor has to do with how the firm's efforts at regulatory compliance are handled, including where in the organization the responsibility for compliance rests. Many firms have found that the difficulties involved with attempting to responsibly control air and water discharges, manage hazardous chemicals and wastes, and ensure occupational safety and health require the services of employees devoting their attention entirely to these subjects. Other firms utilize outside consultants to address specific problems or to develop compliance programs that are then implemented by existing personnel on a full- or part-time basis.

Both of these approaches can be successful under the right circumstances. Success will depend on top management recognizing regulatory compliance as a legitimate and necessary function of the firm, which lessens the liabilities and risks inherent to handling chemicals and wastes. Ultimately, those employees with the responsibility for compliance will need both the authority and resources to achieve this goal. Periodic monitoring of the firm's efforts will be necessary, as will continuing education in the technological advances and regulatory changes.

Periodic monitoring of the firm's efforts becomes especially difficult in multiplant operations. In such situations, regulatory compliance efforts can either be centralized at the corporate headquarters or decentralized at the plant level. Both approaches are common. However, there can be a divergence of concerns between plant management and corporate management, especially regarding the impact of compliance efforts on short-term costs and profits. In any event, the actual level of compliance can be accurately assessed through onsite evaluations by trained and objective personnel. These evaluations, often termed "environmental audits," can be conducted by either qualified employees or consultants. Regardless of who conducts such evaluations, the person must be knowledgeable about the regulatory requirements, the processes and operations conducted, what constitutes "good practices," and a variety of other factors. Observations should be documented for follow-up action or other corrective measures. The audit results should be reviewed by top management to ensure that decisionmakers are aware of the problem situations identified, and the remedial activities undertaken. Environmental audits are more thoroughly discussed in Chapters 16 and 17.

IX. RISK TRANSFER THROUGH INSURANCE

The ninth factor in chemical risk management involves the transfer of risk through the mechanism of insurance. Liability insurance coverage for bodily injury and property damage to third parties as a result of sudden and accidental occurrences is required of all hazardous waste TSD facilities. The required minimum amounts of coverage are $1 million per occurrence, with a $2 million annual aggregate, exclusive of legal defense costs. In addition, those TSD facilities where land disposal is performed must have additional coverage for nonsudden and gradual occurrences, such as contamination of downgradient drinking water wells from offsite migration of hazardous wastes through the subsurface. EPA's regulatory program for USTs, found at 40 CFR Part 280, will require owners/operators to demonstrate financial responsibility for releases and third party damages. Compliance will likely require the establishment of insurance coverage for such risks.

In the past, the standard Comprehensive General Liability (CGL) policy provided coverage for sudden and accidental occurrences, including those involving chemicals and wastes. If the appropriate levels of coverage were purchased, and a properly worded endorsement was provided by the insurance carrier, then compliance with the regulatory requirements was assured for those TSD facilities requiring only sudden and accidental liability coverage. However, recent developments have caused many insurance carriers to totally exclude from their CGL policies all coverage of incidents involving "pollution," broadly defined. This was due to, among other things, heavy losses in general and the "blurring" of the distinction between sudden and accidental occurrences and nonsudden and gradual occurrences. The latter has occurred in some courts when lawsuits to recover losses due to on- and offsite environmental contamination were filed by policyholders.

As a result, firms wishing or requiring coverage for sudden and accidental occurrences that might result in "pollution" or environmental contamination are being forced to also purchase coverage for nonsudden and gradual occurrences. Such policies are often termed pollution legal liability or environmental impairment policies. The availability of such insurance coverage is at present very limited. Prior to issuing such coverage, insurance carriers are requiring applicants to fill out detailed application forms specifying the processes and operations performed at the facilities to be insured, the disposition of all discharges, all by-products and wastes, the nature of the surrounding properties, and environmental compliance strategies of the firm. Often, an onsite evaluation is performed by insurance company representatives in order to (1) verify information provided on the application, (2) identify the extent of exposure to loss, and (3) rate the relative degree of risk.

The implications of this discussion for the management of chemical risk are twofold. First, the transference of risk through insurance for chemical spills and leaks, as well as for offsite property damage and bodily injury to third parties resulting from "pollution," is not automatically a part of existing insurance policies. If such coverage is required, or deemed necessary, it will have to be specified during the negotiation process with an insurance broker or carrier. Levels of coverage in excess of the regulatory minimum amounts may be desirable.

Secondly, in order to obtain the desired amounts of coverage for the lowest possible premium, it will be necessary to demonstrate that the firm's facilities and operations are an "acceptable" risk. Appropriate management responses to the factors discussed in this chapter should go a long way towards this end. Documentation of these and related regulatory compliance efforts will also be viewed favorably during the insurance carrier's evaluation of the firm. The goal in this area is to obtain the desired level of insurance coverage for pollution and other risks at premiums deemed affordable. This can be obtained by the intelligent application of chemical risk management techniques to minimize losses and to control risks and liabilities.

X. MISCELLANEOUS

The tenth factor is actually a combination of elements that are not as universally applicable or of the same level of significance as earlier factors. The importance of thorough, careful investigation of a potential business property transfer is essential to the limitation of liabilities. An environmental audit of past and present chemical and waste handling practices at a site or facility prior to the actual property acquisition is in order in virtually every situation. In some instances, a hydrogeologic investigation including a review of existing data and reports and possible borings and/or installation of groundwater monitoring wells will be necessary to determine the extent of a facility's impact on the surrounding environment.

Related to this aspect and the worker right-to-know issue is the issue of community relations and the extent of the community's "right to know" what types of processes, operations, and chemicals are onsite. The incident at Bhopal, India and those occurring elsewhere have certainly aroused the interest of many public safety officials, environmental activists, concerned citizens, and legislators. Title III of SARA established new authorities for emergency planning and preparedness, community right-to-know reporting, and toxic chemical release reporting. Title III of SARA, also known as the "Emergency Planning and Community Right-to-Know Act of 1986," is intended to encourage and support emergency planning efforts at the state and local level, and provide residents and local governments with information concerning potential chemical hazards associated with fixed facilities in their community.

Regardless of any regulatory requirements, the primary emergency authorities in the area, usually the fire department, should be informed of the nature and amount of different types of chemicals onsite, as well as the various processes and operations conducted. Good communications and an open-door policy with local emergency officials are usually very desirable. Arrangements and agreements with local emergency officials are required by RCRA for firms accumulating, treating, storing, or disposing of hazardous wastes. It is certain that recordkeeping, reporting, and response activities in the area of community exposure to chemicals will become more detailed and wide-ranging, and more damaging to a company's reputation as a result of the media's interest in the spills, leaks, process upsets, emissions, and discharges that are inevitable in industrial operations.

Corporate strategies exceeding the minimum regulatory requirements, or remedial actions in advance of regulatory deadlines, may have beneficial results in all environmental areas of concern, including PCB equipment and asbestos-containing materials (ACM). The goal is to limit the extent and duration of employee and environmental exposure by the removal or enclosure of PCB capacitors and removal or reclassification of such items as PCB transformers. The proper removal or encapsulation of ACM such as friable asbestos insulation also reduces potential human exposures. By achieving this goal, liabilities are minimized, often along with the long-term costs.

CONCLUSION

Ten factors or issues critical to intelligent chemical risk management have been identified. The relevance and significance of each element to an individual firm must be determined. Appropriate responses to the challenges posed must be devised if a firm's exposure to the risks, liabilities, and costs associated with handling chemicals is to be adequately managed. The success of corporate management in meeting the challenges will, to some extent, determine the complexity and extent of future legislative and regulatory initiatives in this area. With these bases covered, corporate management can more confidently respond to the challenges of the marketplace in their efforts to ensure a secure future for their firm.

CHAPTER 16

Environmental Audits as a Management Tool

Environmental audits, sometimes termed "environmental assessments" or "environmental surveys," are becoming increasingly popular as a result of increased corporate sensitivity to environmental problems, as well as the increased complexity of the environmental regulatory programs. Corporate managers want to verify the quality and effectiveness of their firm's environmental management structure and practices. EPA has published a policy statement encouraging the use of environmental audits by regulated entities "to help achieve and maintain compliance with environmental laws and regulations, as well as to help identify and correct unregulated environmental hazards" (51 FR 25003–25010—July 9, 1986.

Environmental audits can be defined as a systematic, documented, periodic, and objective review of facility operations and practices related to meeting environmental requirements. Audits are distinguished from regulatory agency inspections in that they are performed by or on behalf of the regulated entity, and deal with broader concerns than compliance with a single regulatory program.

The definition of environmental audits is somewhat broad. Audits are typically tailored to meet specific needs of specific facilities. For example, audits can concentrate on compliance with one or all environmental program requirements. Table 16.1 identifies some of the environmental regulatory programs of concern for a manufacturing firm. Please remember that these programs can exist at federal, state, and/or local levels of government.

Audits can be modified to achieve objectives in addition to an assessment of environmental compliance. These other objectives can include:

201

Table 16.1.   Environmental and Related Regulatory Programs

Air pollution control
Water pollution control
Solid and hazardous waste management
Hazardous materials transportation
Toxic substances (PCBs)
Community right-to-know
Chemical release reporting
Spill prevention and response
Water withdrawal and supply
Chemical hazard communication
Employee health and safety
Flammable liquids storage

1. identification and evaluation of both regulated and unregulated risks to human health and the environment associated with processes and operations conducted onsite
2. identification and evaluation of opportunities for hazardous waste minimization
3. identification, for planning and budgeting purposes, of those pollution control investments necessary to meet anticipated regulatory developments
4. evaluation of commercial hazardous waste management contractors including landfills, treatment facilities, incinerators, solvent reclaimers, transporters, and laboratories
5. identification and evaluation of risks and liabilities involved with past onsite and offsite disposal practices
6. identification and evaluation of health and safety issues, including chemical exposures and worker right-to-know compliance
7. an assessment of a facility's emergency response capabilities with respect to chemical emergencies
8. identification and evaluation of regulatory requirements and potential liabilities associated with the construction of new facilities, and major modification of existing facilities
9. identification and evaluation of the potential liabilities associated with business property purchases or sales, and with corporate mergers, acquisitions, or divestitures
10. compliance with Securities and Exchange Commission (SEC) requirement 10-K, which requires annual disclosure of environmental compliance problems and anticipated future environmental compliance expenditures
11. compliance with routine application requirements for liability insurance for potential damages from chemical releases
12. compliance with environmental audit requirements that EPA and several states are beginning to include in consent decrees or agreed orders resulting from enforcement actions, particularly for repeat violators

Environmental audits can be performed by qualified employees or consultants. Given the right circumstances, including a management commitment to follow up on audit findings, either approach can work. In certain circumstances, attorneys should be involved to protect the audit report from disclosure to regulatory agencies or plaintiffs.

Regardless of who conducts the audit, the auditor must be knowledgeable about existing and proposed regulatory requirements, the processes and operations involved, what constitutes good practice, and a variety of other factors. Observations and findings should be documented for follow-up action or other corrective measures. Audit reports should be reviewed by top management to ensure that these decisionmakers are aware of the problem situations identified and the remedial measures undertaken.

There are two main criteria of effective audits. *First*, audits should be designed to be as comprehensive as possible, given the specified objectives, in identifying activities and conditions which could have adverse consequences. Regulatory noncompliance certainly has the potential for adverse consequences. However, for some firms other activities and conditions pose greater potential for adverse consequences. *Second*, audits should be seen not as a one-time-only occurrence, but as a part of a rational management process that includes evaluation of any problems identified in terms of potential consequences. Identification and consideration of appropriate responses to the problems identified is also a part of that process. Given limited resources, priorities must be set. For certain risks, both the probability of occurrence and the severity of the consequences should be evaluated when setting priorities and choosing the most appropriate course of action.

Given these two criteria, good audits should be multidimensional. At the first level is determination of the degree of compliance. Manufacturing firms are subject to numerous environmental laws, regulations, and standards of conduct. Many regulatory programs, particularly the hazardous waste program, are in a state of flux. Existing regulations are being revised and new requirements are being developed and implemented. Table 16.2 lists some environmental compliance considerations. Potential adverse consequences of noncompliance include notices of violation, permit revocations, fines, and adversarial relations with regulatory agencies, as well as criminal prosecution of corporate officials.

Table 16.2. Environmental Compliance Considerations

Statutes (federal/state/local)
Regulations (federal/state/local)
Permit conditions
Consent decrees/agreed order provisions
Common law duties
Consensus codes
Best management practices

A second dimension of the audit involves the evaluation of risks posed by the firm's processes or operations with respect to the natural environment and the surrounding community. Manufacturing operations have the potential to impact community air and water supplies, and such firms often have a variety of "neighbors," each with some interest in potential adverse impacts. Once sources of risk are identified, potential pathways and receptors must also be identified and assessed. Table 16.3 lists some environmental risk factors. Potential adverse consequences can include environmental damage, environmental emergencies, citizens' suits over health effects, and bad publicity.

Table 16.3. Environmental Risk Pathways and Receptors

People (employees, neighbors, and local community)
Wildlife and livestock
Water supplies
Waterways
Soil
Air
Surrounding property values and land uses

Finally, audits should address the internal systems set up by the firm to manage and control activities and operations subject to regulation as well as those with the potential for adverse consequences. Organizational and managerial aspects of environmental control can be just as important as the pollution control hardware. This aspect of the audit involves evaluation of the organizational structure, assigned responsibilities and accountability, corporate policies, access to adequate resources (including personnel and capital), training for employees, and internal inspection and reporting systems. Table 16.4 lists several aspects of environmental management systems. Potential adverse consequences of inadequate management systems include noncompliance, inefficiency, unnecessary costs, increased liabilities, lost opportunities, and poor morale.

Table 16.4. Environmental Management System Aspects

Corporate policies
Standard operating procedures
Organizational assignment of responsibility and authority
Internal inspection and evaluation systems
Reporting systems (internal/external)
Internal communications
Training and continuing education
Budgeting and access to corporate resources

To conclude, environmental audits provide corporate managers with a useful tool in determining noncompliance and other risks/liabilities. Audits should review more than just regulatory compliance. Other dimensions, specifically the risks of adverse impacts on the environment and the local community, should be evaluated. The organizational and managerial aspects of responding to environmental requirements and new developments also require some scrutiny. Audits can be utilized in other situations (such as business property transfers) or for other objectives (such as waste minimization). Audits require management support and a commitment for follow-up action or other corrective measures in order to be fully effective as a management tool.

CHAPTER 17

Environmental Auditing Techniques

Richard W. Reynolds

Know ye not, brethren (for I speak to them that know the law), how that the law hath dominion over a man as long as he liveth?

Romans 7:1

INTRODUCTION

Even during Biblical times, there was the sense that laws apply to an individual in perpetuity. The current Congressional climate for the enactment of legislation in the environmental arena has been one of assigning responsibilities for one's actions, as an individual or as a corporate entity, for as long as one exists. Under many of the new laws, once you create or inherit a liability, it is yours forever. How do we, as regulated entities, protect ourselves? In the simplest philosophical sense: don't make mistakes! However, in the real world "mistakes" are a fact of life, partially because the regulatory standards consist of a moving target, subject to change as technology changes and as society demands.

One way to guard against getting caught unexpectedly is to be vigilant about keeping up with technology and regulatory requirements. This is much easier said than done, and it is often very costly in terms of initial expenses. Start-up expenses always seem to be most painful; however, we are now living in a world filled with hidden liabilities which can spell disaster for a company. A relatively inexpensive proactive approach to staying on top of the environmental regulatory onslaught, while keeping current with technological practices, is to institute an internal environmental auditing program. Let's take a look at how it works.

WHAT IS AN ENVIRONMENTAL AUDIT?

An environmental audit is a means of measuring the performance of environmental activities or programs at a facility against some standard or commitment. The word "audit" is used generically here and is often replaced by other words, such as appraisal, assessment, surveillance, inspection, review, etc. The standards and commitments referred to above can be anything from federal, state, and local laws, codes, and regulations to established engineering standards, standard operating procedures, good management practices, corporate or departmental policies, or just plain common sense written as commitments at a local level. The key is that these standards or commitments are generally written documents accepted by or imposed upon program management. These documents then serve as the yardstick for measurement of performance.

Because this definition of environmental audit is broad enough to encompass all manner of environmental audits whether they be internal or external, voluntary or imposed, we'll narrow the focus for this chapter to refer strictly to environmental audits initiated by the management/staff of a facility or program for the purpose of assessing its own activities. The reason for narrowing the scope of this chapter is that if an appropriate effort is self-initiated by a facility, then the facility should be adequately prepared for any external audits.

Table 17.1 shows an abbreviated list of elements which should be included when designing an environmental audit program.

WHY AUDIT?

There are a myriad of sound business reasons for conducting environmental audits of a facility or program. The following list is by no means comprehensive but includes the reasons most frequently cited by the experts:

- to apprise management of current compliance status and environmental performance. This includes the status of operating facilities as well as the environmental media at the facility (air, water, and land), and requires evaluation of permits, SEC reporting requirements, compliance/settlement agreements, etc. (management tool)
- to improve corporate or facility image with regulatory agencies and the public (management tool)
- to provide an early warning device for impending problems (management tool)
- to increase the overall level of environmental awareness (training tool)

- to improve the risk management system by reducing environmental liabilities (risk management tool)
- to control the costs of compliance (management accounting tool)
- to improve overall environmental performance at the operating level (line management tool)
- to develop a proactive program for the optimization of environmental resources (program-wide tool)
- to reduce/eliminate noncompliance conditions and fines due to violations (bottom line tool)

Take note that only the last reason mentioned above relates to meeting "bottom line" compliance conditions. There is a trend to move beyond auditing compliance status to auditing the environmental management control system (the system of controls and mechanisms that are in place to ensure compliance).[1] This type of management audit will be discussed in more detail later.

On the opposite side of the ledger, there is one possibility which continues to send a shiver down an upper-level manager's spine. Once an audit of a facility has been conducted and documented, this documentation can, under certain circumstances, be requested by the EPA. For this reason, do *not* execute an audit of a facility unless top management intends to correct all noncompliance conditions or provide a reasonable schedule for implementation of corrective measures. Failure to implement corrective measures under these circumstances could result in a willful violation of the law and could lead to criminal prosecution of individuals as well as citations and fines against the corporation.

Table 17.1.   Essential Elements of an Effective Audit Program

1. Top management support and commitment to follow up on audit findings
2. Auditing function independent of audited activities
3. Simple and manageable audit process
4. Utilization of in-house personnel where possible
5. Appropriately trained audit staff
6. A standardized approach
7. Specific audit program objectives, scope, resources, and frequency
8. Sensitivity to confidentiality
9. A process that obtains appropriate and sufficient information to achieve audit objectives
10. Procedures for documentation of findings, follow-up on corrective actions, and preparation of schedules for implementation
11. A process that includes quality assurance procedures to ensure the accuracy of environmental audits[a]

[a]Cahill, Lawrence B., and Raymond W. Kane. *Environmental Audits,* 5th Edition (Rockville, MD: Government Institutes, Inc., 1987).

Legal people will tell you about three means of protecting your documents:

1. the attorney-client privilege
2. the work-product doctrine
3. the self-evaluation privilege

Some protection is provided by these techniques, but it is limited and should not be considered a substitute for prudent environmental management. If a company takes the initiative to perform environmental audits, discovers and documents a problem, and promptly corrects the problem or implements a reasonable correction schedule with documentation, then the punitive factor should be removed from any enforcement action the EPA might consider. EPA's policy is to request these types of documents primarily as a part of criminal investigations, consent decrees, or settlement negotiations. Even then, only the salient parts of the audit report can be subpoenaed. EPA also reserves the right to request these documents as provided under various specific statutes (e.g., Clean Water Act section 308, Clean Air Act sections 114 and 208) or in other administrative or judicial proceedings. For federal agencies, such documents are subject to disclosure under the Freedom of Information Act;[2] federal agencies have even less protection than that available to the private sector.

Despite this single negative aspect of conducting and documenting an internal environmental audit program, most companies have accepted environmental auditing as a part of managing a business and the smart ones are using it as a tool to benefit the company in many ways.

WHO PERFORMS THE AUDIT?

Selecting an appropriate auditor or auditing team is one of the most critical steps in the process. Whether a company has its own full- or part-time auditing staff or chooses to use an outside contractor, some of the basic skills that any auditor must have are:

- knowledge of the auditing process
- knowledge of all applicable environmental regulations
- an independent viewpoint
- familiarity with the facility's history and processes
- knowledge of corporate policy
- an understanding of facility organization as well as sensitivity to employee concerns and management philosophy

Many of the above items would most logically suggest a long-standing staff member or members because of the need to understand the facility and its internal dynamics as well as the audit process and all the external requirements. However, if such expertise doesn't exist within the organization (as with many small organizations) it may be necessary to find an expert contractor. Contractors can be found in a multitude of trade journals, in the telephone book under "Environmental," or by calling the local municipal, county, or state boards of health or environmental protection agencies for recommendations. Keep in mind that the knowledge needed and the qualifications mentioned above are the same, regardless of who does the work. Thus, it will require some time (and money) to educate a contractor about your site. Several advantages to hiring a qualified contractor are:

- experience in the auditing process
- current expertise in multiple environmental regulatory programs
- obtaining an impartial survey of facility
- quick turnaround time
- minimal time spent and involvement by facility personnel

If qualified persons are available onsite, then it is crucial to organize the needed expertise either by assigning and training full-time staff or by forming a committee which collectively has the required know-how. The committee should be recruited from among individuals with the following types of responsibilities:

- management
- operations
- maintenance
- legal
- purchasing
- environmental
- facility support engineering
- quality assurance
- emergency response
- occupational health and safety

Corporations sometimes assign separate departments to assess each other's programs. Alternatively, headquarters may carry out this function centrally. If a self-audit is conducted on a single site, be sure that the audit team includes members that are independent of the operation being audited. The team leader or individual should have direct reporting lines to facility-wide management or should be given such privilege during this exercise.

Another critical issue to remember when selecting audit team members is that the audit process is often viewed by subject employees with great suspicion, fear, and even outright hostility. Auditors must be schooled to be sensitive to this perception and exercise good communications and professionalism in the process.

BASIC STRUCTURE OF A WELL-EXECUTED ENVIRONMENTAL AUDIT

A list of elements to be included in any environmental audit follows. Regardless of the size and type of program or organization being audited, each step should be carried out. If the company or facility under consideration is relatively small, many of the steps below may be done quite informally or even verbally rather than in writing. Nevertheless, the time invested in preparation before the onsite portion of the environmental audit is time well spent and is essential to avoid wasted effort, internal organizational strife, or, worse yet, having the audit effort discredited by external parties due to poor quality or improper documentation of environmental conditions.

1. Pre-audit activities:
   - determination of goals and objectives
   - selection of target, subjects, and scope (scoping meeting)
   - plan and schedule (planning meeting)
   - preparation (checklists, questionnaires, etc.)
   - kick-off (previsit exchange of information)
   - review of materials received
2. Field visit
   - inbriefing (start of field visit)
   - review of programs and documentation
   - site tour
   - outbriefing (end of field visit)
3. Audit follow-up
   - report writing
   - corrective measures
   - follow-up (verification)

This list demonstrates three major phases in the process for the auditor. Although it may seem cumbersome, a typical well-run environmental audit involves anywhere from three to ten times as much time spent before and after the audit as during the actual field visit. This may well hold true for the audited party as well, depending upon the required corrective measures resulting from the survey.

It is important, therefore, to keep this time element in mind when scheduling such audits, since valuable time will be taken away from operations people as well as support departments. All parties involved in such an audit must be informed well in advance of the audit, so that plans can be made to free up all appropriate people. It is equally important not to waste the valuable time of these audit participants. Be sure to schedule only the individuals that are actually needed during any phase of the audit. Remember that the audited parties commonly feel that the audit is an invasion of their work privacy and regard it as someone "looking over their shoulders." Most people are uncomfortable in this situation. This is where the careful selection of the audit team members and their level of professionalism really counts.

In the following paragraphs, important accomplishments for each element of the audit will be briefly described.

Pre-Audit Activities

1. Determine Objectives, Strategy, and Frequency

Objectives. Objectives are generally created by top management or presented for approval to top management by the environmental, legal, or other department within the organization. They may also be part of a compliance order, consent decree, or settlement agreement with a regulatory agency. They could be the result of a request by a potential purchaser of the facility or a requirement from corporate headquarters. The assumption, in any case, is that facility management has control over the conduct of the audit. The objective may be to assess the status of some or all aspects of the facility as they relate to environmental compliance and/or performance.

Strategy. The strategy is the guideline to accomplishing the objectives. If the objective is to assess the status of all environmental aspects of the facility, then the strategy might be one of the following:

- assess organizational effectiveness (management audit)
- assess facility operational effectiveness (functional audit)
- assess facility condition (physical inspection)

If we take one of these strategies, for example, the functional audit, there are several sub-strategies to accomplishing this task:

- assess by regulatory program (i.e., RCRA, CWA, TSCA, etc.)
- assess by generic program (air, water, land, etc.)
- assess by a combination of these (water: SDWA, CWA)

Figure 17.1 shows a more detailed description of each type of audit. Also, see the appendix to this chapter for an example of a functional audit employing a combination of generic and regulatory program approaches as used by one Department of Energy (DOE) facility, the Savannah River Plant in South Carolina.

Frequency. Frequency of audits is dependent upon need, internal and external pressures, program condition, and management commitment. Normally an annual audit is sufficient. Many companies choose a two- or three-year cycle during which multiple phases of a total program are covered in succession. (For example, all water and solid waste issues are covered the first year; all air and toxic substances are covered the second year; and all land [including waste sites and solid waste management units] and remaining issues are covered the third year. These issues

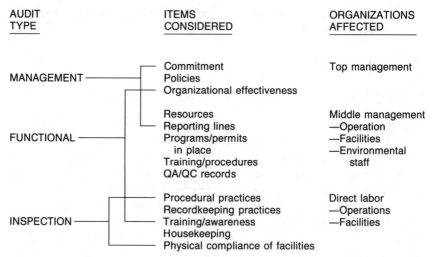

AUDIT TYPE	ITEMS CONSIDERED	ORGANIZATIONS AFFECTED
MANAGEMENT	Commitment Policies Organizational effectiveness	Top management

FUNCTIONAL	Resources
	Reporting lines
	Programs/permits
	  in place
	Training/procedures
	QA/QC records
		Middle management
		—Operation
		—Facilities
		—Environmental
		  staff

INSPECTION	Procedural practices Recordkeeping practices Training/awareness Housekeeping Physical compliance of facilities	Direct labor —Operations —Facilities

Figure 17.1. Levels of environmental audits.

should be put in priority order over the cycle.)

2. *Scope (Select Targets, Subjects, Depth)*

Targets. The target may be a small aspect of a facility or an entire facility. For example, one could choose to audit just the wastewater treatment plant or a stack monitoring system within a facility. Corporate headquarters may request an audit of the entire facility or department.

Subjects. Usually subjects correspond to regulatory programs such as RCRA or the CWA. The subject may be only a narrow aspect of a regulatory program, e.g., NPDES discharges subject to the CWA. It could also be a generic topic, such as water or air emissions, which may cross several regulatory programs. It may even be a comprehensive look at all environmental issues.

Depth. Once targets and topics are selected, a decision is needed as to what depth and breadth of effort is to be expended. A detailed look at all documents, related quality assurance, training, housekeeping, management strategy, and peripheral issues could be undertaken. Perhaps a random selection or sample of a very small percentage of these issues could be evaluated in depth. This depends upon management objectives and strategy.

3. *Plan and Schedule*

This stage is where good communication is critical. All those who will participate in the audit should be informed of the purpose, extent of involvement, the schedule, who is auditing what, and who or what authorizes this audit. Credibility and the subsequent cooperation of participants depend heavily upon timely and thorough communication. Be sensitive to operational schedules so that peak periods can be avoided. Do not waste people's time. Be sure all appropriate people are available. Estimate how much time you will need to visit a site and look over documents, then double it. Finally, stick to your schedule and do not keep people waiting. If additional issues are discovered that will clearly require more investigation, schedule additional time later as needed.

4. Preparation

By this stage, the how, why, what, when, and where have been decided. It's time to design and send out a previsit questionnaire in order to obtain basic information before the interviews. A previsit questionnaire can simply be a list of items to be provided to the auditing team for review during the audit, and may include permits, inspection and training records, analytical data, site plans, etc. It may include a list of questions to be answered concerning organizational structure (organizational chart), environmental resource allocations, assigned responsibilities, physical plant layout, etc. A checklist will also need to be designed for the actual field visit. The checklist will be designed based upon the objectives of the survey. Checklists are normally drawn up in a form such that questions can be answered with simple *yes, no,* or *not applicable* responses. It is also wise to allow a column for comments.

5. Kick-Off

At this stage, previsit questionnaires are sent to the operations to be audited, accompanied by an appropriate cover letter from top management to announce the environmental audit, topics, timing, etc.

6. Review Materials Received

After all appropriate previsit documentation is returned from the field, it must be reviewed thoroughly and critically. Follow-up questions may be required if information is missing or poorly completed by participants. This can be done by letter or telephone if the auditor is not located directly onsite. This information is the key to understanding a facility that may be unfamiliar to the auditor. It should help to identify the key players, key issues, types of facilities, and possible program strengths and weaknesses. It should also help prepare the facility personnel to have all appropriate documentation and people available for the actual audit interview.

Field Visit

7. *Inbriefing*

Regardless of whether the auditor(s) is from the facility, an outside contractor, or from another division, it is necessary to meet at the beginning. This meeting may only be a brief and informal session with each group of participants to explain the purpose, the authority, confidentiality arrangements, and what facilities and documents will be surveyed. Again, this helps to establish rapport with the participants. The object is to put participants at ease.

8. *Review Programs and Documentation*

At the heart of any audit is the actual field work, observations, field notes, interviews, etc. Normally, an interview will begin with a presentation of any previous findings or recommendations to ascertain progress on the resolution of previously identified items. Then the interview will continue with a predesigned questionnaire. The questionnaire should be divided into questions asked of management, questions asked of senior environmental staff, and technical questions asked of environmental professionals or operations people. If practical, separate meetings should be arranged with each category. It is best to ask simple *yes/no* questions with a column for comments. Because timing and punctuality are important, try not to allow conversations to stray too far. If a conversation has strayed but it looks like it could be important or contain relevant information, ask for additional time at the end of the scheduled audit to cover the topic. Be sensitive to confidentiality. If copies of documents are needed, stress that these will be returned or destroyed when the survey is completed. Auditor field notes or working papers are to be closely guarded. Make a note of all follow-up activities that will be required. At the end of each day of a multiple-day audit, summarize the highlights, both good and bad.

9. *Site Tour*

After a good review of documentation and upon completion of the interviews, a tour of the physical facilities is next. This is where many of the facts collected up to this point are either confirmed or denied. It is certainly not unheard of for a facility to be running an excellent paper program while the physical facility is a shambles. The bottom line is that physical programs must be operated and managed well. The true dangers lie not in a poorly documented hazardous waste program, but in a poorly operated program. Both procedural and substantive aspects of the environmental management program are important. Be observant; ask penetrating questions if in doubt, and wander off the beaten path occasionally to ask about suspicious-looking items when possible. Take good notes and summarize highlights at the end of the tour.

10. *Outbriefing*

The outbriefing meeting might be held in two separate sessions. Management may prefer to be briefed first. All remaining key participants should be invited to the general facility outbriefing. An outbriefing should be just that—brief. Normally 15–30 minutes is necessary to review good and bad highlights. Always begin by thanking the participants and reviewing program strengths. At the very end, review areas for improvement. Stress that these are merely general impressions and that more detailed written comments will be forthcoming after an in-depth review of all field notes, documents, etc.

Audit Follow-Up

11. *Report Writing*

The report should follow the order of the outbriefing. After a brief introduction describing the purpose, authority, strategy, etc., it should review program strengths either by regulatory program or by generic program, depending upon the original strategy. It should then describe areas for improvement and finally draw conclusions and make recommendations. Under each area for improvement, be sure to refer to a regulation, code, standard practice, or simply good housekeeping practice, so that the participant has a better understanding of why the recommendation was made.

Any such report should have an executive summary at the beginning where the purpose, authority, and strategy of the project are described briefly, followed by an abbreviated list of conclusions and recommendations.

12. *Corrective Action*

Now that the report is complete, it is up to the top management to decide how to handle the recommendations. If management has been supportive of the venture from the beginning, chances are they will be supportive of making any needed improvements. Small improvements should be completed quickly and documented. Major, more costly improvements should be considered projects which may require additional study, capital expenditure budgeting, and target dates for completion. This should all be documented, if only briefly, so that a regulatory agency would be able to trace the audit/documentation/implementation paper trail and confirm the organization's commitment to environmental compliance.

13. *Follow-Up (Verification)*

A true audit requires confirmation that recommendations were acted upon by the facility. If the audit was conducted by in-house personnel, this confirmation process is relatively simple. A designated audit comptroller might follow up with appropriate operations or management people to check out actions completed and actions in progress with respect to each recommendation. Larger programs computerize such data, since there may be dozens or even hundreds of recommendations with differing implementation schedules. An outside contractor who acted as auditor might need to plan a follow-up visit in six months or a year to confirm actions on recommendations.

OTHER INFORMATION SOURCES

For the true aficionado of environmental auditing, there are many other sources of useful information besides the few mentioned in the reference section at the end of this chapter. One of the best bibliographies available is the seventh edition of the "Annotated Bibliography on Environmental Auditing" (March, 1988, EPA Regulatory Innovations Staff, Office of Policy, Planning and Evaluation, EPA, 401 M Street SW, Washington, DC 20460).

As with any specialized area of endeavor, in addition to a solid literature base one should look to the development of professional societies and standards. Although environmental auditing is a relatively new specialty, there are already several budding professional societies:

1. Institute for Environmental Auditing
   P.O. Box 23686
   L'Enfant Plaza Station
   Washington, DC 20026-3686

2. Environmental Auditing Roundtable
   Gilbert Hedstrom
   Arthur D. Little, Inc.
   35 Acorn Park
   Cambridge, MA 02140

3. Environmental Auditing Forum
   Vinay Dighe
   IT Corporation
   P.O. Box 19724
   17461 Derian Avenue
   Irvine, CA 92714

In addition to literature and private initiatives, the public sector has begun to get involved. Several states, including California, New Jersey, Connecticut, and

Maryland, are considering (or already have instituted) voluntary registration of environmental auditors. The above-mentioned Institute for Environmental Auditing is promoting the idea of a professional certification program on a nationwide basis. These and other developments will undoubtedly hasten the maturation process for the environmental auditing profession.

CONCLUSION

Instituting a proactive environmental auditing system in a company is not a project to be taken lightly. It requires explicit top management support, including the allocation of resources necessary to implement corrective measures. It requires the involvement and cooperation of many different departments and is clearly a facility-wide program. The benefits to such a program far outweigh the potential costs and liabilities associated with the lack of a coordinated program. This has been proven time and again throughout industry from a straight profits-and-losses standpoint.[3]

REFERENCES

1. Arthur D. Little, Inc., Center for Environmental Assurance. "Current Practices in Environmental Auditing," Report to U.S. EPA, EPA 230-09-83-006, February 1984.
2. Environmental Protection Agency. "Environmental Auditing Policy Statement," *Federal Register* Vol. 51, No. 131, Wednesday, July 9, 1986, pp. 25004–25010.
3. Office of Technology Assessment. *Serious Reduction of Hazardous Waste,* OTA-ITE-318, Congressional and Public Affairs Office, OTA, United States Congress, Washington, DC 20510-8025, September 1986.

APPENDIX

DOE Functional Appraisals

Below is an example of a functional audit employing a combination of generic and regulatory program approaches as used by one Department of Energy facility, the Savannah River Plant in South Carolina.

1. *Water Supply and Distribution,* including permitting and monitoring of water systems regulated under the Safe Drinking Water Act. The regulatory guides are derived from the State Primary Drinking Water Regulations (R.61-58); the State Safe

Drinking Water Act (R.44-55); National Primary Drinking Water Standards; Secondary Drinking Water Standards; 40 CFR 16; 40 CFR 122; 40 CFR 141–144; 40 CFR 146; and 42 USC Sections 300f–300j, -10.

2. *Water Pollution Control,* based on elements of the Clean Water Act, which includes point source control (NPDES) and Army Corps of Engineers (COE) permits under Section 404 (Discharge of Dredged or Fill Material). The regulatory guides are derived from the Water Classification Standards System from the State of South Carolina (R.61-68); The Clean Water Act; 40 CFR 122 (NPDES); 33 USC Sections 1251–1326; 33 USC Sections 401–413; Rivers and Harbors Act; and COE 404 Permitting.

3. *Spill Prevention and Control,* including oil pollution prevention, hazardous waste (waste oil) storage facilities, PCB control, hazardous/toxic substances planning, and other elements of the Spill Prevention, Control, and Countermeasures Plan. The regulatory guides are primarily extracted from the Clean Water Act and South Carolina Department of Health and Environmental Control Hazardous Waste Management Regulations (R.61-79), with considerations from 40 CFR 112; 40 CFR 116; 40 CFR 117; 40 CFR 125; 40 CFR 261; 40 CFR 264; 40 CFR 300; and 40 CFR 761.

4. *Nonradioactive Air Pollution Control,* including the areas of permitting and monitoring regulated under the Clean Air Act. The regulatory guides are derived from the South Carolina Department of Health and Environmental Control Air Pollution Control Regulations and Standards (R.61-62); the Clean Air Act; National Ambient Air Quality Standards; New Source Performance Standards; National Emission Standards for Hazardous Air Pollutants (NESHAPS); Prevention of Significant Deterioration; and 42 USC Sections 7401–7642.

5. *Nonhazardous Solid Waste Disposal,* which is based in RCRA regulations controlling sanitary landfills and other nonhazardous solid waste disposal programs. The regulatory guidance is taken from South Carolina nonhazardous waste guides (R.61-6, R.61-46, R.61-59, R.61-60, R.61-66, and R.61-70); the Pollution Control Act; 40 CFR 240–247; 40 CFR 249; 40 CFR 254–257; Solid Waste Disposal Act (RCRA); and 42 USC Sections 6901–6987.

6. *Environmental Radiological Programs,* which include effluent monitoring, environmental monitoring, and facilities appraisals for radiological concerns. State regulations are few, with the primary guidance derived from DOE Order 5480.1A; DOE Order 5484.1; DOE Form 5821.1; DOE/EV/1830-TS (ALARA); Atomic Energy Act; 42 USC Sections 2133–2134 (6); SR Order 5480.1A; Executive Order 12088; DOE Order 5481.1; and DOE Order 5480.2.

7. *Hazardous Waste Management,* which covers programs administered under RCRA provisions, including generation, treatment, storage, and disposal of hazardous wastes, and procedures for the closure and postclosure care of facilities. The regulatory guidance is derived from the South Carolina Hazardous Waste Management Act and South Carolina Hazardous Waste Management Regulations (R.61-79), along with RCRA; 40 CFR 122–124; 40 CFR 260–265; 40 CFR 270; 40 CFR 300; Executive Order 11752; and Executive Order 12088.

8. *Toxic and Hazardous Substances Control,* which includes the management, handling, and reporting requirements for toxic substances, asbestos, and pesticides. The guides used as a basis for these appraisals are TSCA; 40 CFR 712; OSHA

guidelines; NESHAPS; EPA R-61.25; the Federal Insecticide, Fungicide, and Rodenticide Act; CWA Section 307; and Primary Drinking Water Standards.

9. *Environmental Policy,* which covers the programs, assessments, and impact statements under the National Environmental Policy Act (NEPA). Regulatory guidance is derived from NEPA; Council on Environmental Quality regulations; 40 CFR 1500–1508; and 45 FR 20.694.

10. *Ecology and Conservation Policy,* which has a broad scope of issues of ecological concern. Primary areas of concern include archaeological and historical preservation, flood plains management, and ecology/wildlife preservation. The regulatory guides used to support these programs include the Archaeological and Historical Preservation Act; the Endangered Species Act; Executive Order 11593; Executive Order 11988; Executive Order 11990; the Historic Preservation Act; 10 CFR 1022; the Migratory Bird Act; the Fish and Wildlife Coordination Act; and the Bald and Golden Eagle Protection Act.

11. *Laboratory Techniques and Procedures,* which includes sampling, analysis, quality control, and contractor's quality assurance for analytical testing. Guidance is derived from QAMS-001/80; QAMS-005/80; EPA-600/4-79-020; EPA-600/2-80-018; American Standards for Testing and Materials Guides; American Public Health Association Guides; American Water Works Association Guides; and Water Pollution Control Federation (WPCF) Guides.

12. *General Facility Information* is a managerial program that includes organization, administration, communication, planning, and review for SRP facilities. General management appraisal permits are to be used for guidance.

CHAPTER 18

Developing and Implementing the Environmental Management Program

The nature and complexity of environmental laws and regulations present a challenge to those persons responsible for compliance. The hazardous waste regulatory program, in particular, is in a state of flux. Modifications to and changing interpretations of existing regulatory requirements, as well as the addition of new regulatory initiatives, make the management of compliance programs particularly difficult.

The sheer bulk of the regulations regarding air and water effluents and hazardous waste management pose almost insurmountable reading and comprehension burdens on compliance personnel. Further, compliance efforts with respect to one environmental medium (air, water, or land) may adversely affect compliance efforts in another. Finally, regulatory compliance is not a static concept, but is more accurately described as a desired state of affairs. Regulatory compliance is just one aspect of a comprehensive program to manage hazardous chemicals and wastes.

There are a number of different methods or techniques available for managing the solid, liquid, and gaseous residuals (wastes) associated with manufacturing operations. The environmental regulatory programs provide minimum standards for and limitations on those techniques available to the firm. There are also many intricacies and interactions involved in complying with separate sets of regulations protecting air, water, groundwater, and soil resources. The high cost of compliance (and noncompliance) indicates the need for and the advantages of a structured approach. Such an approach would, where possible, take into account the various environmental regulatory programs, and integrate efforts to comply with all such programs.

There is available a framework within which environmental management and regulatory compliance efforts can be placed. This framework is provided by the Waste Management Hierarchy. The hierarchy allows a systematic, methodical approach in thinking about the waste management problem, broadly defined.

This chapter will attempt to adequately describe the Waste Management Hierarchy, and how it can be used to manage the risks, costs, and liabilities faced by manufacturing firms in the current regulatory context. The chapter begins with an overview of the regulatory context, and outlines the organizational factors necessary to effectively implement the Waste Management Hierarchy. The hierarchy is a key element in any comprehensive environmental management program.

CONTEXT

Waste disposal practices have traditionally been predicated on the attitude that residuals from manufacturing operations were to be gotten rid of with a minimum of cost, time, and effort. In the "bad old days," waste (mis)management was a nasty and brutish activity. Many of the environmental problems associated with uncontrolled sites and hazardous waste releases resulted from this mindset. RCRA and CERCLA were the Congressional responses to the problems caused by mismanagement, or more accurately, the lack of management of hazardous wastes.

The perception of waste management functions within manufacturing operations has been changing and continues to change. The proper management of industrial wastes has been elevated to a legitimate concern of corporate management. Resources are being redirected in order to properly and adequately manage the liabilities, costs, and complexities of regulatory compliance. It is now realized that management of hazardous wastes involves more than the legal and proper disposal of any and all hazardous waste generated by the firm. Effective hazardous waste management involves the use of guiding principles and implies the careful consideration of alternatives, as well as the periodic evaluation of program effectiveness. These alternatives include waste minimization, initiatives to reduce the degree of toxicity of those industrial residues generated, and various other risk reduction measures.

The imposition of the RCRA and CERCLA programs began the necessary change in focus and subsequent redirection of resources. The Hazardous and Solid Waste Amendments to RCRA and the Superfund Amendments and Reauthorization Act to CERCLA have accelerated this trend.

RCRA and CERCLA together impose requirements and obligations upon generators of hazardous waste with respect to the disposition and environmental impact of such wastes. Generators of hazardous wastes are ultimately and even perpetually responsible for onsite and offsite damages to human health and environmental quality resulting from their wastes. This responsibility is sometimes termed "cradle-to-grave"; however, it actually extends well beyond the grave.

Industrial hazardous wastes are now a necessary subject of managerial consideration and oversight. It is realized that effective hazardous waste management requires the multidimensional consideration of technical, economic, social, political, regulatory, and legal factors. Managerial efforts are directed toward balancing, within the regulatory framework, the liabilities and costs associated with the generation, storage, transportation, and treatment/disposal of hazardous wastes. Since it is difficult to simultaneously minimize two such interrelated variables as costs and liabilities, the task is to find an optimum balance using a structured approach.

Hazardous waste management is a pervasively regulated field. There are many statutes and regulatory programs. The regulations are lengthy and complex. There are many regulatory agencies involved, at each level of government. There are many nuances in dealing with a variety of regulatory programs, especially when enforced by separate regulatory agencies. Each agency or department within an agency is typically concerned only with compliance with their particular regulatory program. Typically, communication is poor both between and within regulatory agencies.

Legal developments have resulted in the involvement of nonregulatory actors as well, including financial institutions, stockholders, and insurance underwriters. Environmental organizations and neighborhood associations have been taking an active interest in the environmental efforts of certain manufacturing operations.

ORGANIZATIONAL FACTORS

It is obvious that environmental management in general, and hazardous waste management in particular, presents formidable challenges to those individuals responsible for compliance. Environmental management involves more than mere regulatory compliance. Risk reduction and waste minimization are important elements of any corporate environmental management program. The selection of program goals will aid in the development, implementation, and evaluation of a comprehensive corporate environmental management program. Potential goals include:

1. compliance with applicable environmental laws and regulations, common law duties, and standards of conduct
2. minimization of short- and long-term risks and liabilities associated with the use of chemicals and the generation of hazardous wastes
3. minimization of releases and environmental damage from facility operations
4. minimization of the volume and toxicity of waste materials generated
5. minimization of total (short- and long-term) environmental compliance costs (given the constraints involved in attaining the above-listed goals)
6. fulfillment of corporate responsibilities and maintenance of a favorable corporate image

Different corporations have different perspectives on issues such as aversion to risk, centralization of staff functions, and the importance of the corporate image. Given the differences, it is useful to list the organizational factors that are necessary for individuals in a manufacturing setting to investigate, develop, and apply innovative approaches in meeting the requirements of the regulations. These factors are:

1. corporate policy requiring strict compliance with the letter and spirit of environmental laws and regulations. Management support of the corporate environmental management program must be strong and visible.
2. designation of one or more individuals as being responsible for compliance with environmental requirements, and delegation of the necessary authority, technical support, and financial resources to meet that goal
3. recognition of the importance of waste minimization and risk reduction efforts, including flexibility at the plant level to experiment with different inputs, process changes, or operating procedures
4. availability of discretionary capital for measures that are cost-effective, taking into consideration both short- and long-term costs and benefits

Corporate efforts are typically more efficient if the job of environmental management is broken down into specific duties and responsibilities, appropriately divided and assigned to those positions or persons in the firm with the requisite skills and training to properly perform such duties or fulfill such responsibilities. Responsibilities can be divided along environmental program lines (air pollution, water pollution, solid and hazardous waste management) or according to the skill level necessary to carry out the necessary duties. An environmental control committee can be established to develop, implement, and evaluate environmental management programs. However, it is critical that specific duties and responsibilities be assigned to individuals, rather than the committee as a whole.

Those persons assigned the responsibility for specific duties should be given the necessary authority and access to adequate resources, including time, to successfully accomplish those duties. Accountability should be established. Periodic evaluation of individual and collective environmental management efforts is necessary to ensure regulatory compliance and to minimize risks and liabilities in a cost-effective manner. The evaluation of environmental management programs using the technique of environmental auditing is discussed in Chapters 16 and 17.

Concern regarding regulatory compliance has to be translated into the development of plans, programs, and procedures necessary to ensure that all those involved are able to and know how to properly perform their particular duties. Equipment purchases and capital expenditures are typically necessary. Continuing oversight and evaluation are required to verify that the plans, programs, and procedures are consistently followed and carried out.

WASTE MANAGEMENT HIERARCHY

The Waste Management Hierarchy is a concept appearing as early as 1976 in an EPA position statement entitled "Effective Hazardous Waste Management (Non-Radioactive)" (41 FR 35050; August 18, 1976). A 1977 EPA publication entitled "State Decision-Makers Guide to Hazardous Waste Management" expanded upon this concept and gave specific applications. The hierarchy provides a framework for systematic thinking about solutions to the waste management "problem." The hierarchy is a series of seven decisionmaking steps, to be used sequentially in decreasing order of desirability.

The first step (the most favorable and best option) is to prevent the generation of hazardous waste by source control/source reduction measures such as process modifications or input substitutions. The last step (the least favorable or last resort option) is land disposal. The Waste Management Hierarchy is as follows:

1. Prevent the generation of waste (both hazardous and nonhazardous).
2. Minimize the proportion of unavoidable waste that is hazardous.
3. Separate and concentrate waste streams to make further management activities more feasible and economical.
4. Reuse internally or use offsite through waste transfer.
5. Recycle or reclaim material values or utilize the energy value of the waste.
6. Maintain unavoidable and nonreclaimable wastes in a form amenable to treatment (e.g., incineration, neutralization, detoxification). Treat to reduce the volume and degree of hazard.
7. Manage remaining hazardous waste and treatment residues by secure land disposal.

The beauty of the hierarchy is that it can be applied at different levels. It can be used as a guide in determining the preferred disposition of a single waste stream. It can be used to evaluate the practices of an entire manufacturing firm, or it can be used as a framework for making decisions on a state or national level as to how wastes are to be managed.

The U.S. Congress did just that in 1984 when, in the Hazardous and Solid Waste Amendments to RCRA, the basic tenets of the hierarchy were incorporated in the congressional findings and national policy statement. Congress found that ". . . reliance on land disposal should be minimized or eliminated, and land disposal, particularly landfill and surface impoundment, should be the least favored method for managing hazardous wastes." (Section 1002(b)(7))

Congress further declared the national policy of the United States to be that, wherever feasible, "the generation of hazardous waste is to be reduced or eliminated as expeditiously as possible. Waste that is nevertheless generated should be treated, stored, or disposed of so as to minimize the present and future threat to human health and the environment." (Section 1003(b))

Congress felt it necessary to place the core aspects of the hierarchy in statutory findings and the national policy statement because EPA failed to consider or integrate the hierarchy in the pre-HSWA hazardous waste regulations. Indeed, compliance could be achieved with the pre-HSWA hazardous waste regulatory program without even considering the hierarchy. The pre-HSWA hazardous waste regulations revolved around the direct land disposal of hazardous wastes. Land disposal was cheap and easy, compared to the costs of treatment or reclamation. There were few restrictions on what could be landfilled, provided that the waste was not liquid, ignitable, or reactive after solidification. There were no requirements that the generator utilize direct land disposal only as a last resort.

Through the amendments to RCRA, Congress has required EPA to evaluate and restrict or prohibit where necessary the direct land disposal of untreated hazardous wastes. The net result will be that eventually only treatment residuals will be allowed to be managed by secure land disposal. Further, Congressional concern has resulted in EPA requirements for generators to certify on each manifested shipment of hazardous waste that they have a waste minimization program in place, and that the method of treatment or disposal selected is environmentally appropriate. Written statements of waste minimization activities are required with biennial reports of hazardous waste activity.

USING THE HIERARCHY

As mentioned earlier, the hierarchy consists of seven levels, each to be evaluated sequentially, in decreasing order of desirability. The remainder of this chapter will explain the application and uses of the hierarchy, and give examples of the various types of measures applicable at each level of the hierarchy.

Waste materials should be thought of as "resources out of place." Looking at waste materials as resources allows the conception of measures to prevent or minimize the transfer of resources from settings where they have value to places where they have negative value or pose environmental and human health threats. A good example of the concept of waste materials as resources out of place is the hexavalent chromium ion. This ion, which is a necessary and valuable constituent in many electroplating processes, becomes a toxic pollutant if rinsewaters containing it are discharged to sewers or streams. The chromium ion results from the rinsing of excess electroplating solution from the part being plated. If an "end of the pipe" approach is taken, then the rinsewaters are treated to precipitate the chromium as a metal hydroxide. The resulting sludge is then dewatered and buried, while the treated rinsewaters are discharged, still containing trace (<3 ppm) amounts of chromium.

If a waste minimization approach is taken, then less toxic alternatives to chrome plating are evaluated with respect to the necessary aspects of surface finish, durability, and resistance to corrosion of the part. If chrome plating is still deemed necessary, then the feasibility of putting the excess chromium ion back in the plating bath (by, for example, rinsing the part over the plating bath itself) is evaluated. Any necessary chromium rinsewaters are segregated from other rinsewaters containing electroplating metals. Water conservation measures are taken to concentrate the chromium ions in the segregated rinsewater. Reuse as makeup solution to the concentrated bath is evaluated, perhaps after further concentration by evaporation. After reuse opportunities are exhausted, recovery techniques are investigated, including electrolytic recovery and ion exchange. At this point, treatment systems are designed and constructed for the remaining waste materials, which are now greatly reduced in volume. The treatment systems reduce the toxicity of the waste, and extract the contaminant chromium ions for possible recovery, or more likely, for secure land disposal.

Waste Minimization

There are a variety of means of achieving the goal of waste minimization. These source reduction measures include input or feedstock substitution, process modifications, and good operating practices. Often, determining, developing, and implementing the various types of source reduction/waste minimization measures requires the gathering of significant amounts of information.

Efforts are necessary initially to characterize and categorize all the different waste streams by type, composition, and quantity. Determination of waste management costs for each waste stream is also necessary. Accurate records must be kept of this assessment. It is likely that two categories will cover most waste streams: process wastes and pollution control residuals. However, a waste inventory can reveal surprising amounts of off-specification input materials; defective outputs; inadvertent contamination of inputs, process chemicals, and outputs; and obsolete (expired) chemical products. Conducting a waste inventory is discussed in Chapter 3.

Particular attention should be given to problem areas. These include excessive waste amounts per unit of production, excessive process upsets or bad batches, or frequent off-specification inputs. Some effort should be expended to determine the effect of process variables on the waste streams created, and the relationship of waste stream composition to the input chemicals and process methods used. The amount of water used by different processes and the possibilities for reuse could also be examined at this time.

Chemical analysis using proper and uniform sampling procedures and the appropriate analytical and testing methods can accurately determine the nature of the waste and can yield valuable information about the industrial process and the condition of the equipment. Additionally, such information is necessary to determine the feasibility of reuse, recycling, materials recovery, waste transfer, or the proper methods of treatment or disposal.

Input Substitutions

The information gathered by waste stream examination and analysis can suggest or justify the substitution of certain feedstock or process materials or chemicals so that the process wastes and pollution control residuals would be reduced in volume or no longer be of a hazardous nature. Input or feedstock substitutions are often inseparable from process modifications.

Could different (nonhazardous) process chemicals and/or different production processes (yielding nonhazardous residuals) result in equivalent quality of product outputs? Chemical substitution opportunities exist both in production processes and in maintenance operations.

Do purity and concentration requirements for process chemicals bear some relation to process needs? For example, could opportunities for extended life and reuse of process baths be created through more realistic specifications, with no decrease in finished product quality? Alternatively, would improvements in feedstock purity increase mass transfer efficiency or product yield, and/or result in waste streams of lower toxicity? Could spent solutions be reclaimed to sufficient purity and concentration for reuse in the process? Could spent solutions be reclaimed to sufficient purity for reuse by others?

An examination of the market requirements for the finished product may be in order. For example, is the surface finish for appearance or durability? Are less-polluting finish processes available which would meet market demands? Would customers be willing to accept lower quality of appearance as a tradeoff for a lower price?

Specific examples of possible input substitutions include:

1. substitution of alkaline (aqueous) cleaning solutions for chlorinated organic solvents
2. replacement of paints with heavy metal pigments (e.g., cadmium, chromium, or lead) with paints that use alternative pigment systems
3. use of water-based paints instead of solvent-based paints
4. use of noncyanide-based electroplating solutions
5. cartridge filtration in lieu of diatomaceous earth filters
6. use of synthetic coolants in place of emulsified oil coolants
7. use of aluminum in place of chrome-plated steel

Process Modifications

Modifications of production systems with the goal of minimizing hazardous residuals are often suggested or indicated by the information gathered by examination and analysis of waste streams. Increasing the efficiency of production processes or modifying equipment so as to allow the use of alternative inputs can reduce the volume and toxicity of the residuals.

Waste minimization opportunities can exist at several stages in most manufacturing processes. Many individuals are looking at industrial processes as a system, with inputs and outputs. Outputs include both the intended product and the solid, liquid, and gaseous residuals requiring proper management. There is movement away from strictly "end of the pipe" approaches, where treatment and disposal systems are designed and operated for any quantity of waste produced. Front-end approaches such as manufacturing process changes allow evaluation of a variety of waste minimization, reuse, and recycling options. Process changes are also more mindful of the laws of thermodynamics and thus can be much more economical than the "end of the pipe" approach.

The following questions can be helpful in considering process modifications:

Do process steps bear a realistic relationship to need?

How clean is clean enough? How many rinses or how much rinsewater is sufficient?

Do production and process variables have any relationship to the amount of wastes generated per unit of output?

Are "economies" achieved at high production levels or during continuous (24-hr/day) operation, or is there a linear relationship between production and waste generation?

Process changes such as water conservation, segregation, and reuse measures can be introduced to eliminate or reduce the capacity and cost of necessary pollution control equipment and the cost of disposal of the treatment residuals. Such in-plant corrections and modifications can pay for themselves by maximizing the efficiency of the production process in terms of mass transfer, product yield, or some other measure. (Remember that waste materials can be thought of as resources out of place.) The objective is to reduce waste stream quantities or lower their concentration levels to the point where further reduction is more expensive than treatment or disposal of the wastes that remain. Effluent reduction, neutralization, and pretreatment should be used where cost-effective or required.

Short- and long-term costs and benefits should be factored into the decision to modify the process or acquire the necessary equipment. The cost of proper disposal of treatment residuals is a very real element of the operating expenses of any treatment system. Offsite treatment and disposal costs can be expected to increase by at least 10% per year during the foreseeable future.

Specific examples of possible process modifications include:

1. alkaline cleaners instead of solvent degreasing
2. water-conserving rinses (counter-current or spray rinses) in electroplating operations in lieu of rinse baths
3. multiple-stage rinse baths, where contaminant(s) are concentrated in the initial stage for reuse, recycling, or treatment
4. collection and recovery of dragout in electroplating operations

5. electrostatic painting in lieu of spray painting
6. dry filter paint booths instead of waterwall booths
7. powder paint application replacing the use of spray application
8. plastic blast media (mechanical or dry stripping) instead of chemical stripping for removal of paint and other coatings
9. mechanical instead of chemical descaling/pickling

Good Operating Practices

The goal of this strategy is to maximize the useful lifetime of process chemicals as well as to minimize inadvertent releases or wastage. This is accomplished by common-sense practices and methods that minimize unnecessary wastage and those that keep contaminant levels and chemical concentrations within specifications.

Extension of useful life of process chemicals. Activities under this heading include some discussed previously. Maintenance of process baths at the necessary levels of purity and concentration determines just how long solution lifetimes can be extended.

Prevention of contamination is one of the best ways to extend the life of a chemical bath. Rinsewater dilution of process baths can be minimized by spray rinsing or by allowing adequate time for drainage. Depletion of the necessary chemicals can be corrected by periodic additions. Certain contaminants can be removed by filtration. A protocol where the process baths are tested periodically for specific parameters, followed (as needed) by filtration, dewatering, solids removal, oil skimming, bacterial control, precipitation, reclamation and/or replenishment, would prevent any unnecessary waste and extend solution lifetimes to the greatest extent. Portable equipment could be utilized for these purposes.

Contamination of process chemicals can also be prevented by proper process design and layout. This may include the use of rinses, or covered or closed storage tanks and process baths. Dragout can be minimized by allowing proper drainage of dipped parts above process baths, and production lines can be designed to avoid activities above open baths. Cross-contamination can be minimized by proper procedures, container and tank labeling, employee training, and engineering measures.

Prevention of leakage and spills. Activities under this heading include maintenance, inspection, repair, and replacement to avoid leaky pumps, flanges, and valves or ruptured process or storage tanks. Spills can be minimized by adequate freeboard in open tanks, and by proper startup, shutdown, maintenance, chemical addition, and material transfer procedures.

Specific examples of good operating practices include:

1. filtration of plating baths, process tanks, etc.
2. replenishment of plating baths
3. dragout reduction with drain boards or dragout collection tanks
4. use of lids, sideboards, and chillers on solvent cleaning tanks and vapor degreasers, and maintaining adequate freeboard in such units
5. preventative maintenance on tanks, pumps, piping, etc., to prevent leaks or spills
6. material handling improvements, including proper chemical transfer equipment (such as dry disconnect valves) and procedures
7. waste stream segregation
8. production scheduling improvements, especially in batch processes where equipment cleaning is necessary after each batch or change in color
9. chemical inventory control, including rotation of inventory
10. employee training in effective response to spills
11. good housekeeping

As the measures potentially available to minimize waste generation are reviewed, thought should be given to minimizing the proportion of unavoidable waste that is considered hazardous, under the existing regulatory definitions. Many of the same measures are appropriate to both Level 1 and Level 2 of the Waste Management Hierarchy.

Separation/Concentration

At this point, it is necessary to ensure separation and concentration of the waste, where appropriate. Separation (also termed segregation) and concentration in the manufacturing process, a "front-end" initiative, can be much more effective and efficient than "end of the pipe" treatment and disposal approaches. Separation of the hazardous components of a waste stream, or of different types of waste streams, prevents normally nonhazardous wastes from becoming characteristically hazardous or hazardous by virtue of being mixed with a listed hazardous waste. Separation also allows for management of concentrated waste streams, rather than large volumes of mixed and diluted wastes.

Separation includes two distinct aspects. The first aspect involves the separation of different waste streams. The second aspect involves the isolation and concentration of the hazardous or toxic components from the mixtures in which they occur. Remember the "resource out of place" concept. Separation and concentration of waste stream components early in the production process, along with segregation of different waste streams in different containers, provides the greatest opportunities for subsequent reuse, recycling, and/or economical treatment.

Separation typically makes higher-level activities such as reuse or reclamation much more feasible technically, as well as less costly. Mixing of waste streams is not recommended unless justified as a treatment measure, or where mixing is compatible with the highest feasible level of management activity under the hierarchy. Obtaining offsite TSD acceptance of mixed waste streams can become extremely difficult and very expensive.

Reuse/Waste Transfer

The next step is to evaluate the use or reuse potential of the waste material. Use or reuse implies that such activities are feasible without prior processing. If some processing is necessary, the activity is more properly classified as reclamation for reuse.

Reuse includes using waste materials both for their original purpose and for a different purpose. Reuse can either occur onsite or offsite. The materials involved can be wastes, manufacturing by-products, or off-specification, obsolete, or surplus raw materials. Waste transfer is the physical movement of residual materials from firms (or operations within a firm) that find no further use or value in them to other operations or firms that can use such materials as effective substitutes for materials or chemicals otherwise purchased in a virgin state.

Specifications for chemical quality, purity, concentration, and quantity vary widely among different users; thus, transfer is possible to users that have lower specifications or that can use a lower-quality chemical. Off-specification products such as inks or paints can sometimes be reworked into future batches of the same product, or used as an ingredient or feedstock in batches of products of differing color (i.e., black), end use, or quality requirements. Spent pickling liquors from steel finishing operations are used in some sewage treatment plants as flocculant aids and for phosphorus removal.

Reuse is typically possible in a sequential manner, whereby water, solvents, cleaning agents, etc., go from processes with high quality and production rate requirements to those with lower quality/purity/concentration requirements. For example, in some printed circuit board manufacturing operations, electronics-grade 1,1,1-trichloroethane is used in a final cleaning step, in order to remove dust particles and any traces of moisture. Spent 1,1,1-trichloroethane solvent from this application would meet all of the specifications for technical-grade 1,1,1-trichloroethane typically used in vapor degreasers to remove oil from stamped metal parts. In general, transfers can take place from large companies with continuous processes to smaller companies using batch processes, from basic chemical manufacturers to chemical formulators and users, and from industries with high purity requirements to those with lower purity requirements.

Several organizations have evolved to promote the concept of waste transfer. Generally called "waste exchanges," they publish anonymous or "blind" listings of wastes or by-products available and wastes or by-products desired. These organizations provide a mechanism whereby generators and users of industrial by-products can be placed in contact. Waste exchanges operate in many states. Appendix H consists of a list of waste exchanges operating in North America.

Recycling and/or Material and Energy Recovery

Recycling is similar in concept to waste transfer, except that some processing (e.g., filtration, distillation, or re-refining) is necessary before the material can be used again. Wastes generally recognized as having components of potential value include:

1. solvents
2. flammable and combustible liquids (for fuel)
3. oils
4. drosses, slags, and sludges having high concentrations of recoverable metals
5. precious metal wastes
6. acids and alkalies
7. catalysts

In the commercial recycling market, wastes can be reprocessed for reuse by the original generator, or are reprocessed for sale to a third party by the recycling firm. Additionally, wastes can be processed to meet the purity specifications of the virgin chemical, or to be useful only in another, lower-grade application. For example, a spent solvent can be reclaimed to be reused in the process from which it was taken, or it can be reclaimed sufficiently to serve only as a cleanup solvent or supplemental fuel.

An entire industry has developed around the concept of processing industrial wastes so that they may be reused again in the same or a different application. These commercial facilities extract the components of value from the wastes, and either return the recovered materials to the generator, or sell them to another user. Often the component of value is the energy content of the waste stream. Processed petroleum-based waste streams are often used as supplemental fuels for industrial boilers or furnaces.

Specific examples of activities possible under the category of recycling (onsite or offsite) for reuse (onsite or offsite) include:

1. solvent distillation
2. filtration of oils
3. re-refining of used oils
4. regeneration of spent activated carbons
5. use of spent oils or solvents as supplemental fuels
6. ion exchange for recovery of metals from wastewaters
7. evaporation of rinsewaters to recover process chemical "dragout" for reuse
8. electrolytic metal recovery
9. reverse osmosis
10. ultrafiltration

Treatment

Even with process modifications and material substitutions, there may still be portions of some waste streams requiring some type of onsite treatment before discharge to a municipal sewer or a waterway. Additionally, some treatment (performed either onsite or offsite) is necessary before certain waste streams can be managed by land disposal. Dewatering of waste streams has assumed additional importance now that special requirements and restrictions on the disposal of liquid wastes (and wastes containing free liquids) in hazardous waste landfills are in effect (40 CFR 265.314). Dewatering will also reduce quantities needing transportation and disposal.

It is important to understand federal, state, and local pollution control regulations before purchasing any particular type of treatment or pollution control equipment. Air and water pollution control equipment should be chosen knowing the types and amounts of pollution control residuals that will result from the operation. Ideally, the cost of disposal of such residuals should be factored into the decision to purchase. Finally, innovative technologies for waste treatment and control should be investigated.

Specific examples of possible treatment activities include:

1. neutralization of corrosives
2. hexavalent chromium reduction
3. hydroxide precipitation of heavy metals
4. alkaline chlorination of cyanides
5. separation of liquids from solids—filter presses
6. evaporative concentration of wastewaters using waste heat
7. reduction of wastewater treatment sludge volume using sludge driers
8. removal of oils from water (including emulsion breaking)
9. incineration

Secure Land Disposal

Secure land disposal should be utilized as a last resort option for those waste materials not amenable to volume reduction, reuse, recovery, treatment, or destruction. Ideally, only residues from treatment processes will be managed by secure land disposal. EPA's program of Land Disposal Restrictions, found at 40 CFR Part 268, will likely prohibit the direct land disposal of all untreated hazardous wastes.

SUMMARY

The complexity of the ever-growing body of environmental regulations and the high costs of compliance (or noncompliance) indicate the advantages of a strong managerial approach to environmental compliance. Such an approach would use the Waste Management Hierarchy in determining actions toward this goal, recognizing that waste management takes place in a complex statutory and regulatory context, and is a more involved undertaking than merely meeting each requirement separately, without perceiving the total effect of the environmental regulations on the manufacturing process or the effect of changes in the manufacturing process on compliance with the regulations. Interactions include the effect of compliance methods for one set of environmental requirements on the applicability of and compliance with other environmental regulations.

An integrated management approach toward compliance could require the talents and efforts of more than one individual within the organization. An environmental control or waste management committee could be established, involving representatives from various corporate functions. Responsible persons from purchasing, plant engineering, production, environmental, safety/industrial hygiene, and finance functions could be included on such a committee. A cooperative labor-management approach, along the lines of quality circles, might yield surprising results.

The committee would be responsible for developing an environmental management strategy capable of meeting agreed-on goals, including regulatory compliance, at an acceptable cost to the company. The committee should examine alternatives, develop programs, and implement the necessary measures. Individuals on this committee, or their subordinates or designees, would be assigned specific tasks and would be held accountable for the successful completion of those tasks. This chapter may provide a starting point for discussion.

It is hoped that this information may be of value to generators who have not taken the comprehensive environmental management approach. The diligent application of intelligence, creativity, and ingenuity in this effort may make the imposition of strict controls on hazardous waste more palatable to both the employees involved and their employers. RCRA is a law conceived with good intentions, but it will require trained professionals with an innovative approach to enable these good intentions to be transformed into environmentally sound and cost-effective actions.

CHAPTER 19

Chemical Hazard Communication

Steven R. Ball

Developing and implementing a chemical hazard communication program is a task often assigned to persons with responsibilities for hazardous waste management. Proper handling of the purchased chemicals used in production and maintenance operations is just as important as the proper management of hazardous wastes. Many of the same principles and precautions apply. However, the Hazard Communication Standard (HCS) is administered and enforced by OSHA, while the hazardous waste management program is an EPA program. These different federal regulatory agencies have somewhat different purposes. OSHA was established to protect the safety and health of the nation's work force, while the EPA's mission is to protect public health and environmental quality. These and other reasons have resulted in different focus and terminology between the HCS and RCRA. Both OSHA and EPA allow for delegation of authority to state regulatory agencies.

Several of the tasks necessary to effectively implement a hazard communication program are similar to those tasks involved with hazardous waste management. These tasks include developing an inventory of potentially hazardous materials, performing the hazard determination, labeling, employee training, and documentation. The HCS is a performance-oriented regulation. There are wide variations in methods companies can use to achieve compliance with the HCS. The hazardous waste management program under RCRA is much more prescriptive than the OSHA HCS.

To achieve compliance, a company's primary goal with respect to the HCS should be to ensure that employees handling chemicals understand the hazards of chemicals or materials they may be exposed to when performing their job. In order to meet this goal, a hazard communication program must be implemented that allows each employee to learn *and* practice proper and safe work practices and principles. Therefore, employees need to know a chemical's potential health and physical hazards as a framework in following safe work principles. Employers should enable employees to achieve these safe work principles through positive methods of information dissemination and through effective training techniques. The information and training employees receive should reinforce any existing work rules and safety policies, as well as effectively modify behavior with respect to chemicals within the workplace.

The purpose of this chapter is to outline the required elements of a chemical hazard communication program at a manufacturing firm. Guidelines and recommendations are provided. However, for a more complete understanding of this topic, the reader is advised to read the Hazard Communication Standard itself, OSHA's directive to its inspectors and state agencies regarding the HCS (CPL 2-2.38B; 8/15/88), and other sources of information. One useful reference is *Lowrys' Handbook of Right-to-Know and Emergency Planning* by George and Robert Lowry. This book is available from Lewis Publishers (800/525-7894).

BACKGROUND

Though chemical hazard communication programs are relatively new to the regulated community, the concept actually was part of the Occupational Safety and Health Act of 1970. The OSH Act included language that directed OSHA to promulgate standards that prescribed methods in which employees could be apprised of all hazards to which they are exposed. This was to be accomplished through the "use of labels or other forms of warning as necessary to provide employees with the hazard information, signs and symptoms of exposure, emergency treatment, and proper conditions and precautions of safe use or exposure" (48 FR 53280–53281). The chemical-specific regulations OSHA has promulgated (i.e., for asbestos, benzene, lead, vinyl chloride, etc.) have been developed with hazard communication elements. However, there are far too many chemicals used in the workplace for OSHA to develop specific regulations for each. For that reason, in the mid-1970s OSHA began action to develop the framework of the current Hazard Communication Standard, found at 29 CFR 1910.1200.

Several states and communities have developed their own hazard communication or "employee right-to-know" regulations. However, the OSHA HCS specifically preempts both the state and local *occupational* right-to-know laws at 29 CFR 1910.1200 (52 FR 31860–31861). Unless hazard communication regulations are established under the authority of an OSHA-approved state plan, state or municipal governments are not permitted to regulate in the area of occupational chemical hazard communication. This is a very controversial topic and several disputes concerning preemption have been or will be decided in the courts.

As a result of litigation, OSHA has expanded the application of the HCS. Initially, the HCS only applied to employers in the manufacturing sector (Standard Industrial Classification [SIC] Code prefixes 20–39). Amendments to the HCS expanded its application to all employers having hazardous chemicals in the workplace. An effective date of May 23, 1988 was specified for expansion of the HCS to nonmanufacturing employers. The HCS was effective with respect to chemical manufacturers, chemical importers, and chemical users in the manufacturing sector on May 25, 1986.

METHODS TO ACHIEVE COMPLIANCE

Applicability

There are two basic determinations a company must make in order to establish whether it is regulated by the Hazard Communication Standard. The first determination relates to the firm's usage of chemicals within its operations. The HCS applies to all chemical manufacturers, importers of chemicals, chemical distributors, *and* employers that use, handle, generate, blend, formulate, manufacture, or repackage hazardous chemicals or materials, as defined, in their processes or operations. If a firm does not use chemicals to perform any of the above-listed activities, then it is not regulated by the HCS. If chemicals are used or if the above-listed activities are performed, then a second determination is necessary.

The second determination is whether any of the chemicals or materials used are "hazardous chemicals," as defined by OSHA. Normally the hazard determination for OSHA purposes is performed by the chemical manufacturer.

The chemical manufacturer or importer is required by the HCS to assess the hazards of any chemical produced or imported. OSHA has specific criteria which the chemical manufacturer or importer must use to evaluate a chemical's physical and health hazards (see Appendices A and B of the HCS). This hazard information must then be forwarded to all companies using the chemical or material by means of labels, MSDSs, or other forms of warning. If hazardous chemicals (per the chemical manufacturer) are not used, then the company is not subject to the HCS, and no compliance program is necessary.

The Hazard Communication Standard has limited application to certain operations. The standard applies to laboratories only as follows:

1. Labels on incoming containers of hazardous chemicals cannot be removed or defaced.
2. MSDSs received with incoming shipments of hazardous chemicals must be

maintained and made available to employees.
3. Employees must be informed of chemical hazards through information and training efforts.

Operations that only require handling of hazardous chemicals in sealed containers (i.e., retail sales and warehousing operations) have only the following
requirements:

1. Labels on incoming containers of hazardous chemicals cannot be removed or
defaced.
2. MSDSs received with incoming shipments of hazardous chemicals must be maintained.
3. Upon employee request, MSDSs must be obtained for sealed containers of
hazardous chemicals when such MSDSs are not already on file.
4. All MSDSs on file must be made available and accessible to employees during
each work shift.
5. Employees must receive training and information concerning the chemical hazards they may be exposed to during their work activities.

Chemical Inventory

Once the applicability of the HCS is verified, the next task involves an inventory of all chemicals used in the workplace. An inventory is the most important
and logical starting point for efforts to achieve compliance with the HCS. Complete compliance begins with knowing just how many different chemicals are used
within the operation, as well as the locations or departments where the chemicals
are used. The HCS requires that a list of hazardous chemicals used in the company be developed and maintained. The inventory is the primary means of preparing the hazardous chemicals list. The list can cover the entire facility, or separate
lists can be prepared for each department or work area.

There are some chemicals and materials that are exempt from the HCS and
need not be inventoried. These include hazardous wastes; tobacco products; wood
or wood products (wood dust is not exempt); food, drugs, cosmetics, or alco-
holic beverages in a retail establishment and/or consumed by employees; consumer products used in the same manner as normal consumer use; medicines;
office products (adhesive tape, correction fluid, etc.); and articles. An article is
a manufactured item formed to a specific shape, has end-use functions, and does
not release or result in exposure to a hazardous chemical under normal conditions of use. One example of an article is a sheet of aluminum to be attached
to a wall for wall covering. In this application, it has end-use functions. However,
if a sheet of aluminum is welded or sanded upon within a manufacturing or production process, it would not have end-use functions and could result in exposure
to a hazardous chemical. In this instance, it would be considered a hazardous
chemical (or material) and must be inventoried.

The best (but most time-consuming) approach to completing a chemical inventory is performing a physical tour of the plant and searching for suspect products or materials. The inventory preparer must walk through each work and storage area of the facility while recording specific data about chemical products or materials found. Data necessary to record includes the identity (from the label) and common name of the material, manufacturer's name and address, and location where the material is stored or used. Upon completion of the inventory, purchasing records should be reviewed as a check system to ensure that all materials were recorded. A review of purchase orders over a specific time period (e.g., one year) could be performed. Any chemicals found in the purchasing records and not on the inventory list should be added. It is also a good idea to estimate quantities of chemicals or materials in inventory and quantities purchased annually, since community right-to-know regulations (discussed in Chapter 20) require chemicals onsite above certain thresholds or used in certain quantities to be reported to state and local emergency response agencies. As a general rule, the following types of materials should be included in an inventory:

1. cleaning agents—acids, caustic solutions, solvents, or other cleaning solutions
2. process chemicals specific to the operations (i.e. inks in a printing industry)
3. industrial oils—cutting, hydraulic, lubricating oils; coolants; heat-treating and quench oils and greases
4. surface-finishing agents—electroplating chemicals, rust inhibitors, primers, and paints
5. water and wastewater treatment chemicals—neutralizing, oxidizing, or reducing agents; water softeners; biocides; flocculants

When performing the inventory, do not include those materials exempt from the regulation (articles, tobacco products, etc.). However, in the case of consumer products, an evaluation of how the product is used in a manufacturing setting must be performed. The consumer product may or may not be included in the inventory. For example, a corrosive cleanser purchased from the supermarket and used one or two hours once per week by a part-time janitor may not be included in the inventory since the cleanser is used as a consumer would at home. However, if a full-time janitor is using the same product on a daily basis, four to eight hours per day, the product must be included in the inventory. Also, if the consumer product is used in industrial quantities or at industrial concentrations, then the product must be inventoried. If there is doubt as to whether a certain product is to be inventoried, the best approach is to include the product in the inventory.

The purpose or reason that a product is present in the facility should also be evaluated. If it is found that a product is no longer used or necessary to a process, it should be returned to the supplier, listed in a waste exchange, or properly disposed of. The product should be included in the inventory until it is removed from the facility, at which time it may be removed from the inventory.

An evaluation of a chemical's potential by-products or decomposition products must also be made. For example, chemicals such as carbon monoxide generated from the incomplete combustion of fuels or ozone generated during arc welding processes must be included in a chemical inventory. Wood dust generated by woodworking operations is also regulated under the HCS. MSDSs for gasoline or welding rods should provide information as to the hazards of decomposition products or by-products. If not, these by-product or decomposition compounds should be included in the inventory.

Hazard Determination

Once the inventory has been completed, manufacturers or vendors supplying the chemical or material to the company should be contacted to provide information as to whether the product has been determined to be hazardous, per the HCS hazard definitions and procedures. Users of the chemical are not required to perform a hazard determination unless they choose not to rely on the determination performed by the manufacturer or importer of the chemical.

The best way to obtain the hazard information is to request a copy of an MSDS for each product. This request can be made by telephone or written correspondence. If contact is made via telephone, be sure to document the date, time, company and person contacted, and product for which the MSDS request was made. Form request letters can also be developed where mass mailing may be more convenient to obtain several MSDSs from different manufacturers or vendors. Written correspondence is best, because the chemical manufacturer is placed on notice that the user is relying on them to provide a complete, accurate, and up-to-date MSDS.

A manufacturer's response to the MSDS request may be either of the following:

1. An MSDS that documents the material's physical and health hazards will be provided.
2. An MSDS or some other document will be provided stating that the material is not hazardous according to the hazard determination methods specified at Appendices A and B of the HCS.

Please note that the OSHA definition of a hazardous chemical is very broad and inclusive. Most industrial chemicals and materials are considered hazardous by OSHA, for purposes of the HCS. Hazardous chemicals can be in liquid, gaseous, powdered, granular, or solid forms.

If the material is not hazardous, it is important to maintain the manufacturer's statement that the material is not hazardous. OSHA may request to review this documentation during inspection activities. MSDSs obtained for hazardous chemicals must be maintained and made available for employee review. Proper MSDSs and MSDS files will be discussed in the next section.

MSDSs

An MSDS is similar to a product specification sheet. The MSDS will provide specific health and physical hazard information about a chemical or product. Figure 19.1 shows an example of OSHA Form 174, Material Safety Data Sheet. It is not mandatory that this specific form be used by the chemical manufacturer. However, the basic requirement is for the chemical manufacturer to provide an MSDS with the same information elements as OSHA Form 174.

MSDSs are to address each of the following items:

1. identity (name) of the product and appropriate chemical and common names
2. physical and chemical characteristics
3. physical hazards
4. health hazards
5. routes of entry (inhalation, ingestion, skin/eye contact)
6. exposure limits
7. carcinogenicity
8. precautions for handling and use
9. control measures
10. emergency and first aid measures
11. date of preparation
12. name, address, and telephone number of a responsible party

An MSDS must be obtained for every hazardous chemical or product used or present within a company. Once an MSDS is received, it should be evaluated to determine that it contains all of the necessary information. MSDSs, under the HCS, cannot have any blank spaces in any of the sections. If no relevant information is available for any given category, the MSDS preparer is to mark the MSDS indicating that no applicable information was found. Designations such as Not Applicable, Not Available (NA) or Not Determined (ND) may be found in some of the MSDS sections.

MSDSs must be collected, maintained, and filed. Once all MSDSs are received and reviewed, they should be filed in some organized fashion to facilitate retrieval of specific MSDSs. The MSDS filing method can be developed in conjunction with the development of the hazardous chemicals list.

Material Safety Data Sheets must be provided to chemical purchasers prior to or with an initial shipment of a hazardous chemical. It is the employer's responsibility to ensure that MSDSs are obtained if not supplied with the shipment. MSDSs must be maintained for each hazardous chemical onsite and be readily accessible for employee review. It has been interpreted that "readily accessible" requires that the MSDSs be available for employee review without any involvement by or with members of management. In other words, employees should not need to ask permission or obtain approval to review MSDSs. This can be accomplished by providing additional copies of MSDSs in a lunch room, break room, or general work area that is easily accessible to employees.

Material Safety Data Sheet
May be used to comply with
OSHA's Hazard Communication Standard,
29 CFR 1910.1200. Standard must be
consulted for specific requirements.

U.S. Department of Labor
Occupational Safety and Health Administration
(Non–Mandatory Form)
Form Approved
OMB No. 1218–0072

IDENTITY *(As Used on Label and List)*	Note: Blank spaces are not permitted. If any item is not applicable, or no information is available, the space must be marked to indicate that.

Section I

Manufacturer's Name	Emergency Telephone Number
Address *(Number, Street, City, State, and ZIP Code)*	Telephone Number for Information
	Date Prepared
	Signature of Preparer *(optional)*

Section II — Hazardous Ingredients/Identity Information

Hazardous Components (Specific Chemical Identity; Common Name(s))	OSHA PEL	ACGIH TLV	Other Limits Recommended	% *(optional)*

Section III — Physical/Chemical Characteristics

Boiling Point		Specific Gravity ($H_2O = 1$)	
Vapor Pressure (mm Hg.)		Melting Point	
Vapor Density (AIR = 1)		Evaporation Rate (Butyl Acetate = 1)	
Solubility in Water			
Appearance and Odor			

Section IV — Fire and Explosion Hazard Data

Flash Point (Method Used)	Flammable Limits	LEL	UEL
Extinguishing Media			
Special Fire Fighting Procedures			
Unusual Fire and Explosion Hazards			

(Reproduce locally)   OSHA 174, Sept. 1985

Figure 19.1. Material Safety Data Sheet (front).

Section V — Reactivity Data

Stability	Unstable		Conditions to Avoid
	Stable		

Incompatibility (*Materials to Avoid*)

Hazardous Decomposition or Byproducts

Hazardous Polymerization	May Occur		Conditions to Avoid
	Will Not Occur		

Section VI — Health Hazard Data

Route(s) of Entry:	Inhalation?	Skin?	Ingestion?

Health Hazards (*Acute and Chronic*)

Carcinogenicity:	NTP?	IARC Monographs?	OSHA Regulated?

Signs and Symptoms of Exposure

Medical Conditions
Generally Aggravated by Exposure

Emergency and First Aid Procedures

Section VII — Precautions for Safe Handling and Use

Steps to Be Taken in Case Material Is Released or Spilled

Waste Disposal Method

Precautions to Be Taken in Handling and Storing

Other Precautions

Section VIII — Control Measures

Respiratory Protection (*Specify Type*)

Ventilation	Local Exhaust		Special	
	Mechanical (*General*)		Other	
Protective Gloves		Eye Protection		

Other Protective Clothing or Equipment

Work/Hygienic Practices

☆ U S.G.P.O. 1986–491–529/45775

Figure 19.1, cont'd.   Material Safety Data Sheet (back).

MSDSs must be maintained as long as the chemical is used or present within the facility. When a chemical is no longer present or used, the MSDSs need not be retained for any specified time period. However, some record of the identity of the material or its specific chemical ingredients, where the material was used, and when it was used must be retained for at least 30 years. If an MSDS is made part of an employee's medical record, the MSDS must be retained for the same length of time as the medical record (duration of employment plus 30 years).

Labels and Other Forms of Warning

Chemical manufacturers must label, tag, or mark containers of hazardous chemicals supplied to a purchaser for use within the workplace with the following information:

1. name of product (as it would be identified on the MSDS)
2. appropriate hazard warning(s)
3. name and address of the chemical manufacturer, importer, or other responsible party

Chemical manufacturers must ensure that labels do not conflict with labeling required by the DOT.

Labeling requirements do not apply to:

1. any pesticide subject to labeling requirements and labeling regulations issued by the EPA
2. any food, food additive, color additive, drug, cosmetic, or medical or veterinary device, including materials intended for use as ingredients in such products
3. any distilled spirits, wine, or malt beverage intended for non-industrial use
4. any consumer product or hazardous substance regulated by the Consumer Product Safety Commission

Companies using hazardous chemicals within the workplace must ensure that in-plant containers are labeled, tagged, or marked with the identity of the hazardous chemical and appropriate hazard warning. Signs, process sheets, batch tickets, or posted operating procedures are alternate methods for labeling stationary containers such as tanks or manufacturing process units. For example, hazards of paints and solvents used within a room containing paint mixers could be communicated by means of a sign or placard in the immediate work area. This is one alternative to labeling each individual process vessel.

Individual piping systems do not need to be labeled. However, in cases where a hazardous chemical is transferred by piping systems to a remote dispensing station, a sign or placard with the chemical name and hazard warning could be located at the dispensing station.

Portable containers for immediate use by an employee or group of employees do not need to be labeled. However, each individual using the chemical must have an opportunity to read the label on the larger chemical container that is the source of the chemical in the portable container.

Good safety practice prescribes that at least the name of the chemical or material be provided on any portable container or piping system. This is necessary to prevent intermixing of incompatible chemicals and provides the employees with basic information about the container or pipe contents.

Labels or other forms of warning must be in English and prominently displayed on the container. Companies that have non-English-speaking employees must provide labels or warning information in the languages spoken, in addition to the English labels.

Employee Training

An effective hazard communication training program is a key factor to providing employees a safe and healthy working environment. As stated at the beginning of this chapter, the primary goal of hazard communication is for employees to understand the potential hazards of workplace chemical exposures. Container labels and Material Safety Data Sheets are just part of the information transfer process. Employee training programs should be structured in such a way that employees can interpret the information provided on labels and MSDSs, then develop and maintain safe work practices based upon the information provided.

Since employee education levels will vary, information provided in a training program must be able to be understood by all participants. Explanation of some chemical hazards could also be alarming to some individuals. The information must be presented in a manner to alleviate any fears. One method is to compare hazards of chemicals at work to hazards of chemicals commonly used at home. Another method is to identify and explain the engineering controls, work practices, personal protective equipment, and administrative measures that act to minimize actual exposure to the hazardous chemical.

If employees are only exposed to a small number of chemicals, training could be limited to a discussion of just those chemicals. However, in most operations, employees may have potential exposures to a variety of chemicals. In this instance, training could be accomplished by informing employees about hazards of chemicals in general hazard categories.

Common hazard categories include fire hazards, corrosivity hazards, reactivity hazards, toxicity hazards, and sudden-release-of-pressure hazards. An explanation of the characteristics of each of these hazard categories should be provided. Training can be enhanced by giving examples of chemicals used in the workplace that fit into these hazard categories.

The training program must also include a discussion of the company's written hazard communication program, where MSDSs are located, and container labeling systems. Emphasis should be placed on understanding and using the information provided by container labeling systems, as well as how to read and interpret MSDSs.

Employees will undoubtedly have questions about specific chemical hazards and hazardous situations. Be sure to provide time to field questions during a training session. Again, the goal is to make sure employees understand the potential hazards and how to handle the chemicals safely, so as to minimize the actual hazards. Providing this information in a positive manner will act to alleviate undue fear.

Written Hazard Communication Program

The written hazard communication program is a document which identifies and explains the methods the company will use to achieve compliance with the HCS. The written program should begin with a policy statement emphasizing management support for hazard communication compliance.

Another vital portion of the written program is the hazardous chemicals list. The HCS requires that a list of hazardous chemicals present onsite be made a part of the written program. The hazardous chemicals list may actually be the list of chemicals recorded during the chemical inventory. The list should contain the name of the product, manufacturer, hazard category, and location where the chemical may be used or stored in the facility. The list should be compiled in any format convenient for the company's operations. For example, the list could consist of separate listings for different work areas. It is recommended that the hazardous chemical list be used as a table of contents for files of MSDSs. This should expedite retrieval of MSDSs, when necessary.

The remainder of the written program should basically be a brief discussion of how the company will achieve compliance with each segment of the HCS: chemical inventory, hazard determination, MSDSs, labeling, and employee training. The written program must also specify methods the company will use to inform employees of the hazards of nonroutine work tasks or emergency situations.

The final portion of the written program should be a discussion of how the company will inform contractors about chemicals to which their employees may be exposed during work activities while on the company's property. The company should arrange for an exchange of information with the contractor concerning potential chemical hazards. Chemical hazards can be posed by the operations of either party. For example, contractors often use hazardous chemicals in their activities. Should company employees be exposed to chemicals brought onsite by the contractor, additional information and training efforts may be necessary for any new or different hazards of exposure during contractor activities.

In concluding this discussion of HCS compliance, three points should be kept in mind. The first point is to remember why this regulatory program is being implemented. Its goal is to reduce the incidence of chemical-related occupational illnesses and injuries. I think we can all agree that this is a noble and highly desirable goal.

The implementation of a program to achieve this goal will result in several identifiable benefits to the employer, as well as the employee. They include:

- reduction in lost work days and lost production due to chemical-source injuries and illness

- reduced medical costs and worker's compensation insurance premiums

- reduced employee turnover

- improved employee morale

- reduced probability of fires, spills, and associated property losses

- improved work practices and increased use of personal protective equipment

The second point concerns the importance of documentation in compliance with this regulation. It is no longer sufficient to merely inform and train workers regarding chemical hazards. It is also necessary to document compliance. Documentation for this particular regulation involves preparation of plans, procedures, programs, and lists, keeping good records of training, and ensuring that program elements are complete and up-to-date. Changes in chemicals used or in personnel will require additional activity.

The third point is that chemical hazard communication, or "employee right-to-know," has established itself as an issue of continuing concern to corporate management, regulatory officials, and employees. The HCS has raised standards of acceptable practice in chemical handling. Employers must develop and implement effective compliance programs in response to the regulatory requirements. These efforts, if properly structured, should have positive effects on hazardous waste management and community right-to-know compliance.

CHAPTER 20

Community Right-to-Know

Charles W. White

Responding to community right-to-know initiatives is a project often assigned to those persons already responsible for hazardous waste management. The community right-to-know program is administered by EPA and newly established state and local entities. However, the focus of this program is much more broad than industrial and hazardous wastes. Emergency planning and community right-to-know efforts are concerned with hazardous chemicals, as defined by OSHA, and those subsets of hazardous chemicals termed "extremely hazardous substances" (EHS) and "toxic chemicals."

The purpose of this chapter is to provide an overview of the statute involved, as well as a brief explanation of the major regulatory requirements. Responsible parties are well advised to delve into the regulations themselves, instructions supplied with reporting forms, and EPA guidance manuals. EPA is operating a toll-free hotline for those parties requesting information or having questions regarding this program. The telephone number of the EPA Chemical Emergency Preparedness Hotline (Title III Hotline) is 800/535-0202. The toll number is 202/479-2449.

Effectively and efficiently responding to the massive data collection and reporting requirements discussed in this chapter will require the involvement of company purchasing and accounting personnel. Their information systems, particularly with respect to chemical inventory levels, volumes purchased, and production levels, will need to be revised to ease regulatory compliance burdens. The gradual reduction of reporting thresholds over time will make regulatory compliance impossible without comprehensive, systematic, computer-aided data collection and reporting systems.

The statutory authority for federal, state, and local efforts in this area was provided by Title III of the Superfund Amendments and Reauthorization Act of 1986 (SARA), appropriately entitled the Emergency Planning and Community Right-to-Know Act of 1986 (EPCRA). This statute, often referred to as EPCRA, SARA Title III, or simply Title III, was established to encourage emergency planning efforts at the state and local levels and to provide public and local emergency response agencies with information concerning potential chemical hazards associated with fixed facilities in their communities. The emergency planning requirements of this act were developed to meet the need for well-planned response procedures at the local level. The emergency plans would focus on chemical releases that might threaten public health and the surrounding environment. The various reporting provisions of EPCRA are intended to provide the information necessary to facilitate appropriate emergency planning and emergency response activities.

A summary listing of compliance activities with respect to EPCRA is found in Table 20.1.

Table 20.1.   Title III Reporting Requirements

1. Notify state emergency response commission (SERC) that the facility is subject to the emergency planning provisions of EPCRA. Deadline is May 17, 1987, or within 60 days after facility becomes subject to the requirements.

2. Notify local emergency planning committee (LEPC) of facility representative who will participate in the local emergency planning process. Deadline is September 17, 1987.

3. Submit MSDSs or a chemical list organized by hazard category for all OSHA hazardous chemicals at the facility in quantities greater than or equal to 10,000 lb and for all extremely hazardous substances (EHS) at the facility in amounts greater than or equal to the threshold planning quantity (TPQ) or 500 lb, whichever is less. Deadline is October 17, 1987, or three months after the facility becomes subject to this requirement. The MSDS set or chemical list must be updated as necessary due to changes in chemical inventories.

4. Facilities subject to Requirement 3 (above) must submit an inventory form (Tier I) to the SERC, LEPC, and the local fire department. Deadline is March 1, 1988, and annually thereafter.

5. Facilities in Standard Industrial Classification Codes 20–39, with 10 or more full-time employees, that manufactured or processed toxic chemicals above the regulatory threshold, or that used over 10,000 lb of any toxic chemical listed at 40 CFR 372.65, must submit Form R ("Toxic Chemical Release Inventory Reporting Form") for each toxic chemical manufactured, processed, or otherwise used. Deadline is July 1, 1988, and annually thereafter.

EPCRA is organized into three subtitles. Subtitle A establishes a framework for state and local emergency planning programs. Subtitle B requires covered facilities to submit information concerning the hazardous chemicals stored at those facilities to state and local governments. Subtitle C contains general provisions concerning the protection of trade secrets, enforcement of the act, citizen suits, and public access to the information provided by EPCRA. The remainder of this chapter is concerned primarily with the reporting requirements contained in Subtitle B and portions of Subtitle A.

SUBTITLE A REQUIREMENTS

Section 302 of Subtitle A requires the owner/operator of a covered facility to notify the state emergency response commission (SERC) that it is subject to the emergency planning provisions of EPCRA. This notification was to have been made by May 17, 1987, or within 60 days after a facility becomes subject to the requirements, whichever is later. Section 303 requires the owner/operator of a covered facility to designate a facility representative (facility emergency response coordinator) to participate in the local emergency planning process. Typically, a given facility will already have designated an emergency coordinator in the contingency plan and emergency procedures document developed for potential hazardous waste releases. The owner/operator of the facility must provide the local emergency planning committee (LEPC) with the name of the facility representative. This notification was required on or before September 17, 1987. Regulations implementing the statutory language of Sections 302 and 303 are found at 40 CFR Part 355.

In order to determine if a facility is required to comply with the emergency planning and facility emergency coordinator notifications, the owner/operator must first inventory all chemicals at the facility. This inventory should then be compared to the List of Extremely Hazardous Substances and their Threshold Planning Quantities (TPQs). This list is found in 40 CFR Part 355, Appendix A. If the facility inventory contains any chemical listed as an Extremely Hazardous Substance (or a chemical that contains an EHS) in an amount that is equal to or in excess of the corresponding TPQ, the facility must comply with the notification requirements described above. The amount of each extremely hazardous substance must be calculated in pounds to allow comparison to the TPQ for that substance. The amount of an extremely hazardous substance relevant for reporting purposes is the amount present at the facility at any one time. Mixtures, blends, or proprietary (trade name) chemicals containing an EHS at concentrations greater than 1% by weight (regardless of location, number of containers, or method of storage) must be included in quantity calculations. If a mixture contains an extremely hazardous substance in a weight percentage of 1% or greater, that percentage must be multiplied by the mass of the total mixture to determine the actual amount of the extremely hazardous substance. More specific guidance for powdered solids, solids in solution, and molten solids is provided at 40 CFR Part 355.30(e).

Section 304 of Subtitle A requires owners/operators of facilities at which hazardous chemicals are produced, used, or stored to immediately notify the community emergency coordinator for the local emergency planning committee and the SERC in the event of a release of a reportable quantity of any extremely hazardous substance or CERCLA hazardous substance. Various notification requirements are specified at 40 CFR Part 355.40(b) and are further discussed in Chapter 9 of this book.

SUBTITLE B REQUIREMENTS

Subtitle B of EPCRA establishes reporting requirements that are intended to compel covered facilities to provide specific information to state and local officials concerning chemicals produced, stored, imported, or used at the facilities, as well as potential emissions of those chemicals. This portion of EPCRA will have the greatest impact on manufacturing firms using hazardous and toxic chemicals, as defined. Following is a discussion of the basic requirements of Subtitle B.

Section 311

Section 311 of Subtitle B applies to any facility that is required to maintain or prepare an MSDS for a hazardous chemical according to regulations promulgated under the Occupational Safety and Health Act of 1970. OSHA has specified procedures to be utilized in determining whether a chemical is a hazardous chemical, and thus subject to MSDS requirements, in the HCS, which is found at 29 CFR 1910.1200. The determination as to whether a chemical is hazardous for the purposes of the HCS is typically performed by the chemical manufacturer. Please note that OSHA's definition of hazardous chemical for purposes of the HCS is very broad and inclusive. Most chemicals are considered hazardous chemicals by OSHA. Hazardous chemicals can be in liquid, gaseous, powdered, granulated, or solid form.

Owners/operators of covered facilities are required to submit an MSDS for all hazardous chemicals at the facility in amounts greater than or equal to 10,000 pounds, and for all extremely hazardous substances present at the facility in amounts greater than or equal to the TPQ or 500 pounds, whichever is less. The appropriate MSDSs must be submitted to the local emergency planning committee, the state emergency response commission, and the local fire department with jurisdiction over the facility. Owners/operators of covered facilities were to have submitted the appropriate MSDSs by October 17, 1987, or three months after the facility first becomes subject to the requirements. Reporting thresholds are scheduled to be lowered by EPA, likely resulting in a revised Section 311 submission in October of 1990.

Alternatively, an owner/operator of a covered facility may fulfill the Section 311 reporting requirement by submitting a list of the hazardous chemicals and/or extremely hazardous substances present at the facility in amounts greater than the thresholds described above. Chemicals in this list must be grouped in one or more of five specified hazard categories. The EPA-specified hazard categories, along with examples of chemical hazards corresponding to each category, are found below:

1. immediate (acute) health hazard—highly toxic agent, toxic agent, irritant, sensitizer, corrosive, blood toxin, eye hazard, skin hazard, nervous system toxin
2. delayed (chronic) health hazard—carcinogen, kidney toxin, liver toxin, lung toxin, reproductive toxin
3. fire hazard—flammable liquid, combustible liquid, flammable aerosol, flammable gas, flammable solid, oxidizer, pyrophoric
4. sudden-release-of-pressure hazard—compressed gas, explosive
5. reactive hazard—unstable reactive, organic peroxide, water-reactive

EPA regulations under Section 311 of EPCRA are codified at 40 CFR Part 370. The above-referenced hazard categories are defined at 40 CFR Part 370.2. Note that toxic agents can be categorized as chronic or acute health hazards. In the list above, the various aspects of chemical toxicity have been placed in the category in which they most often occur. Also note that any one chemical may be placed in more than one category. For instance, a solvent that presents an acute health hazard may also be a flammable liquid, and thus a fire hazard as well.

The owner/operator of a covered facility is responsible for updating the Section 311 chemical list annually as necessary. Revised lists must be sent to the LEPC, SERC, and the local fire department. Different provisions for providing current information apply to facilities supplying MSDSs instead of chemical lists. The covered facility is required to submit copies of MSDSs for any new hazardous chemical that is introduced into the facility's inventory, or a copy of a revised MSDS whenever new information on an existing chemical is discovered. Updates must be sent to the local emergency planning committee, the state emergency response commission, and the local fire department within three months of the receipt of new MSDSs or revised MSDS information. Another Section 311 submission will be necessary in October of 1989 for hazardous chemicals present at or above the final (lower) reporting threshold, which at this writing has not yet been determined by EPA.

Section 312

Section 312 of Subtitle B builds on the information required by Section 311 by requiring the owner/operator of a covered facility to submit an inventory form entitled the Emergency and Hazardous Chemical Inventory or "Tier I." Tier I contains specific information about the hazardous chemicals present at the facility during the preceding calendar year. Only chemicals in amounts greater than the minimum thresholds described above are reported on the Tier I. Some chemicals are exempt from Tier I reporting. The exemptions include hazardous waste; consumer products or hazardous substances that are used in the workplace in the same manner as normal consumer use; food additives; color additives; substances used in research laboratories or medical facilities; and substances used in routine agricultural operations. EPCRA Section 312 provisions are codified in EPA regulations found at 40 CFR Part 370.25. A complete list of exemptions is found at 29 CFR Section 1910.1200(b) and Section 311(e) of EPCRA.

The Tier I form is organized into four sections. The first section requests general information about the facility. This information includes the name and address of the facility as well as the facility's primary SIC code and Dun and Bradstreet number. This section also requires the name, mailing address, and telephone number of the facility owner/operator.

The second section of the form requires the name, title, work telephone number, and 24-hour telephone number of at least one emergency contact for the facility. Space is provided for similar information for a secondary contact as well.

The third section is the most important section of the form. This section requires specific information concerning the hazardous chemicals present at the facility. The chemicals that must be reported in this section are the same chemicals that were included in the MSDS inventory described above. However, the quantities are aggregated according to the five hazard categories described previously. For each hazard category the owner/operator must provide the maximum amount, average daily amount, and the number of days onsite.

The best way to complete this section of the form is to calculate the amounts of OSHA hazardous chemicals exceeding 10,000 pounds and the amounts of extremely hazardous substances exceeding 500 pounds or the TPQ, whichever is less. (*Note:* The 10,000-pound minimum threshold is scheduled to drop to a final threshold of zero on March 1, 1991, for all hazardous chemicals and all extremely hazardous substances present at the facility during the preceding calendar year.) These quantities must be calculated in pounds based on the preceding calendar year's inventory. If a chemical is part of a mixture, the owner/operator has the option of reporting either the weight of the entire mixture or only the portion of the mixture that is a hazardous chemical. Whichever option is chosen, it must be consistent with the way the chemical was reported in the MSDS inventory. Once the weights of the individual chemicals are calculated, they can be grouped into the five hazard categories described above. Remember that any one chemical can fall into more than one hazard category. Each chemical must be counted in all applicable hazard categories. Also, maximum amounts and average daily amounts for each hazard category are reported on the Tier I form using designated "Reporting Ranges" for each category. The 10 designated Reporting Ranges are given in Table 20.2.

The general location within the facility where each chemical hazard may be found must also be entered alongside hazard category information. For each hazard type, descriptions, including names or identifications of buildings, tanks,

Table 20.2. Tier I Designated Reporting Ranges

Range Value	Weight Range in Pounds	
	From	To
00	0	99
01	100	999
02	1,000	9,999
03	10,000	99,999
04	100,000	999,999
05	1,000,000	9,999,999
06	10,000,000	49,999,999
07	50,000,000	99,999,999
08	100,000,000	499,999,999
09	500,000,000	999,999,999
10	1 billion	>1 billion

storage areas, etc., should be listed for all applicable chemicals. A simple way to complete this information is to attach a facility diagram to the Tier I form. A list of numbered locations as indicated on the diagram may then be included on the Tier I form.

The last section of the Tier I form is a certification statement which must be signed and dated by the owner/operator or the owner/operator's authorized representative.

As an alternative to completing a Tier I form, an owner/operator may elect to submit a Tier II form instead. This is generally not done, however, as Tier II requires much more detailed chemical information and would involve much more time and effort to complete than the Tier I form. Owners/operators of covered facilities must submit Tier II if requested to do so by the state emergency planning commission, local emergency planning committee, or local fire department.

Tier I must be submitted by March 1, 1988, and annually thereafter. It is important to remember that the Tier I is completed annually for the *preceding calendar year*. Thus, a facility which begins operations in 1988 will not be required to report until March 1, 1989. Also note that the minimum threshold for determining whether the reporting requirements apply to a facility are scheduled to decrease in 1989.

Section 313

Section 313 of Subtitle B of EPCRA is perhaps the most complex section for the facilities to which it applies. Although it will not be covered here in great detail, this discussion should give readers a basic understanding of Section 313. Section 313 requires owners/operators of facilities that "manufacture, import, process, or otherwise use" certain chemicals to annually report the emissions to the environment of those chemicals. EPCRA Section 313 requirements are found in regulatory form at 40 CFR Part 372. EPA requires that reports be filed by owners/operators of facilities that meet the following three criteria:

1. The facility is included in SIC Codes 20–39.
2. The facility has 10 or more full-time employees.
3. The facility used over 10,000 pounds of any toxic chemical, or manufactured or processed over 75,000 pounds (in 1987) of any toxic chemical listed in specific Toxic Chemical Listings found at 40 CFR 372.65.

Facilities subject to the Section 313 reporting requirements must complete a Form R ("Toxic Chemical Release Inventory Reporting Form") for each toxic chemical manufactured, processed, or otherwise used in excess of the applicable reporting threshold. Covered facilities are to submit the completed form(s) to the EPA and to the state emergency response commission. Form R must be submitted on or before July 1 of the year following the calendar year reported. For example, a Form R covering January to December, 1989 must be submitted by July 1, 1990. Form R will require determination and disclosure of toxic chemical emissions to air, surface water, groundwater, soil, and offsite disposal facilities.

Form R consists of four parts. Part I is entitled "Facility Identification Information." This part requires the owner/operator of the facility to provide much of the same general information about the facility that Tier I required, although in more detail. Part II of Form R is entitled "Off-Site Locations To Which Toxic Chemicals Are Transferred In Wastes." This part requests specific information about treatment, storage, or disposal facilities to which wastes containing toxic chemicals were transferred during the previous calendar year. Part III, "Chemical Specific Information," is the most important part of Form R. Part III must be completed separately for each toxic chemical at the facility. This part is discussed further below. Finally, Part IV, "Supplemental Information," provides additional space for information that would not fit in the preceding parts of the form. Part IV must be included when Form R is submitted, even if it is left blank.

As with Tier I, there are minimum reporting thresholds for Section 313. Unlike Tier I, however, there are two different thresholds, depending on the activity the particular chemical is involved in and the year for which the report is completed. If a listed toxic chemical is "manufactured" or "processed" at the facility in amounts greater than the following quantities, it must be reported.

- 75,000 pounds in calendar year 1987
- 50,000 pounds in calendar year 1988
- 25,000 pounds in calendar year 1989 and subsequent years

A chemical is "manufactured" if it is produced, prepared, or imported. The definition of manufacture also includes toxic chemicals that are produced as by-products. The term "process" means the preparation of a listed toxic chemical, after its manufacture, for distribution. Any listed chemicals that do not fall into one of the two definitions above are considered "otherwise used." If a listed toxic chemical is "otherwise used" in quantities greater than 10,000 pounds, it must be reported on Form R. When making threshold determinations, the three activities defined above must be compared separately to the appropriate thresholds. More specific definitions of the three activities are found at 40 CFR Part 372.3.

Owners/operators can make threshold determinations based on a mass balance, using purchasing records and year-end chemical inventory levels. This approach is allowable if chemical use records are not readily available. The previous calendar year's inventory plus the amount purchased during the year minus the year-end inventory will give the amount of a particular chemical used during the reporting year. For example, a facility's December 1986 inventory contained 5,000 lb of acetone, a listed toxic chemical. The total amount of acetone purchased during 1987, derived from purchasing records, was 12,000 lb. The December 1987 inventory was 6,000 pounds. The total amount of acetone used in 1987 was $5,000 + 12,000 - 6,000 = 11,000$ pounds. Thus, the facility used more than the 10,000-pound threshold and is required to submit a Form R for acetone.

Listed toxic chemicals that are contained in trade name products or mixtures must also be factored into both the threshold determinations and Form R. Amounts of toxic chemicals in mixtures or trade name products should be calculated using the best available information. MSDSs are good sources of information for this evaluation. If only the maximum concentration of a chemical is known, that maximum must be used when calculating the total amount of the chemical. It is important to note that if a listed toxic chemical is present in a mixture at or below 1% by weight (0.1% for carcinogens), or if the specific concentrations or a maximum concentration of the chemical is not known, it does not have to be factored into the threshold determination or reported on Form R. There are specific threshold determination procedures for metal-containing compounds, compounds containing more than one metal, and recycled or reused toxic chemicals. These procedures may be found in the instructions for completing Form R. Additional information is provided in the EPA guidance manual entitled "Estimating Releases and Waste Treatment Efficiencies for the Toxic Chemical Release Inventory Form" (EPA 560/4-88-002).

Once threshold determinations have been made for the listed toxic chemicals used at the facility, a Form R must be completed for each chemical exceeding the threshold. In Part III of Form R, the owner/operator must calculate releases of each toxic chemical to the environment from various sources of emissions. These calculations must be as accurate as possible. However, owners/operators of covered facilities are not required to perform monitoring or measurements other than what has already been performed or is otherwise required by law or regulation. Emissions may be calculated using the following methods:

1. based on monitoring data
2. based on mass balance calculations, such as the difference between the amount of a chemical that enters a process and the amount that leaves
3. based on published emission factors
4. based on best engineering judgement

The method used must be indicated for each release estimate in Part III of Form R. Part III also requires information on transfers of the reported chemical to off-site locations, and waste treatment methods and their efficiency.

Section 313 provides a specific exemption when determining thresholds or estimating releases for any listed toxic chemical contained in an "article." An "article" is a manufactured item that does not release a toxic chemical under normal conditions of processing or use. The term "article" is more completely defined at 40 CFR Part 372.3. It is important to note that if a release of a toxic chemical occurs during processing or use of an item, it is not considered an article. The appropriateness of this exemption should be carefully considered.

Section 313 also provides exemptions for listed toxic chemicals that are structural components of the facility; used in routine maintenance and janitorial operations; personal use products; used for vehicle maintenance; or used, processed, or manufactured in laboratories. The exemption for laboratory chemicals does not include specialty chemical production or pilot plant studies. These exemptions are more thoroughly described at 40 CFR 372.38.

Section 313 also sets forth provisions for maintaining trade secrets. Owners/operators who wish to claim a chemical as a trade secret must submit two versions of Form R to EPA. One version specifically identifies the chemical. This version will be kept by EPA and will not be available to the public. The second or "sanitized" version will not identify the chemical specifically, but gives it a generic name.

Some important distinctions between Section 312 Tier I reporting and Section 313 emissions reporting should be made. First, Tier I information concerns *chemical inventory* during the previous calendar year. Form R, on the other hand, is concerned with *chemical use and the associated releases* during operations of the previous calendar year. Second, there are two different sets of chemicals employed in determining thresholds. Section 312 reporting concerns OSHA hazardous chemicals and that subset of hazardous chemicals identified as extremely hazardous substances. Section 313, on the other hand, is concerned only with that subset of hazardous chemicals identified as toxic chemicals. Finally, while the Tier I reporting requirement applies to any facility that stores or uses an OSHA hazardous chemical, a Form R must be submitted only by facilities in SIC codes 20–39 with at least 10 full-time employees that manufacture, process, or use toxic chemicals at levels above regulatory thresholds.

This chapter by no means describes every aspect of the Emergency Planning and Community Right-To-Know Act of 1986. It does, however, provide a discussion of the basic framework of Title III. It is hoped that the information presented in this chapter clarifies what can be a confusing and complex set of regulations. A more detailed discussion of Title III is found in *Lowrys' Handbook of Right-to-Know and Emergency Planning,* available from Lewis Publishers (800/525-7894).

CHAPTER 21

Documentation

The importance of a formal, well-organized environmental management program cannot be overstressed. It is no longer sufficient that a manufacturing firm merely manage its wastes properly. It is now necessary to document compliance. The purpose of this chapter is to provide a better understanding of the need for and benefits of the documentation portion of an environmental management program.

A corporate environmental management program necessarily implies documentation. Documentation includes preparation of plans; implementation of management systems and programs and waste-handling standard operating procedures; keeping good records and compiling information in useful and reportable formats; documenting decisions, determinations, and actions; and filing reports. Sections I and II outlined suggested and required documentation necessary for a good environmental management program.

These suggestions and requirements should be used by environmental management personnel at the firm to begin implementation of the program, or to modify any existing procedures or programs. Of course, each manufacturing firm is unique in its production processes, layout and physical plant, organization of functions among employees, and other factors. The suggestions and examples given can and should be modified to meet the unique circumstances of each firm. However, the minimum regulatory requirements for each regulatory category must be met. Even so, waste minimization activities may make it possible to alter the applicable regulatory category.

Broadly speaking, the documentation associated with an environmental management program should not be viewed as unnecessary "makework." It is useful to look at the documentation associated with a good program as a risk management

261

measure. In other words, by preparing plans and implementing procedures, by documenting the decisions, determinations, and actions of the firm with respect to its hazardous waste management activities, and by collecting appropriate data, the risks and liabilities inherent in such activities can be effectively managed. Efforts to reduce such risks can be developed and implemented upon this necessary foundation.

A formal, well-organized environmental management program will certainly make it easier for regulatory agencies to ascertain compliance. Violation letters or compliance orders, and the potential fines associated with noncompliance, can be avoided. The reporting aspects of such a program can limit liabilities under RCRA (as amended) and CERCLA (as amended). Documentation requirements can encourage better decisions and make employees more conscious of how they handle waste materials.

There are other advantages to an environmental management program. A program that uses the guidance and suggestions found in Chapter 18 can actually reduce waste management costs, or at least mitigate increases resulting from treatment/disposal price hikes. Further, a "house in order" is a prerequisite to doing business with the more reputable commercial hazardous waste management firms. Disposal of "unknowns" is at best more expensive and difficult than disposal of segregated hazardous wastes of known composition, and at worst is impossible. Finally, insurance companies will base the premiums (and availability) of comprehensive general liability and environmental impairment liability insurance at least partially on the quality of environmental management practices at the manufacturing facility under consideration.

The hazardous waste management regulations seem to be all paperwork to waste management personnel at manufacturing firms. There are waste analysis plans, contingency plans, operating records, inspection logs, personnel training records, closure plans, manifest procedures, and of course, recordkeeping and reporting. That there is a substantial amount of paperwork associated with compliance is beyond dispute. The reasoning behind the paperwork is less obvious.

Many firms are willing, even anxious, to manage their industrial wastes in a proper manner. They have complied with the intent and spirit of the law and some of the substantive aspects of the regulations, i.e., analysis of wastes, notification and permit application, inspections of the storage area, proper packaging, transportation under manifest, and treatment/disposal at a facility under interim status. However, some of these companies have not produced the correct types and amounts of documentation associated with compliance. Although they are on the right track, they have yet to comply with both the spirit and letter of the requirements.

The rationale behind the procedural requirements listed earlier is to document that some thought has gone into the sampling and analysis of wastes, the necessary reactions to emergency situations, the content and frequency of inspection, the content and extent of training, etc. The procedural requirements also document that certain activities have been carried out or followed through.

The inspector has no recourse other than to rely on such documentation in the evaluation of the extent of compliance with certain of the regulatory requirements. Indeed, a review of the documentation (including records) is the only way an evaluation can be made for some of the requirements.

In summary, the documentation required by regulation is indeed necessary and vital to the manufacturing firm interested in developing and implementing a comprehensive environmental management program. Instead of viewing such requirements as busywork, they can perhaps be more profitably perceived as opportunities to rethink old methods of waste management, to reduce risks and liabilities, to plan for emergencies, to develop written standard operating procedures, to eliminate substandard practices, and to pursue new directions in industrial processes and waste management methods that will help the firm achieve compliance in a least-cost fashion. Viewed in this manner, these "paperwork" requirements are not nearly as burdensome. Indeed, they become part and parcel of enlightened industrial management.

SECTION IV

Selected Considerations in Implementing the Environmental Management Program

Onsite accumulation or storage of hazardous wastes before offsite treatment or disposal is largely a managerial exercise. As a result, there are many aspects of industrial waste management that cannot be described in regulations. An understanding of some of these aspects usually is gained only through experience. The purpose of this section is to provide the reader with some guidance in these areas. It is hoped that this guidance will prove useful to environmental managers, and will help them avoid some of the more common problems and pitfalls.

CHAPTER 22

Selecting an Offsite Treatment or Disposal Facility

Finding and selecting a treatment or disposal facility that (1) is competently managed, (2) is in compliance with the applicable regulatory standards, (3) poses manageable and acceptable risks to human health and the environment, and (4) possesses the permits, equipment, and expertise necessary to handle a particular waste is crucial to corporate compliance and limitation of liability. However, the regulations do not provide much guidance to that end, nor will regulatory agencies typically recommend any one firm over others.

Matters are further complicated by the fact that many commercial hazardous waste management facilities offer only one waste management technology (e.g., secure land disposal, chemical treatment, solvent reclamation, or incineration), thus limiting the types of wastes they will accept. This poses a problem when there are several types of wastes at a manufacturing facility, each requiring different waste management methods or a different sequence of methods. Additionally, some wastes can be managed properly by more than one method. Finally, some commercial hazardous waste management facilities will accept all types of hazardous wastes, but can perform only one (or a few) waste management operation(s) onsite. The wastes they cannot handle onsite will be "brokered" to another commercial hazardous waste firm. The facilities initially accepting the waste can legally accept such wastes and sign and return a copy of the manifest to the generator. However, the return of the manifest to the generator does not, in this case, signify that the material has been managed as promised (e.g., by land disposal, incineration, or recycling). The return of the manifest may only mean that the wastes have been accepted for storage. The wastes represent a potential liability should the TSD broker accepting the waste go out of business, or should the wastes be brokered to a substandard treatment/disposal facility.

The purpose of this chapter is to provide some guidance in the choice of whether to do business with a particular commercial hazardous waste management firm. It is assumed that there has been a decision as to which method(s) are appropriate for the waste stream in question. The decision between waste management technologies is waste-specific and largely technical. However, the decision to deal with a particular commercial TSD facility is only partially based on the equipment and techniques or methods available there. The nontechnical elements of selection are to be discussed shortly.

One further preliminary warning is necessary. There are commercial TSD facilities under interim status that have little chance of meeting the standards necessary to obtain a full RCRA permit. These facilities will close or change hands shortly before termination of interim status by either statutory deadline or after Part B of their RCRA permit application is requested by the EPA or the state. It is necessary to keep this in mind when dealing with commercial TSD facilities, and to be especially cautious until all such facilities have full RCRA permits.

The key component of the hazardous waste management system is the responsibility of the generator for the disposition and environmental impact of his hazardous wastes. Given this responsibility, it is imperative to do business only with commercial TSD facilities that will minimize the risk of future liability claims on the generator. The legal liabilities and statutory and common law responsibilities of manufacturing firms as hazardous waste generators are discussed in Chapter 26.

There are several areas by which the "quality" of a particular commercial hazardous waste management firm can be evaluated. These areas are permits, physical plant, management and personnel, housekeeping and operations, insurance, compliance status, environmental impact, and community relations. In addition, the issue of assessing potential liabilities in the event of premature facility closure should be addressed.

PERMITS

This element includes appropriate federal and state hazardous waste permits (or interim status) as well as other environmental permits for discharges to waterways, sewers, or air. Permits are a necessary condition of doing business with a commercial facility, but in themselves they are not a sufficient condition.

There are many questions to be asked, and items to be verified. Is the facility in possession of a valid EPA identification number? Ask to see a copy of the "Acknowledgement of Notification of Hazardous Waste Activity" (EPA Form 8700-12A) sent to the facility by the EPA. Has the facility filed Part A of the federal consolidated permit application (EPA Forms 3510-1 and 3510-3)? Ask to see a copy and verify that the application includes both the specific type(s) of wastes you wish to be managed at the facility and the process code(s) for the waste management method(s) you desire to use. Most commercial TSD facilities have been required to submit Part B of the RCRA permit application. Determine the status of the application in the permit process. Has the application been deemed complete by the regulatory agency? Has a draft permit or Notice of Denial been issued?

If the facility involves a wastewater discharge, ask to see a copy of its NPDES permit (for discharge to waterways) or municipal pretreatment permit or letter of approval (for discharge to municipal sewer systems).

Inquire as to possession of all necessary state permits (e.g., state sanitary landfill permit or state hazardous waste management facility permit). Review such permits for design life (important for landfills), operating capacity, and any permit conditions or restrictions on capacity that could affect the ability of the facility to promptly process your wastes. In addition, verify which wastes can and cannot be handled at the facility. Of importance are the specific restrictions (e.g., no wastes with free liquids, no chlorinated wastes, or concentration limits on cyanides or heavy metal wastes).

Get copies of all permits, including any air pollution and wastewater discharge permits. Does each waste stream or waste shipment need prior written approval (a "permit letter" or "approval letter") from the state regulatory agency?

PHYSICAL PLANT

Tour the facility. Look at and verify the existence and working condition of any processing equipment described in promotional literature. Observe the general housekeeping. Determine the types and amounts of security, emergency, and spill control equipment. Are emergency instructions posted? Are there sufficient alarms and communication equipment?

Look at waste storage areas. Is all drum storage conducted on concrete pads with curbs and collection sumps? Ask what is done with the spills, leaks, and precipitation collected in the sump. Is there impermeable secondary containment around all storage tanks? Look for nearby spill control equipment at the areas where drums are off-loaded, where drums are pumped, and where any drums are staged for shipment offsite.

Landfills and facilities with surface impoundments require special attention. Does the facility meet the HSWA minimum technology standards? For landfills this means dual synthetic liners on a compacted clay base. There should be a leak detection system between the liners and a functional leachate collection and removal system above the upper liner. What is done with the collected leachate? Does the facility have the required groundwater monitoring system? Has the facility submitted the monitoring results to the appropriate regulatory agencies? What is the disposition of stormwater runon and runoff? Does the discharge from any stormwater holding pond have an NPDES permit?

For landfills it is necessary to inquire as to the disposal method. Are hazardous wastes commingled, or are different cells or subcells utilized for different classes of hazardous wastes (i.e., organics, metals, acids, etc.)?

Is there a laboratory onsite? A modern, well-equipped laboratory is a necessity for hazardous waste treatment facilities. Hazardous waste landfills should have sufficient laboratory capabilities to "fingerprint" wastes (i.e., to verify by a few tests that the wastes received are the same type as they agreed to accept). Properly staffed laboratories at treatment facilities should not only have "fingerprinting" capabilities, but should be able to completely analyze the wastes, the treated effluent, and any residues and sludges from treatment. Inquire as to the existence of quality control programs for the laboratory procedures and equipment.

MANAGEMENT AND PERSONNEL

It is necessary to evaluate the qualifications of facility personnel and their ability to properly operate a business in the complex and highly regulated field of hazardous waste management. Credentials (university degrees) are nice, but training and experience in hazardous waste management or related fields (chemical manufacture, wastewater treatment) must also be considered. An intimate knowledge of regulatory requirements is a must. Inquire as to the quality of the training program for employees, and its regularity. Does training include emergency control and spill response training? Is there a formal safety program? Are workers provided with safety equipment? Do employees follow safe procedures and actually wear the protective clothing and equipment provided? OSHA has regulations specific to hazardous waste facilities and emergency response operations. Determine the facility's compliance with 29 CFR 1910.120 ("Hazardous Waste Operations and Emergency Response").

Observe the familiarity of key personnel with the actual day-to-day operations of the facility. Are they effectively overseeing operations, or merely letting things slide? Is the facility adequately staffed, at both the operations and managerial levels? The elements of this category are not easily evaluated, but the quality of the firm in terms of management and personnel is one of the more important factors in the choice of which firm to deal with.

HOUSEKEEPING AND OPERATIONS

This category is relatively easy to observe and evaluate. Although it applies primarily to treatment, recycling, and incineration operations, the suggestions and questions below have some applicability to landfills. Observe the general level of housekeeping, especially around off-loading and waste transfer (pumping) areas. Watch the workers' waste handling practices. Are the workers wearing protective clothing and equipment? Do they appear cautious? Are they causing sloppy conditions? Are small spills promptly cleaned up? Are there written material handling and spill prevention and control procedures?

Although the larger, well-run treatment and recycling operations use bulk (tank) storage and transportation almost exclusively, there will likely be some drum storage. Observe the conditions of drum storage. Are the drums on a concrete pad? Are they stacked more than two high and two wide for a row? Do they appear excessively aged (i.e., are the drums weathered, rusty, leaking, or damaged; have the labels and markings faded or fallen off)? Does there seem to be an excessive inventory for facility operations? Does the inventory consist of wastes that cannot be handled onsite (either "brokered" wastes or wastes generated by the facility itself)? To which facilities do "brokered" wastes go? How are wastes generated by facility operations (process residues, sludges, cooling water, wastewaters) managed, and where? Does the inventory consist of "product" (reclaimed solvents or oils) awaiting sale? Does the firm have a steady market for any reclaimed products?

INSURANCE/FINANCIAL RESPONSIBILITY

Under the broad heading of insurance come the considerations of the financial stability of the firm, and its ability to pay for the responsibilities associated with a hazardous waste management facility. These responsibilities include:

1. proper closure of the facility at any time
2. postclosure monitoring and maintenance (for disposal sites)
3. the ability to compensate injured parties for claims arising from facility operations that accidentally injure persons or property

All facilities must have compensatory capabilities with respect to sudden and accidental occurrences (fires, explosions, or releases). Surface impoundments and disposal facilities must, in addition, have such capabilities for gradual and accidental occurrences (such as groundwater contamination).

A firm that has the financial strength and the financial instruments (such as trust funds and insurance policies) to meet the costs associated with the closure and accidents will protect the generators using the site from claims against the commercial TSD facility. Compliance with RCRA regulations does not eliminate CERCLA liabilities for either generators or the commercial TSD facilities they select. Remember that generators are ultimately and perpetually responsible for the disposition and environmental impact of their hazardous wastes. Evaluate whether the HWM facility is sufficiently stable to continue operations over the long term. The legal liabilities of manufacturing facilities as generators and users of offsite TSD facilities are discussed in Chapter 27.

The federal regulations (40 CFR 265 Subpart H) require owners or operators of hazardous waste management facilities to estimate the cost of closure (and post-closure care, if applicable). Owners or operators of such facilities are to establish financial responsibility for these costs through:

1. trust fund
2. letter of credit
3. surety bond guaranteeing payment into a trust fund
4. financial test
5. insurance policy providing funds for closure and postclosure care
6. corporate guarantee

The federal regulations (40 CFR 265 Subpart H) also require hazardous waste facility owners or operators to demonstrate liability coverage for bodily injury and property damage to third parties resulting from facility operations. For sudden accidental occurrences resulting from facility operations, liability insurance in minimum amounts of $1 million per occurrence and $2 million annual aggregate (exclusive of legal defense costs) is required. A certificate of insurance or other evidence of coverage for sudden accidental occurrences must be submitted to EPA and/or the relevant state. It must be noted that standard comprehensive general liability policies have exclusions for pollution events, and therefore cannot be used to meet RCRA financial responsibility requirements.

Surface impoundments and disposal facilities are, in addition, required to have coverage for gradual accidents. For gradual accidental occurrences, coverage for claims is required in the amounts of $3 million per occurrence, $6 million annual aggregate, again exclusive of legal defense costs. A company may demonstrate compliance with the liability requirements by obtaining the required insurance policy(ies) or by meeting the specified financial test, or using the corporate guarantee, or some combination of insurance and either the financial test or corporate guarantee.

This brief outline of the federal financial responsibility requirements makes possible a knowledgeable review of commercial TSD documentation. It is important to determine whether and how the facility has provided for closure (and postclosure, if applicable) financial burdens. Find out which of the mechanisms have been used for this requirement.

With regard to the liability insurance policy, it will be necessary to get a copy of the facility's certificate of insurance for sudden occurrences. Verify that coverage with the issuing insurance firm. Determine the amount of the deductible for the policy. For landfills or surface impoundments, inquire as to the existence of coverage for gradual occurrences. Most existing insurance policies for environmental impairment do not distinguish between sudden and nonsudden events, in that both types of occurrences are covered.

An evaluation of whether the commercial facility is sufficiently stable financially is also in order. The goal is to assess whether the firm is capable of continuing operations over the long term. This involves their credit rating, and access to capital for necessary and desirable improvements. A Dun & Bradstreet report may be of some value in this assessment, as can be annual reports for publicly held companies.

COMPLIANCE STATUS

The evaluation of a commercial facility's compliance status begins at the facility itself, and ends with a verification of those findings with the appropriate federal and state regulatory officials. It is necessary to review the facility's compliance programs and documentation.

While on a tour of the facility, ask to see a copy of the facility's contingency plan. Does it cover the types of emergencies most likely to occur at the facility? Ask the whereabouts of the primary emergency coordinator listed in the contingency plan.

Ask to see a copy of the facility's past inspection reports by federal or state hazardous waste facility inspectors. Review the inspection form and the accompanying letter (sometimes termed a violation letter or compliance order) for major deficiencies noted by the inspector. Ask to see the facility's written response to the inspection, which should note the actions the facility took to come into compliance. Verify that such actions were actually performed.

At this point it is necessary to review the history of the site, first with facility personnel and then with regulatory officials. Are there any "notice of violation" letters or compliance orders for which a written response and appropriate remedial actions have not been performed? Are there any enforcement actions pending against the facility? Has the facility ever been taken to court to obtain compliance with environmental regulations? Is the site operating under an agreed order or consent decree? Is the facility in environmental receivership? Are there any court-issued orders relating to a lack of progress in meeting the terms of compliance orders, agreed orders, or consent decrees?

After completion of the facility tour, it is necessary to verify your findings and the completeness of any statements by facility personnel. This is best done by a visit to the state regulatory agency's office. Make an appointment with the field inspector for the facility in question. Arrive early and examine the agency's file(s) on the facility. Remember that the facility may have an NPDES (water pollution) permit also. After reviewing the files, meet with the facility inspector. Verify that the facility is in possession of or has applied for the necessary permits. Explore any questions or problem areas that were raised as a result of your facility tour. Review the compliance status and history of the site, as told you by the facility personnel.

Other items need to be verified through conversation with the inspector or through review of the files. Is the facility "brokering" any unprocessed wastes from generators to another TSD? Where are "brokered" wastes sent? Do these facilities have the necessary permits? What is the disposition of any leachates, cooling water, or process waters? Do any discharges need permits? Are there restrictions on the types of wastes the facility may accept? What types of wastes is the facility permitted to accept? Do waste streams or waste shipments need any approval from the state regulatory agency? Has the facility filed evidence of compliance with state financial responsibility requirements for closure and post-closure, and sudden and gradual liability insurance? If the facility is a landfill or a surface impoundment, has it submitted the required groundwater monitoring results? Do these results indicate any groundwater contamination?

The final subject to broach with the state inspector is one for which you may not get a direct answer. Inspectors are generally instructed not to recommend facilities. However, ask their opinion of the facility. Is it well run compared to similar facilities? What is its reputation with the various state environmental program departments? Do not expect a recommendation to use any one facility over others, or any glittering evaluations. Read between the lines, remembering that it is an inspector's job to find faults and ensure that the facility meets the regulatory standards.

ENVIRONMENTAL IMPACT

It is very possible for a commercial TSD facility to be in possession of all necessary federal, state, and local permits, while at the same time causing adverse environmental impacts. Pre-RCRA waste handling and disposal practices have resulted in soil and groundwater contamination at many RCRA TSD facilities. Such contamination often results from releases from disposal cells (termed solid waste management units [SWMUs]) abandoned or closed prior to the effective date of RCRA, and thus never subject to RCRA.

EPA has the statutory and regulatory authority to require owners/operators of TSD facilities to investigate for possible contamination and then take any necessary corrective actions for releases from existing, closed, or other SWMUs on their property. As a result, it is necessary to evaluate the facility's impact on air, groundwater, and surface water resources. Review the possibility of direct contact with wastes, as well as fire and explosion. A risk assessment scenario can be utilized, with identification/evaluation of sources, pathways, and receptors. Determine whether the facility has been issued any corrective action orders by the EPA or state regulatory agency.

Current operations should be evaluated, as well as any pre-RCRA disposal activities that may cause adverse environmental impacts. Both the probability of releases and the severity and magnitude of the effects should be taken into account. Potential receptors should be identified, and ranked in terms of potential adverse effects. It is important to remember that CERCLA and common law liabilities still apply to facilities which have achieved compliance with RCRA requirements.

COMMUNITY RELATIONS

This element is often difficult to assess, unless the facility has an adversarial relationship with its neighbors and the community. Responding to the fears, concerns, and complaints of ''neighbors,'' however far away, is an important aspect of maintaining good community relations.

Once relations with the community deteriorate, it is often impossible to restore trust and confidence. Disgruntled neighbors, effectively organized, can disrupt operations and attract media and regulatory scrutiny to the facility. Frequently, organizations are established with the sole purpose of closing an existing facility or preventing a new facility from being built. Local politicians, once summoned to the cause, can create many legal roadblocks to continued operations.

Facility representatives should be questioned as to their relationship with neighbors, any neighborhood associations, and the local community. It is important that they be sensitive to the special concerns of such groups. The facility should be making efforts to keep on good terms with such groups.

RELATIVE CONTRIBUTION OR "OTHER DEEP POCKETS"

This aspect of the facility evaluation has pessimistic overtones. A scenario is posed involving several assumptions. First, it is assumed that the TSD facility will be closed, either through permit revocation or denial or as a result of significant groundwater contamination. It is also assumed that the resources of the facility owner/operator, and any financial instruments such as insurance policies and trust funds, are insufficient to fund corrective actions and to compensate injured third parties. Finally, it is assumed that the facility poses a significant level of risk, triggering federal enforcement actions under CERCLA, or similar state actions under state cleanup statutes.

Here it is necessary to estimate how deeply the generator might be involved in the resulting governmental enforcement action, including the relative share of cleanup costs.

A conscious attempt should be made to find out who the customers of the facility are, and their financial strength and relative contribution to the amounts of waste being processed. CERCLA cleanup costs are typically assessed by EPA on the basis of the relative volume of wastes shipped by a generator to the site. In other words, if a generator shipped 10% of all wastes present at a Superfund site, then the generator would be responsible for 10% of the total cleanup costs.

Potentially responsible parties always find it better to be a small fish in a big pond, or to be a small fish in with much larger fish, when it comes to CERCLA liabilities.

TSD FACILITY REQUIREMENTS OF GENERATORS

It is important to note that well-run firms will be as selective in dealing with you as you should be with them. These commercial facilities will not accept any and all hazardous wastes, as doing so could put them in violation of their permit as well as potentially causing immediate (fires, explosions) or future (groundwater contamination) problems at their facility. Well-run commercial firms will require you to fill out forms giving detailed information about the waste and the processes that generated it. They will require analyses and samples of the wastes, and certifications by the generator regarding the accuracy of the information and the nature of the wastes shipped to the site. Keep a file of your dealings with a particular firm, including the results of the facility tour.

TSD FACILITY REEVALUATION

Once you begin doing business with a commercial firm, it is necessary to periodically reevaluate the facility and verify the actual disposition of the manufacturing firm's wastes. It is important to find out what actually happened to each shipment of waste to the facility. Was the waste immediately processed (treated, incinerated, or landfilled)? Has the waste been placed in storage (in the original container or in bulk [tank] storage)? Has the waste been brokered to another firm, or is it awaiting approval (from the state or from the other TSD) to be brokered? Determine the processing or "turnaround" time for your type of waste. This is the time the wastes sit in storage before treatment or disposal. In general, the shorter the processing time, the better the facility is run, and the smaller the generator's exposure to liability is from incidents that may occur during storage.

Many firms offer a certificate of disposal verifying the time and date that the waste was actually processed (treated, incinerated, or landfilled). This certificate of disposal is in addition to a signed manifest. The receipt of a signed manifest from the designated TSD facility really only signifies that the facility has received and accepts the wastes specified on the manifest. The receipt of a signed manifest does not imply that the designated facility actually processed the wastes as agreed on before shipment. Please note, however, that certificates of treatment or disposal do not relieve the generator of CERCLA liabilities.

BROKERS

Any discussion of the evaluation of commercial hazardous waste management vendors must include brokers and transporters. Brokers and transporters pose potential liabilities out of proportion to their involvement in offsite management of hazardous wastes. This is because brokers and transporters are typically less sophisticated than TSD facility personnel, are less capitalized than firms owning TSD facilities, and are underinsured or not insured for the risks they impose on generators.

Firms brokering hazardous wastes have established a niche in the commercial hazardous waste services market. This niche often is the result of a lack of sophistication on the part of the generator regarding waste management facilities that could accept his wastes directly. Often, brokers are involved in transportation, offering small quantity pickups or "mixed load" pickups, where different wastes are delivered to different facilities.

There are different types of brokers. As mentioned earlier, TSD facilities will broker to other TSD facilities those wastes that cannot be treated/disposed of with their equipment and technology. Other brokers own the transportation vehicles, and invoice the generator for both transportation and treatment/disposal services. Many brokers act as nothing more than independent salespersons for facilities and transporters. They subcontract both the transportation and treatment/disposal.

Tremendous potential liabilities are involved with the use of brokers, particularly with those brokers that own neither the transportation vehicles nor the TSD facility. First, brokers typically select the treatment/disposal (T/D) facility for the generator. The generator often has no knowledge prior to shipment of the particular disposal facility to be utilized. Often, disposal facilities change from shipment to shipment. Allowing a broker to select the T/D facility usually results in selection of a facility according to the broker's priorities and perspective. The broker does not share the same perspective as the generator, particularly with respect to potential long-term liabilities.

Brokers often select substandard facilities, because they offer the broker the lowest price, the easiest waste approval procedures, and most favorable credit terms. This allows the broker the maximum possible markup on costs to the generator. These facilities are often out-of-state, making it more difficult for the generator to conduct an independent evaluation. Distant facilities are often selected by the broker, even though suitable facilities are closer, because the broker has arrangements with the more distant facility. This results in higher transportation costs to the generator.

Because brokers that are not transporters or TSD facilities are not a regulated party under RCRA, there are tremendous opportunities and incentives to mismanage a generator's waste. Brokers typically are not potentially responsible parties in CERCLA 106/107 actions, as they are neither generators, transporters, nor owners/operators of TSD facilities. Brokers typically have no liability insurance that would cover such risks. Brokering operations have low capital requirements; in fact, a brokerage can be run with a leased office, business phone, typewriter, and automobile. There are no assets to be won, should the generator decide to sue.

Brokers involved in selecting the transporter and who bill for both transportation and disposal can pull some exceedingly nasty "tricks" on the generator. Selecting a substandard facility has already been mentioned. The broker can invoice the generator and be paid, yet neglect to pay the transporter and disposal facility. The broker can also direct the transporter to illegally dump the generator's waste, then sign the manifest for the designated TSD facility and mail the original copy back to the generator. The broker then invoices for both transportation and disposal, with disposal charges being pure profit. As indicated above, extreme caution has to be exercised when utilizing the services of a broker. Dealing directly with the TSD facility has some advantages in this regard.

Generators that choose to utilize the services of brokers should determine the ultimate disposition of their wastes, evaluating the facility or facilities involved as if they were dealing directly with them. After each shipment, the ultimate facility should be contacted to determine whether the waste actually arrived. The facility where treatment/disposal ultimately occurred may not be the designated facility on the generator's manifest, as some brokered wastes are shipped to other TSD facilities.

TRANSPORTER SELECTION

Selection of a transporter (often termed a carrier or a hauler) also requires some care. TSD facilities often offer transportation services, so evaluation of the facility can be performed jointly with the evaluation of the transporter. There are several criteria necessary to evaluation and selection of a hazardous waste hauler. These criteria include:

1. the condition of the transportation vehicles, including safety and personal protective equipment, tires, and permanent, flip-type placards on all four sides of the vehicle
2. the level of driver training with respect to hazardous materials, as well as the other aspects of driver qualifications. Other aspects are the driver's physical qualifications, including medical certifications, as well as the driver's record with respect to accidents and traffic violations.
3. the existence of adequate emergency response plans for traffic accidents or spills in transit or at transfer facilities
4. whether state hazardous or industrial waste hauling permits have been obtained, as necessary, for the states in which the transporter operates. Such permits can be required in addition to a federal EPA identification number.

5. the adequacy of existing insurance coverage for liability for bodily injury/property damage as well as for environmental restoration. A copy of form MCS-90 should be requested. Minimum coverage should be demonstrated in the amount of $5,000,000.
6. standard operating procedures to ensure secure loads. These procedures should address loading, blocking, and bracing. Appropriate types and amounts of equipment to secure containerized loads are necessary on each semitrailer or box van.
7. compliance with the Federal Motor Carrier Safety Regulations of the Department of Transportation, Federal Highway Administration (49 CFR Parts 390–399)

CONCLUSION

There are many elements to evaluate in selecting an offsite TSD facility to do business with. This chapter has only brought up some of the more relevant questions to ask and things to determine or evaluate. No weights have been assigned to these matters, and thus the decision to use a particular facility cannot be as simple as assigning each possible facility a ''score.'' Further, no mention has been made of the price of a commercial firm's services. To a great extent, the old adage that ''you get what you pay for'' is true in this decision. Waste management personnel at manufacturing firms will need to apply their own decision rules, assign their own priorities, and use good judgment and common sense in making the very important decision of which offsite firm(s) to use.

CHAPTER 23

Principles of Container Handling and Storage

Storage and handling of containerized hazardous waste in a safe and proper manner is not an intrinsically difficult task. The principles are fairly simple. Hazardous waste storage is often conducted improperly due to ignorance of the minimum regulatory requirements and a lack of forethought regarding how to perform storage. In the past, this lack of thought has resulted from the secondary nature of waste materials in a manufacturing operation. Wastes were primarily viewed as unwanted and inevitable by-products of production, to be disposed of with a minimal cost, time, and effort. The container storage principles to be discussed in this chapter include:

1. containment
2. protection from weather
3. storage area location
4. storage area arrangement
5. spill control
6. labeling and marking of containers
7. housekeeping
8. empty containers
9. storage levels

The regulations that came about partially as a result of past improper storage (and disposal) practices have been discussed earlier in this book. This chapter will only refer briefly to the standards as they apply to onsite accumulation (Chapter 9) and storage (Chapter 11). Appendix K provides a summary of the regulatory requirements for containerized hazardous wastes. The bulk of this chapter is an attempt to bring together these regulatory requirements with the common sense, basic chemical knowledge, consensus codes, and good practices involved in the safe and proper storage of hazardous wastes. At a minimum, this involves handling wastes with the same degree of care and consideration given to the input (virgin) chemicals used at the manufacturing facility. Of course, these virgin chemicals are often major constituents of manufacturing process wastes. With this in mind, please note that many of the same principles and concerns in waste storage also apply to the storage of containerized chemical products.

CONTAINMENT

The first principle of hazardous waste container storage is to place such containers on an impermeable base. This will make cleanup of any spills or leaks much easier and cheaper, since there should be no contaminated soil to dispose of, and the possibilities of soil and groundwater contamination are minimized. The storage base or pad is usually poured concrete. The pad should be constructed of a material that will not be dissolved or deteriorated by spills or leaks. Chemical-resistant coatings are often applied to minimize degradation, as well as to prevent permeation into the base material. There should be provisions for collecting any spills and leaks as well as for preventing uncontrolled drainage to soil or to plant or city sewers. The pad is often sloped to a collection sump for this purpose. Any accumulation of liquids in the sump should be promptly removed.

Ideally, the storage area should be curbed or diked to collect spills, leaks, and precipitation. A recessed floor or grated concrete drainage trench are sometimes utilized in lieu of sills or curbing, especially in areas of heavy forklift traffic. Any trench, curb, or dike should be "keyed" into or otherwise sealed to the floor so as to adequately contain liquids. Floors and joints should be liquid-tight. Cracks or expansion joints, if unsealed, provide a conduit for migration of releases to the subsurface. The containment mechanism should be sufficient to contain a volume of liquid equal to 10% of the volume of the maximum inventory of containers, or the volume of the largest container, whichever is greatest. Collection of precipitation or sprinkler/fire control water should also be taken into account when sizing containment sumps. The base should be sloped (1% minimum) to a collection area or sump, to minimize contact between liquids and containers. Separate collecting sumps may be necessary to prevent incompatible materials from mixing. If an existing drum accumulation or storage area is to be modified to meet these guidelines, it is suggested that the 40 CFR 264 Subpart I permit standards for storage in containers be consulted. Although the Part 264 permit standards do not apply to accumulation areas, they can serve as guidelines. At some time in the future, circumstances may make it necessary to obtain a permit for what is now an accumulation area.

PROTECTION FROM WEATHER

The second principle of storage is that the wastes should be protected from the elements. In other words, storage should be under a roof, either inside the manufacturing facility, in a separate building for the storage of input chemicals and/or wastes, or within a pole barn or three-sided shed. A few firms are marketing prefabricated, movable structures especially designed for secure storage of containerized chemicals. These structures typically have secondary containment, security, and fire protection features as integral parts of their design and manufacture.

The primary concern here is to minimize the possibility of contamination of surrounding areas from drainage of precipitation. There are other concerns involved. The movement of waste containers becomes nearly impossible when they are surrounded by mud or by snow and ice. Further, some wastes should not be exposed to direct sunlight (especially in black drums) or should be protected from freezing. Exposure to the elements will weather the waste container and the required markings and labels.

Remember, however, to provide adequate ventilation of any enclosed areas. Flammable liquid vapors are heavier than air, so ventilation measures should be within 12 inches of floor level to take this property into account. Mechanical ventilation for any indoor flammable liquids storage area should meet NFPA and/or OSHA standards with regard to ventilation rate. Electrical equipment, including wiring, blowers, and lighting systems, must be suitable for use in explosive atmospheres. OSHA regulations (29 CFR 1910 Subpart H), state fire codes, and NFPA standards should be consulted regarding requirements for flammable and/or hazardous material storage. Requirements for such items as sprinkler systems, mechanical ventilation, electrical wiring and equipment, aisle space, stacking limitations, and total quantity limitations should be followed.

STORAGE AREA LOCATION

The third principle deals with the location of the storage area(s) within the property controlled by the manufacturing facility. The primary constraints have to do with the safety of workers, equipment, and structures should there be a release, fire, or explosion, or should incompatible wastes come into contact with each other. Ignitable and reactive hazardous waste storage areas must be located at least 50 feet from the facility's property line (40 CFR 265.176). Ignitable and reactive wastes must be protected from processes or operations which could act as a source of ignition (e.g., cutting and welding, or operations involving molten metal). Flammable liquids should not be stored in basements or other inadequately ventilated low-lying areas, or near elevator shafts, stairways, or other pathways leading to lower areas where vapors can accumulate.

Storage areas where ignitable, reactive, or incompatible wastes are held should be readily accessible to internal emergency personnel (fire brigades or spill response teams) as well as to outside organizations (e.g., fire department or spill response contractors) that might be called on to provide emergency services. Storage areas for such wastes are often separated from plant operations. Sometimes specially constructed chemical storage rooms are necessary to protect workers and equipment. Security precautions should be taken to prevent unauthorized entry into the storage area. Often, areas are locked to prevent uncontrolled access by employees or trespassers.

Security precautions are necessary to reduce the potential for theft, sabotage, or accidental discharge. Designated storage areas should be prominently marked as to their purpose and the principal hazards of the chemicals stored therein. "Danger—Unauthorized Personnel Keep Out" signs are required for interim status or fully permitted hazardous waste storage areas. Such signs are advisable for other designated storage areas. Warning signs should be posted at all entrances to chemical storage areas.

Changing the physical location of waste accumulation or storage areas within the facility's premises may have insurance consequences, so consult with the insurance carrier and its safety experts. Storage area locations are often regulated by state and local fire codes, building codes, or zoning ordinances. Exterior storage areas are often required to be located a minimum distance from buildings, property lines, streets, alleys, public ways, or exits to a public way. NFPA codes are sometimes applied by these regulatory agencies as well as by insurance carriers. Inside storage areas for flammable liquids may require automatic sprinkler and mechanical ventilation systems, as well as explosion venting meeting certain specifications.

The storage area location must also be accessible by the type of transportation vehicle utilized to remove the waste material offsite. A nearby loading dock makes loading container shipments much easier than using a lift gate or ramp. If wastes are removed in bulk (i.e., container contents are pumped into a tank truck), the distance between the storage area and a road surface capable of supporting a tank truck becomes important. Remember the regulatory constraint that containers of ignitable or reactive hazardous wastes cannot be stored closer than 50 feet to the property line (265.176).

STORAGE AREA ARRANGEMENT

The fourth principle has to do with the arrangement of the waste storage area itself to provide accessibility. Accessibility is required for various purposes. Since many offsite TSD facilities can accept only one type of waste (e.g., solvents, oils, or heavy metal sludges), drums of each waste type at the manufacturing facility must be easily identified and accessed for removal to the offsite TSD facility. In addition, the drums must be inspected weekly for leaks or corrosion (265.174), and adequate aisle space must be maintained for unobstructed movement of personnel and equipment (265.35). Aisle space is necessary for inspection of containers and to permit access for emergency response purposes.

These two regulatory requirements must be met while at the same time ensuring the segregation of wastes with incompatible chemical properties. Chemicals can be segregated by waste type or hazard classification. For example, flammable and combustible liquids should be separated from corrosives and oxidizers, as well as other sources of ignition. Such sources include cigarette or cigar smoking; welding, grinding, and cutting operations; unprotected electrical equipment; open flames; hot surfaces; and static electricity; as well as oxidizing chemicals. It is critical to separate acids from cyanide- and sulfide-containing wastes. Strong acids should be separated from strong bases (also called caustics or alkalies) in the corrosive category. Poisons should be segregated from other chemicals, including flammable and combustible liquids.

Incompatible drummed wastes are to be separated by means of a dike, berm, wall, or other device (265.177). Physical separation of containers of incompatible wastes by an appropriate distance (not less than 20 feet) is often utilized to comply with this requirement. Drainage from areas where incompatible materials are stored should be separated, in separate sumps, for example. Table 23.1 identifies common combinations of incompatible hazardous wastes, and incompatible hazardous wastes and materials.

Table 23.1. Common Combinations of Incompatible Hazardous Wastes, and Incompatible Hazardous Wastes and Materials

1. Acids and cyanides

2. Flammable or combustible materials and oxidizers

3. Strong acids and strong alkalies

4. Acids and water

5. Solvents and corrosives (acids and alkalies)

6. Flammable liquids and ignition sources

7. Strong corrosives (acids and alkalies) and aluminum, magnesium, and zinc alloys

When a manufacturing facility has two or more waste streams that are incompatible, the drums should be very clearly marked, perhaps even color-coded. Employees should be informed of the potential dangers of inadvertent mixing. Additionally, the specific areas within the storage facility where such wastes are to be stored should be clearly designated, by signs and/or markings on the floor. Appendix V of 40 CFR Part 265 provides examples of waste categories that are incompatible, as well as the potential consequences should such wastes come in contact with each other. Appendix V of 40 CFR Part 265 is reprinted in Appendix G of this book. Further information is found in the EPA publication "A Method for Determining the Compatibility of Hazardous Wastes" (EPA-600/2-80-076). Please note that segregation of waste streams is typically advisable even for compatible waste materials. Offsite (commercial) treatment and disposal firms are very particular about the wastes they accept, and are required by regulation to inspect incoming loads to ensure that the waste received is what they had agreed to accept. Unauthorized mixing of wastes can result in rejected loads, off-specification surcharges, and termination of future shipments by the commercial facility.

In determining the arrangement of any chemical product/waste storage area, the following suggestions and considerations are offered:

1. Containers of virgin chemicals should be stored in separate aisles from containers of waste chemicals.

2. Containers should be marked or labeled as to their contents and hazards. These markings or labels should be visible during storage and handling. Inconsistent or inaccurate markings should be removed from containers being reused for a different chemical.

3. Chemical properties and incompatibilities should be taken into account when determining which chemicals should be stored in adjacent aisles. For example, acids should not be stored next to caustics, nor should flammables be stored next to oxidizers.

4. Containers should be stored in rows no more than two containers wide, with three feet of aisle space between rows. When forklifts are utilized, main aisles of eight feet in width should be maintained. It is recommended that four feet of aisle space be maintained between rows of palletized containers. Rows could be designated by floor striping, signs posted on walls, and/or markings painted on floors designating the specific chemical to be stored in that row.

5. Chemical containers should not be stacked more than two containers high, unless rack storage is utilized or the containers are structurally able to support such loads. Ignitable wastes should not be stacked more than one container high.

6. Chemical containers should be stored closed, except when chemicals are added to or removed from them. Tight-head containers must have all bungs in place, tightly secured. Open-head containers must have the lid with gasket in place, secured by a lever ring or ring-and-bolt assembly. Keeping containers closed is a regulatory requirement for hazardous wastes and is generally considered good practice. It minimizes potential employee exposure hazards and fire hazards from volatilization of contents.

7. "NO SMOKING" signs should be posted in chemical storage areas. Other sources of ignition should be kept at least 25 feet away from the chemical storage area. Such sources include cutting, grinding, and welding; open flames; hot surfaces; frictional heat; static, electrical, and mechanical sparks; lightning; radiant heat; spontaneous ignition; and heat-producing chemical reactions. Internal combustion forklifts should not be parked in chemical storage areas.

   Trash, garbage, debris, and other combustible material not incident to storage should be kept out of chemical storage areas. A minimum distance of eight feet should be maintained between chemical containers and ordinary combustibles, such as stacks of pallets.

8. Chemical storage areas should be adequately ventilated. Continuous mechanical exhaust ventilation of inside areas is recommended. Ventilation rates should not be less than 1 ft^3/min/ft^2 of floor area in the chemical storage area. OSHA specifies at least six air changes per hour for flammable liquids storage areas (29 CFR 1910.106(d)(4)(iv)). Exhaust ventilation should take into consideration the density of any potential gases or vapors. Flammable liquid vapors are typically heavier than air, so exhaust should be taken from a point within 12 inches of the floor.

9. Grounding and bonding techniques and devices sufficient to prevent static electrical charges are necessary where flammable liquids are poured into containers or transferred from container to container. (See OSHA regulations at 29 CFR 1910.106 and NFPA 77 "Recommended Practice on Static Electricity.")

10. Adequate types and amounts of emergency and spill control equipment and supplies should be readily available. Such equipment and supplies should include alarms or emergency signal devices, telephones, fire extinguishers, personal protective equipment, and absorbents. These should be readily available, but not so close to the containers that a fire or chemical release would restrict their accessibility or endanger those attempting to reach them. Decontamination equipment such as plumbed eyewash stations and deluge showers is recommended, especially when corrosive liquids are present. Emergency and spill control equipment and supplies should not be blocked by containers in storage or other equipment.

11. Chemical/waste storage areas should have a sufficient number of exits, strategically located to allow personnel to safely evacuate in case of emergency. Emergency exits should not be blocked by containers in storage or other equipment.

12. Chemical/waste storage areas should be included in emergency plans ("Contingency Plans") developed for dealing with chemical emergencies. Local emergency authorities should be made aware of the location of such

areas within the overall property or specific buildings. A single-page summary of emergency response procedures, including plant emergency coordinators and emergency telephone numbers, should be posted near chemical/waste storage areas.

Other considerations and practices related to container storage and management cannot be elevated to the level of first principles, but are provided here in the form of suggestions.

Spill Control

Because waste-handling operations are often accompanied by spills and leaks, it is important to have spill control equipment and supplies readily available. These supplies include absorbents, spill booms, and drain covers. Commonly used absorbents include calcined clay, sawdust, corncob meal, and lime. Some of the synthetic absorbent media are especially effective and easy to use. Because of land disposal restrictions on biodegradable absorbents, selection of spill absorbents should be made in conjunction with the selection of the probable treatment/disposal method and facility for the spill cleanup residues.

The selection of the type and amount of emergency and spill control equipment is highly dependent upon the physical/chemical properties and hazards of the chemicals/wastes in storage, as well as the total volume and number of containers. The degree of containment, potential pathways such as storm sewers, nearby waterways, etc., and the proximity of potential receptors (residential neighborhoods or manufacturing employees) must be considered. Also, the quality of existing emergency procedures and the presence and resources of emergency response organizations, both internal (fire brigade, spill response team) and external (fire department, public sector Hazardous Materials [Haz Mat] Teams, spill response contractors), also enters into selection of the types and amounts of on-site emergency and spill control equipment to be procured. The level of expertise and training of both internal and external spill responders also must be considered.

Some general recommendations can be made as to minimum acceptable types and quantities of emergency and spill control equipment for well-ventilated container storage areas where relatively small numbers (<30) of containers are held.

The following is a list of such items:

1. type ABC fire extinguishers with minimum ratings of 10A 40BC. The fire extinguishers could be mounted at opposite ends of the chemical storage area, preferably just outside the exits from the chemical storage area. It is recommended that extinguishers be located not less than 10 feet and no more than 50 feet from flammable liquid storage areas.
2. alarms, such as manual pull stations, hand-held compressed gas alarms, or telephones with intercom capabilities, to be mounted next to the fire extinguishers

3. spill control supplies, such as bags of clay absorbents or pillows, tubes, or pads of synthetic absorbents. Neutralization agents, such as lime, are indicated where containerized acids are stored.

4. shovels, brooms, squeegees, mops, etc., to clean up spill residues and saturated absorbents. Nonsparking tools are recommended where flammable liquid spills may occur.

5. empty containers to repackage leaking containers or to containerize spent absorbent materials. Both open-head and closed-head drums are necessary, preferably two drums of each type. Consideration should be given to purchasing 85-gallon overpack drums, into which can be placed an entire 55-gallon container.

6. personal protective clothing and equipment, such as chemical splash goggles or face shields; chemical-resistant gloves, aprons, and boots; and saran-coated tyvek coveralls. Respiratory protection needs in emergency situations should be evaluated by qualified personnel in accordance with a written Respiratory Protection Program per OSHA (29 CFR 1910.134).

7. DOT-specification 17-H open-head containers with a lever-ring type of closure, or an 85-gallon overpack or salvage drum, where spill control supplies could be kept in the storage area. Personal protective equipment, such as goggles and gloves, would be placed on top of absorbent materials. The drum could be painted yellow or red, and stenciled to indicate the contents. The drum could be placed on a drum dolly to maximize mobility. In case of a spill or leaking drum, the protective equipment is donned by the spill responder, and the absorbents are removed from the container and used to contain or absorb any liquids. The now-empty open-head container is available to containerize the saturated absorbents and any other spill residues, pending appropriate disposal arrangements.

Labeling and Marking of Containers

Container labeling problems, with both chemical products and wastes, are common in industry. Chemical container labels get damaged, deteriorate, or fall off. Chemicals are transferred from one container to another, and empty containers are often reused for different chemicals than they originally held. Chemical labeling requirements apply to both chemical products and wastes. A commitment to, and a system for, container labeling is critical to regulatory compliance and cost minimization.

The absence of labels or misleading labels poses safety/health hazards and regulatory compliance problems (OSHA and EPA), and can be expensive. Expenses can result from:

- good material going unused
- good material being inadvertently disposed of
- "bad" material being unknowingly substituted for good material
- laboratory analyses to establish the identity and quality of chemicals in unmarked containers
- laboratory analyses to determine hazardous waste characteristics and make disposal arrangements for materials in unmarked containers

- disposal charges that could have been avoided
- any fines or legal expenses resulting from enforcement actions

Unlabeled containers of hazardous chemicals are a violation of regulations issued under the OSHA Hazard Communication Standard (29 CFR 1910.1200). Unlabeled/unmarked containers of hazardous waste violate several hazardous waste regulatory requirements.

Use of some form of internal labeling system can be very helpful in complying with the regulations and keeping a handle on various types of wastes generated at the facility. Generators who accumulate must, of course, mark each drum of hazardous waste with the words "hazardous waste" and the accumulation start date. (See Figure 9.5.) Satellite accumulation containers can be marked either with the words "hazardous waste" or other words that identify the contents of the container. The accumulation start date is the date that the container was filled (for satellite accumulation areas) or the date the first drop of waste is placed in the container (for those containers in areas not qualifying as satellite accumulation areas). Although not strictly required prior to offsite shipment, other information could be marked or labeled on containers. Such information includes (1) the generator's name; (2) the waste's common name; (3) the hazard(s) posed by the waste; (4) the department, process, or machine generating the waste; and (5) a reminder to workers to keep the container closed when wastes are not being added. Labels are available that will identify drums containing nonhazardous wastes, as well as "empty" containers (Figure 23.1).

In-plant labeling, in conjunction with proper supervision, can minimize the use of hazardous waste containers for plant garbage. The presence of gloves, rags, trash, soda cans, etc. mixed with waste materials can result in off-specification charges from the treatment/disposal facility. Management supervision is also necessary to ensure effective segregation of different waste materials, as well as to prevent half-full drums from being shipped offsite. Shipping mixtures of waste streams to a treatment/disposal facility, or partially full drums to landfills, can result in rejection and return, or off-specification surcharges.

A system of color-coded drums could be used to supplement the internal labels. Each waste stream would be assigned a specific color drum. White or light-colored drums should be used for ignitable wastes or volatile (high vapor pressure) liquids. The use of light-colored drums for such wastes is important if wastes are stored outside. Placing ignitable or flammable wastes in black drums that are subsequently exposed to direct sunlight is not recommended. In order to avoid container deformation or rupture, flammable liquids packages must not be completely filled. Sufficient head space (outage) must be available to allow expansion of the contents, to ensure that the container will not be liquid full at 130°F. This translates into 2 inches of outage below the bung for DOT 17-E 55-gallon containers.

Figure 23.1. Labels to identify barrels containing nonhazardous and nonregulated waste, as well as empty containers.

Overfilled containers cannot be shipped offsite, per 49 CFR 173.116(b). Deformed or otherwise deteriorated containers cannot be shipped offsite. They must be repackaged into another DOT-specification container or overpacked.

One of the primary reasons for in-plant labeling and marking is to give the workers handling the waste information regarding the type of material they are handling, and its hazard classification. This information is necessary for them to be safety-conscious while handling the wastes. This will also give them basic information necessary to respond to a spill or fire.

The author contends that the labeling and marking necessary for in-plant purposes and offsite transportation is best done as the wastes are initially placed in containers, prior to being moved to accumulation areas or onsite storage facilities. This will prevent the container from becoming an "unknown" (especially if an internal labeling system is not used), requiring costly analysis to re-establish the identity of the contents. This practice will also allow the segregation of wastes and maintenance of an operating record (required for TSD facilities at 40 CFR 265.73). Please remember that many commercial TSD facilities require that each container be marked (typically on the lid) with a unique waste product number or code assigned by the TSD to the particular waste.

If the drums are stored outside a building, there can be problems with markings/labels fading, peeling, or washing off. This is not an insurmountable problem. Weather-resistant labels (usually vinyl) are available (Figure 9.5). Indelible markers or spray paint and stencils also work well.

Housekeeping

Housekeeping in and around chemical and waste storage areas is often neglected by manufacturing facility personnel. Housekeeping, in the author's opinion, is a legitimate area of concern for environmental managers. The level of cleanliness and orderliness accepted by the firm both inside and outside the plant sends a strong signal to both employees and regulatory agencies as to the importance management attaches to proper handling, storage, and disposal of all types of materials, including hazardous wastes. As a former state hazardous waste facility inspector, the author can speak to the use of the general level of housekeeping as an indicator of the degree of scrutiny required of the waste handling and disposal practices at any particular firm. Housekeeping problems include open and unmarked containers (both input chemical and waste containers), empty containers, discolored soils and chemical residues in and around storage areas, and garbage, trash, pallets, scrap metal, and equipment in and around chemical storage areas.

Many manufacturing operations have particularly difficult problems with partially full chemical containers. These drums are often scattered throughout the plant, as well as in outside areas. At the opposite extreme are the chemical storage rooms or buildings filled wall-to-wall with partially filled containers. The presence of multiple partially full drums of the same chemical is not uncommon. Problems arise when these containers are left open, and when labels are damaged or absent. Outside storage of chemical product containers can result in precipitation entering the container. This can contaminate the product inside, or result in contaminated rainwater. In either case, a disposal problem results.

Some housekeeping problems have both safety and regulatory compliance implications. Many hazardous chemicals, when spilled, become hazardous wastes. Damaged or deteriorated containers should be repackaged or overpacked before any release of contents. Any spills or releases that do occur should be promptly cleaned up. Labeling of containers of hazardous chemicals is required, whether the contents are usable material or hazardous wastes. Containers of hazardous chemicals must be labeled per the OSHA Hazard Communication Standard. Containers of hazardous waste must be labeled per RCRA regulations. Additional labeling may be necessary to satisfy DOT requirements once the container is ready for offsite shipment.

The potential problems posed by discolored soils also deserve mention. Many of these areas result from leakage or spillage of oil. Presently, oil to be recycled is not considered to be a hazardous waste, except under unusual circumstances. EPA had proposed regulations that would cause waste oil of all kinds to become listed hazardous waste. As a result, oil spills and leaks would become hazardous waste discharges, subject to reporting and cleanup requirements. In any case, many regulatory agencies have an informal "discolored soils policy," where they assume any areas of discolored soils resulted from hazardous waste releases. The generator is required to determine whether the discolored soil is a hazardous waste through sampling and analysis. The discolored soils are often required to be removed, regardless of whether they are a hazardous waste. It is often more cost-effective to remove areas of discolored soils and properly dispose of them in advance of an inspection.

Empty Containers

Manufacturing operations typically have an excessive inventory of "empty" chemical containers. Often, these empties (with or without labels) are scattered throughout the plant and outside areas and become a housekeeping problem. If checked, many of these containers will *not* be found to meet the EPA definition of "empty," and thus unregulated, container. These drums will typically contain either unused product or contaminated rainwater.

"Empty" containers, unless thoroughly rinsed, retain product residues and should be appropriately managed. Empties can pose safety (fire), environmental, and housekeeping problems if allowed to accumulate. They should be stored closed, with the bungs in place, or with the lid securely fastened. Empties stored outside can collect rainwater, which becomes contaminated with product residues. It is recommended that an appropriate area be designated for the temporary accumulation of empty containers. Some chemicals are sold by the vendor in deposit containers, with deposits ranging from $10 to $50. These should be promptly returned for credit when empty. Many chemicals, however, are sold in no-deposit containers, or are not accepted by the chemical supplier for reuse. Containers not to be reused internally can be sent to a drum reconditioner or scrap dealer, or can be deheaded and crushed for sanitary landfill disposal. The possibility of purchasing chemicals in deposit containers or requiring suppliers to accept empty containers for proper management could be explored. Bulk purchase and storage of high-volume chemicals, such as paints and oils, could be explored as a cost-cutting measure that would also reduce empty drum problems. Portable "Tote" tanks are becoming more common as an intermediate alternative between 55-gallon containers and fixed bulk storage tanks.

Many chemical suppliers and drum reconditioners are refusing to accept empty containers that have any residues remaining. A few require that such containers be rinsed prior to acceptance. Regulatory requirements on drum reconditioners are such that many are unwilling to purchase empties unless there are significant numbers or they have been cleaned. Disposal charges for empty containers in the $5-$15/drum range are not uncommon. Because of liability considerations, "empty" containers should not be given to employees or others for use as trash containers or burn barrels.

Some empty virgin chemical containers would be suitable for reuse as waste containers. This is particularly the case with empty steel drums. However, the residues in empty containers must be chemically compatible with the wastes to be stored, and all previous labeling must be removed. The containers should be promptly relabeled as appropriate. The empty drums should be visually inspected prior to use to ensure that they are in good condition, and are capable of being closed. Both bungs must be present, and the bung gaskets must be intact. Open-head drums should have ring-and-bolt closure assemblies and a lid gasket. Inspection procedures would include looking at the drum's bottom surface to see to which DOT specification the drum was constructed. DOT specifications 17-E, 17-H, and 17-C should be appropriate for many types of noncorrosive wastes. As a general rule, fixed-lid ("tight-head" or "bung type") DOT-specification containers should be used for solvents and oils, while removable lid ("open-head") DOT-specification containers can be used for solids, sludges, and high-viscosity material. DOT regulations regarding reuse of containers are found at 49 CFR 173.28(m), and are discussed in Chapter 10.

Storage Levels

The final consideration to be covered is that of the amounts of wastes to be kept in storage. Waste storage should be kept at "reasonable" levels. This presents little problem to the small quantity generator or the generator who must move wastes offsite within 90 days. For generators who have an onsite TSD storage facility, "reasonable" levels could be thought of as less than truckload-quantity levels. The number of "truckloads" present will depend on how many different commercial treatment and/or disposal sites are necessary for the types of wastes generated at the manufacturing facility. It should be emphasized that the prices of offsite treatment or disposal will continue to rise faster than inflation, so stockpiling will not save money. Stockpiling ignitable or reactive wastes increases risks and will cause concern with the firm's insurance carrier.

CONCLUSION

It is hoped that the principles and common-sense hints contained in this chapter will prove useful to manufacturing firms currently using improper storage practices. In the past, wastes have all too often been placed in deteriorated containers without bungs, lids, or markings. These containers were then placed on bare soil or gravel areas for indefinite periods. Subsequent releases contaminated soil and groundwater. Great progress has been made in upgrading storage practices, but some improvement is still necessary for wastes to be accumulated or stored in a safe and proper manner.

CHAPTER 24

Considerations in Waste Sampling and Analysis

Charles W. White

Proper sampling and appropriate analysis of waste materials is often critical to regulatory compliance. Sampling and analysis of waste materials for the four hazardous waste characteristics is usually necessary for waste materials not meeting the listing descriptions. However, there are several other possible reasons for devising a waste sampling and analysis plan. Of course, the analytical parameters of concern will vary with the specific goal of the sampling and analysis plan. This chapter discusses some of the main reasons for analyzing waste samples, as well as some basic elements of a sound waste sampling and analysis program.

As mentioned above, there are several possible reasons for performing laboratory analyses on waste samples. The types of analyses necessary to fulfill each purpose will differ. In other words, the parameters most appropriate for achieving the given purpose will vary. Reasons for analyzing a given waste include, but are not limited to, the following:

1. determining whether the waste material is hazardous (or nonhazardous)
2. obtaining approvals from commercial treatment/disposal vendors, after their evaluation of the waste's suitability for treatment/disposal at their TSD facility
3. determining the technical and economic feasibility of use, reuse, recycling, reclamation, or treatment of the waste
4. determining the waste's physical properties and chemical characteristics with respect to the various DOT Hazard Class definitions, in order to assign a proper DOT shipping description
5. determining whether the waste is subject to any restrictions on land disposal, per 40 CFR Part 268

6. preparing delisting petitions, per 40 CFR 260.22
7. preparing Material Safety Data Sheets based on total composition data
8. determining the nature and extent of environmental contamination resulting from chemical releases

Remembering the purpose of a particular sampling and analysis program is critical to obtaining valid data necessary to achieving the program's goal(s). Both sampling and analysis must be given proper consideration. Regardless of the purpose, there are several elements that are important to include in every sampling and analysis program. These elements are:

1. ensuring the use of method(s) and equipment recommended or required by regulatory agencies for the type of sample to be taken
2. use of strategies, methods, and sample collection equipment sufficient to ensure that the sample is representative of the waste material as a whole
3. use of appropriate (and clean) sample containers
4. use of appropriate preservatives (including temperature control)
5. sample container labeling (including the name of material sampled, date and time of sampling, name of sample collector, and sampling method). An example of a sample container label is illustrated in Figure 24.1.
6. maintenance of a chain-of-custody record (see Figure 24.2) from the time the sample is collected until the time all analyses are completed. This includes the ability to definitively match a particular laboratory result with a specific sample of a specific waste material.
7. ensuring that recommended sample holding times are not exceeded by the laboratory prior to analysis
8. ensuring that the techniques or methods recommended or required by regulatory agencies for the parameter(s) of concern are utilized in the analysis of those parameter(s), and that the detection limits achieved are appropriate to the purposes of the analysis

---

Date Sampled _____ **Time Sampled** _____

Sample Collector _____

Sampling Method _____

_____

Sample Description _____

Sample Location _____

Figure 24.1 Sample label.

Project Location   Name of Client   Client Telephone No.   Project No.

Item No.	Sample No.	Number & Size of Containers	Description	Transfer Number and Check						
				1	2	3	4	5	6	7

Person Responsible for sample	Affiliation	Date	Time	Transfer No.	Item No.	Transfers Relinquished By	Accepted By	Date	Time
				1					
				2					
				3					
				4					
				5					
				6					
				7					

Purpose of analysis

Figure 24.2. Chain-of-custody record.

Consideration of these elements is important for all of the potential purposes of sampling and analysis programs. However, their consideration and documentation become critical when making determinations on waste materials that may be *non*hazardous, and in situations involving ongoing or potential enforcement action by regulatory agencies. The documentation of proper methods and procedures can prevent resampling and additional analyses should nonhazardous or other analytical results be questioned by regulatory agency personnel.

Table 24.1 lists analyses typically performed for hazardous waste determination purposes, along with EPA method numbers for the required method/technique for that parameter.

Table 24.1   Typical Analyses for Hazardous Waste Determination Purposes

Characteristic	Parameter	Test Method
Ignitability	Flash point	ASTM Standard D-93-79 or ASTM Standard D-3278-78
Corrosivity	pH	SW-846 9045
Reactivity	Reactive cyanide	SW-846 9010
	Reactive sulfide	SW-846 9030
EP Toxicity	Extraction procedure	SW-846 1310
	Arsenic	SW-846 7060
	Barium	SW-846 7080
	Cadmium	SW-846 7130
	Chromium	SW-846 7190
	Lead	SW-846 7420
	Mercury	SW-846 7470
	Selenium	SW-846 7740
	Silver	SW-846 7760

A number of factors are affected by how a waste is sampled and analyzed. Perhaps the most important factor from the perspective of the firm is the analytical costs. Laboratory charges for many parameters of concern in the RCRA regulatory program are substantial. Some analyses, such as EP toxicity for the eight specified heavy metals, can cost anywhere from $150–$450 per sample. Evaluation by TCLP of F001–F005 listed wastes for land ban purposes can cost $1000–$1400 per sample. Analysis of samples that are not representative of the waste stream as a whole, or the use of nonapproved methods or techniques, can result in inaccurate and/or invalid results. Inaccurate results differ from the "true" value for the parameter of concern. Invalid results are those that are unacceptable to regulatory agencies for compliance purposes. In either event, unnecessary expenses are incurred from reanalysis, and/or the regulatory compliance status of the firm is adversely affected.

For analyses utilized for regulatory purposes, or those with regulatory implications, the primary reference to be utilized is EPA's "Test Methods for Solid Waste: Physical/Chemical Methods" (SW-846). SW-846 is currently in its third edition. SW-846 methods are mandatory in the RCRA regulatory program.

Appendix I of 40 CFR Part 261 requires that SW-846 methods be used through their incorporation by reference. Other reference manuals of value for environmental analyses include:

1. *Standard Methods for the Examination of Water and Wastewater*, 16th edition (Washington, DC: American Public Health Association, 1985)
2. "Manual of Methods for Chemical Analysis of Water and Wastes" (EPA-625/6-74-003a)
3. ASTM Standards (For further information, write to the American Society for Testing and Materials, 1916 Race Street, Philadelphia, PA 19103.)

SAMPLING METHODS/TECHNIQUES

The most appropriate sampling method(s) in a particular situation will depend upon a variety of factors. These factors include:

1. the physical state of the waste (e.g., solid, semisolid, single-phased liquid, multi-phased [layered] liquid, powder or dust)
2. the expected degree of uniformity or homogeneity of the waste with respect to space and time
3. the type of containment, if any (e.g., 55-gallon drum, bulk tank, roll-off box, waste pile, surface impoundment)
4. the type of analyses to be performed (e.g., GC/MS volatile organics, metals, flash point)
5. the purpose to be served by the sampling and analysis program

As previously discussed, it is important to obtain a representative sample of the waste. A representative sample is a sample that can be expected to exhibit the average properties of the entire waste. A representative sample is important in order to accurately determine the physical/chemical properties of the waste. The appropriate sampling method(s) to be utilized primarily depends on the physical characteristics of the waste. There are many references that can be consulted to help determine the appropriate sampling method for a particular waste. Three such references are:

1. SW-846 (mentioned above)
2. 40 CFR Part 261, Appendix I (also mentioned above)
3. "Samplers and Sampling Procedures for Hazardous Waste Streams" (EPA-600/2-80-018), January, 1980

Some examples of appropriate sampling methods for different waste types are discussed in the following paragraphs.

Liquid containerized wastes are often sampled using a small-diameter (0.5- or 0.75-inch) glass or polyvinyl chloride (PVC) tube. The tube is inserted vertically into the waste until the bottom of the container is reached. The upper opening of the tube is then closed, and the tube is removed from the drum and emptied into an appropriate sample container. In this manner, a representative vertical profile of the waste is obtained. If the waste is stratified (layered), each layer will have been sampled.

Another liquid sampling method utilizes a coliwasa (*Composite Liquid Waste Sampler*) in the same manner described above. A coliwasa is essentially a tube with a small internal rod connected to a closure device at one end (see Figure 24.3). The tube is inserted into the waste and the bottom end is sealed. The coliwasa may then be removed from the waste, and the contents drained into a sample container. Coliwasas are commonly made of glass, stainless steel, or PVC.

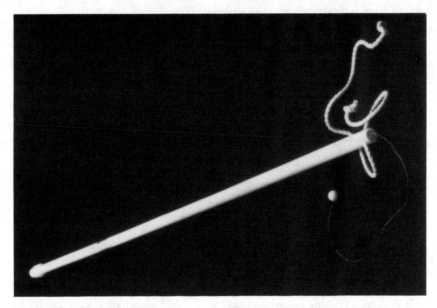

Figure 24.3. Composite liquid waste sampler (Coliwasa). (Photo courtesy of Soiltest, Inc., Lake Bluff, IL)

Wastes such as solids, sludges, viscous liquids, powders, or dusts may also be sampled using variations of the methods described above. If vertical homogeneity of the waste is relatively uniform, then a clean plastic or stainless steel scoop or trowel may be used to obtain a sample (see Figure 24.4). Material should be taken from various surface locations and composited to ensure a representative sample is obtained. The number of locations sampled and the number of composites created will depend on the amount of the waste and/or number of waste containers.

If a solid waste is not vertically homogeneous, various related methods may be used to obtain a representative sample. The simplest method is to drive a 2-inch-diameter PVC or clean steel tube through the entire depth of the waste, remove the tube, then extract the contents into an appropriate sample container. Solids may also be sampled using a trier. This is essentially a tapered trowel which is inserted into the waste, turned in a circular motion, then extracted, removing a small coring of the waste. Sludges and viscous materials are often sampled using a modified coliwasa. Dusts and powders may be sampled using a device called a grain thief. This piece of equipment consists of a slotted tube with a tight-fitting inner sleeve. The thief is inserted into the waste and the powder enters the tube through the slots. The inner sleeve is twisted to cover the slots and retain the sample as the thief is removed.

Figure 24.4. Stainless steel scoops. (Photo courtesy of Soiltest, Inc., Lake Bluff, IL)

Soil samples may be obtained from an area of potential contamination using common excavating tools. Depending on soil conditions, the desired size of sample increments, and the desired final boring depth, soil samples may be obtained with anything from a hand-operated bucket auger to a large-scale drilling rig. Regulatory agencies often recommend that soil sample increments be obtained from 0–6-, 6–12-, 12–18-, and 18–24-inch increments. Below 24 inches, samples are taken in one-foot increments (i.e., 2–3 feet, 3–4 feet, etc.). Data obtained from analysis of these samples will provide a vertical profile of soil concentrations for the parameter(s) of interest. It also allows for sequential analysis of increasingly deeper soil samples until a specified cleanup standard is reached. In this manner, the amount of overexcavation is reduced during remediation because the depth of contamination may be more accurately pinpointed using this method.

As a side note, it must be pointed out that the EP toxicity analysis for metals is not applicable to soil samples unless disposal approval for contaminated soil is being sought. For contaminated site soil investigations, metals should be analyzed on a total dry weight basis. It is also important during these types of investigations to obtain background soil samples using the same methods and in the same increments described above. These background samples must be obtained from an area where no disposal, storage, or manufacturing operations have taken place. A facility's front yard or a neighboring empty undisturbed field are common areas from which site-specific background samples may be obtained. Data from these background samples may be used to set a cleanup standard for the parameter(s) of interest.

It is very important to thoroughly clean sampling equipment between samples, no matter what type of waste is sampled or which method is used. Dirty or contaminated sampling equipment can lead to erroneous analytical results that could cost money and time. Cross-contamination between samples can be minimized conveniently by using disposable sampling devices. For some methods, however, disposable equipment is not an available option.

There are other specific sampling methods that may suit particular wastes. The most important considerations when choosing the proper method are the five factors discussed above. These factors will indicate the most appropriate sampling method to be used.

STATISTICAL SAMPLING

In some instances, the analysis of a single sample is insufficient to meet the intended purpose of the sampling and analysis program. Examples of such instances include:

1. heterogenous waste streams, or combined waste streams of variable composition
2. large-volume waste streams
3. multiple unmarked drums of unknown contents
4. waste materials considered hazardous wastes by virtue of analytical results slightly above the regulatory threshold
5. evaluation of soil in areas of known or suspected contamination

Statistical sampling involves initially evaluating the variability of past data (if any exists) in order to determine the appropriate number of samples to be analyzed. A statistical evaluation of the resulting new data, such as determination of the mean and standard deviation and establishment of upper and lower confidence intervals, may then be completed. With sufficient data, it can be said with a specific level of certainty that the upper confidence interval number is above (or below) the regulatory threshold; thus, the waste material is a hazardous (or nonhazardous) waste with respect to that parameter. Statistical sampling often involves selecting one or two parameters of concern, and focusing on their further evaluation.

Statistical evaluations may be used to compare sample data to background data, as well as a regulatory threshold. This type of evaluation is often used during contaminated site soil investigations. Background data are used to establish a cleanup standard by calculating a mean, variance, and upper confidence interval (CI) for background. If a particular sample exceeds the CI, it does not meet the cleanup standard and is considered contaminated.

A statistical sampling program will typically involve simple random selection of an appropriate number of sample locations or waste containers. The data from these samples may then be statistically compared to a regulatory threshold or some type of cleanup standard using a simple Student's t-test, for example. There are various statistical references that may be used to select appropriate statistical approaches to sampling and analysis, including SW-846. In addition, some state regulatory agencies recommend and/or require the use of specific statistical methods for some situations. Guidance documents are usually available from these agencies.

The regulatory and monetary implications are key to whether statistical strategies are appropriate, since laboratory invoices can become exceedingly large. Costs of further analyses must be balanced against potential savings realized by analyzing fewer samples and making worst-case assumptions. The transportation and disposal cost differences between hazardous and nonhazardous wastes are typically one order of magnitude. There are also long-term liabilities associated with hazardous waste disposal. All of these factors must be carefully weighed when deciding the appropriate sampling strategy. For example, statistical evaluations can be justified for large-volume waste streams, or where the waste marginally exceeds a regulatory threshold (RT) for one or two parameters (e.g., EP toxic lead at 5.2 ppm vs the RT of 5.0 ppm). It may be possible to statistically demonstrate in this instance that the waste as a whole is actually nonhazardous. The regulatory and cost implications in this example serve to further emphasize the necessity of giving appropriate consideration to representative sampling and use of approved test methods, as well as the other elements of a sound sampling and analysis program.

SELECTING A LABORATORY

The selection of a laboratory to perform the analyses on waste samples is also of critical importance. Laboratories tend to specialize, so it is important to deal with a laboratory that specializes in environmental analyses. However, even among environmental labs there are wide variations in quality. Although reliable, high-quality data is the name of the game, there are other considerations as well, including:

1. laboratory turnaround time
2. service (e.g., provision of sample containers, pickup of samples, advice regarding necessary preservatives, etc.)
3. use of appropriate analytical procedures and methods
4. maintenance of a proper quality assurance/quality control (QA/QC) program
5. pricing

EPA and the various state regulatory agencies are becoming increasingly cognizant of the importance of high-quality data, placing emphasis on both the sampling and the analytical portions of the equation. Laboratories can create problems in this regard by using nonapproved methods, by exceeding maximum sample holding times, by having detection limits higher than regulatory thresholds, and by failing to establish QA/QC programs. Increased scrutiny of analytical data and environmental laboratories has resulted from perceived deficiencies. Some states, such as New York, go so far as to certify environmental laboratories. A laboratory's qualification for EPA's Contract Laboratory Program (CLP) is a form of indirect certification. Regardless, environmental laboratories are expected to have a formal QA/QC program. Elements of QA/QC programs include equipment calibration using reference standards, analysis of blank, spiked, and replicate samples, and round robin analysis of blind samples, as well as extensive documentation.

Once an environmental laboratory is selected, it is necessary to know how to deal with the laboratory for maximum effectiveness. Normally, the laboratory will need to know the purpose of the sampling and analysis effort, and/or have the parameters of interest and appropriate methods specified for each sample. A list of common laboratory acronyms and abbreviations used for such purposes is included in this chapter as Table 24.2. The laboratory may need to be informed of any requirements for low detection limits (e.g., 1 ppm or 1 ppb), and labs typically like to know what the expected concentration ranges are, so they can dilute or concentrate the sample as necessary for the "range" of the analytical instrument.

The detection limit is the lowest analytical measurement which can be accurately attained for a particular parameter. It is important to note that the detection limit may vary depending on the capabilities of the analytical instrument used, the analytical parameter specified, and the type of waste being analyzed. Many wastes that are particularly "dirty," such as highly contaminated waste, or opaque materials, such as paint or ink, may cause interferences, thus raising the detection limit.

It is important to be able to match a laboratory report to a particular sample. Information necessary to adequately identify the sample on the laboratory report must be indicated on the sample container (or a label), or should be in a chain-of-

Table 24.2   Common Laboratory Acronyms and Abbreviations

AA	atomic absorption
A/B/N	acid/base/neutral
AIHA	American Industrial Hygiene Association
APHA	American Public Health Association
ASTM	American Society for Testing and Materials
BOD	biochemical oxygen demand
Btu	British thermal unit
CLP	(EPA) Contract Laboratory Program
COD	chemical oxygen demand
DL	detection limit
ECD	electron capture detector (GC)
EP	extraction procedure
FID	flame ionization detector (GC)
GC	gas chromatography
GC/MS	gas chromatography/mass spectrometry
HECD	Hall Electroconductivity Detector
HOC	Halogenated Organic Compounds (Appendix III of 40 CFR Part 268)
HPLC	high-performance liquid chromatography (often high-pressure liquid chromatography)
HSL	Hazardous Substance List (40 CFR Part 302)
ICAP	inductively coupled argon plasma emission spectrophotometer (often ICP)
IR	infrared spectrophotometer
MDL	method detection limit
mg/kg	milligrams per kilogram
mg/L	milligrams per liter
ND	not detected
NPD	nitrogen-phosphorus detector
OVA	organic vapor analyzer
PAH	polynuclear aromatic hydrocarbons (often PNA)
PCB	polychlorinated biphenyls
PID	photoionization detector
ppb	parts per billion
ppm	parts per million
PQL	practical quantitation limits
QA/QC	quality assurance/quality control
SVOA	semi-volatile organic analyses
SW-846	"Test Methods for Evaluating Solid Waste: Physical/Chemical Methods"
TCDD	tetrachlorobenzo-p-dioxin
TCL	target compound list
TCLP	toxicity characteristic leaching procedure
TOC	total organic carbon
TOX	total organic halogen
TSS	total suspended solids
TTO	total toxic organics
TX	total halogen
μg/kg	micrograms per kilogram
μg/L	micrograms per liter
VOA	volatile organic analyses
VOC	volatile organic compounds
Appendix VIII	"Hazardous Constituents," found at 40 CFR Part 261
Appendix IX	"Ground-Water Monitoring List," found at 40 CFR Part 264

custody form accompanying the sample. The chain-of-custody form should reference some type of identification on the sample container (i.e., a sample number). Typically, report information includes:

1. the name of the material sampled
2. company name
3. date and time of sample collection
4. sample location and sampling method
5. name of sample collector

A good laboratory report will reference some (or all) of this information, as well as provide the sample results. Other aspects to a good laboratory report include:

1. identification of parameter (e.g., total cyanide)
2. identification of result, with an appropriate number of significant digits (e.g., 8.25)
3. identification of units (e.g., mg/L)
4. detection limit achieved (and units) (e.g., 0.25 mg/L)
5. method number utilized (e.g., SW-846 9094)
6. QA/QC documentation
7. date of analysis
8. analyst (initials)

The above items should be reported to provide a complete and accurate "analytical picture" of the waste.

This chapter has touched on some basic concepts that are involved in the development of a complete and appropriate waste sampling and analysis program. There is certainly much more room for discussion on this topic. However, the information provided here should be sufficient to place the reader on the right track as far as what to consider when developing a plan for the sampling and analysis of a particular waste. A well-designed plan will minimize unnecessary expenditures, and provide reliable data for effective decisionmaking.

CHAPTER 25

Dealing with Regulatory Personnel

Jack Cornpropst

This chapter discusses dealing with people from governmental regulatory agencies, from the viewpoint of an electroplater and licensed industrial wastewater treatment operator.

Electroplating and other surface-finishing operations seem to be the subject of close government scrutiny and regulation. Much of this attention is understandable, due to the nature of surface-finishing operations and the types of materials and process chemicals used in such operations. Part of being a good electroplater is proper management of the solid and liquid wastes resulting from electroplating and industrial wastewater treatment operations. In other words, one is not only an electroplater, but also an industrial garbage man. By properly managing a company's wastes, one fulfills an important corporate and personal responsibility.

It is obvious that for all government regulatory programs to succeed, there must be a high degree of voluntary compliance on the part of the regulated industries. There simply are not enough government employees or enough jail cells available, were there to be any significant degree of resistance to such programs. For the most part, there is not basic disagreement between government and industry on the goals of clean air, clean water, a safe workplace, and control of toxic and hazardous materials and wastes.

The best way, and perhaps the only way, that the national hazardous waste program can be as successful as we wish it to be is through development of a spirit of cooperation between industry and government authorities. This cooperation is necessary for voluntary compliance. Of course, enforcement and fines should be reserved for those who fail to comply with substantive aspects of the regulations and/or fail to make good faith efforts to comply with paperwork requirements. Industry wants fair and reasonable enforcement; otherwise, the companies in compliance compete against firms that do not bear the very real costs of compliance.

A cooperative approach will alleviate the stresses associated with compliance efforts during a confrontation with a government agency. Cooperation, though, is a two-way street. There are certain things that industry can expect of the state environmental agency and there are those things that the state agency can expect of industry.

Industry should be able to expect some advance notice of new environmental regulations, and the opportunity to comment on such regulations before they take effect. This is far better than to have to try to get unworkable regulations changed after they are finalized. Also, industry should be able to readily obtain interpretations of federal and state environmental regulations. It is not always perfectly obvious what is required, even after several readings of the *Federal Register*. Thirdly, the regulations should be consistently and fairly applied. Industry can reasonably expect the state environmental agency to make an attempt to understand general (industry-wide) problems and specific (one plant or firm) problems or situations. Finally, there should be some way or mechanism available for a company to discuss compliance problems with the state agency without fear of a citation or reprisal.

Again, cooperation is a two-way street. For state environmental personnel to be able to understand the compliance problems of a certain industry or firm, they must be allowed to become familiar with the particular processes and operations involved. Company personnel should cooperate with inspectors. This includes access to the property, an explanation of production processes and operations, complete answers to questions, and access to relevant records and information. This does not, however, entail doing the inspector's job for him/her. Also, the state agency should be able to expect diligent, good faith efforts to comply with the requirements. Finally, problems that may affect others, such as spills or leaks to waterways, should be reported immediately. In some cases, this may mean bringing "skeletons out of the closet" (e.g., old company disposal areas) so that the situation can be addressed before any serious environmental impacts result.

Company waste managers should read the applicable regulations and make an attempt to understand them. Questions on the applicability of certain requirements or on the correctness of a particular interpretation of a "grey area" should be resolved before any deadlines have passed. Ignorance of the law should not be used as an excuse; nor should the lame response that "others are doing it."

Cooperation usually begins and develops at the lowest involvement level, i.e., between a company's hazardous waste manager and the state inspector. This cooperation is based on the realization that both parties have a job to do and that both are accountable to superiors. Further, the company's waste manager and the inspector are not (or should not be) on opposite sides of the fence. In fact, they have similar objectives. The objective of the hazardous waste manager at the company is to properly handle and dispose of the hazardous wastes resulting from company activities. The objective of the hazardous waste inspector is to verify that proper waste management is actually being performed at the facility, in accordance with the applicable laws and regulations.

With a realization of these similar objectives, each party must make an effort to understand, assist, and offer encouragement to the other. Both the inspector and the waste manager possess specialized information, knowledge, and experiences that could be beneficially used by the other.

Of course, there is always the possibility of encountering an individual with whom the development of a working relationship is impossible. Fortunately, such persons are rare. All one can do in such a situation is to make sure that all statements made (both oral and written) are absolutely correct, be firm in one's convictions, and hope for the best.

To summarize, the hazardous waste management laws and regulations are lengthy and complex. Whether any company could ever be in complete compliance, at all times, with the letter of the regulations is questionable. Given these premises and the accountability of industrial waste managers to corporate management, it is believed that cooperation with regulatory agencies is the only workable alternative. The hazardous waste management system is still relatively new. Both industry and government are learning to implement and fine-tune their respective programs. With the right attitude and recognition of their respective responsibilities, a workable and effective system can result.

CHAPTER 26

Legal Responsibilities and Liabilities

John M. Kyle III
Barnes & Thornburg
Indianapolis, Indiana

The topic of legal responsibilities and liabilities associated with handling hazardous wastes and other hazardous substances is complex and constantly evolving, and forms the entire subject matter of several recent books and multivolume treatises. Because of the depth and bredth of this subject, I have tried simply to cover some of the major issues and to provide a flavor of this area of the law. This chapter is not intended to be, nor should it be considered, a definitive discussion of the topic.

Previous chapters in this book have discussed the statutory and regulatory requirements for handling hazardous and solid wastes pursuant to the federal Resource Conservation and Recovery Act (RCRA) and the states' analogs. However, as suggested earlier in the book, RCRA and RCRA-spawned regulations are not the only source of legal responsibilities applicable to handling hazardous waste or other hazardous materials. Additional obligations may arise from other federal statutes such as the Comprehensive Environmental Response, Compensation and Liability Act (CERCLA or Superfund); common law (i.e., "judge-made" law); and state statutes and regulations unreleated to RCRA. Because state requirements vary widely from state to state, this chapter cannot address those requirements in any detail. A company doing business in any state must be familiar with that state's laws and regulations, as well as local laws and regulations, to determine what additional obligations exist.

This chapter will first discuss potential liabilities for violating RCRA or RCRA regulations and then address various responsibilities and liabilities that may arise under Superfund and common law. As will be seen, noncompliance with these various laws and regulations can result in serious civil penalties and, in some cases, jail terms for responsible corporate officers.

RESOURCE CONSERVATION AND RECOVERY ACT

RCRA establishes a wide variety of standards of conduct applicable to generators, transporters, treaters, storers, and disposers of hazardous wastes. Many of these requirements are "self-executing"—for example, compliance with the generator or transporter regulations is required without any further action by EPA or a delegated state. Other regulations, such as those for hazardous waste treatment, storage, and disposal (TSD) facilities, are specifically applied to a particular facility through an EPA or state permit (initially, interim status and eventually, Part B permits). Compliance with such permits constitutes full compliance with RCRA Subtitle C permit regulations (but not with §7003 actions for imminent hazards, Superfund, or common law), and erects at least a temporary shield from evolving permit standards under RCRA until the permit expires or is reopened and reissued or modified.

Section 3008 of RCRA (42 USC §6928) provides very serious civil and criminal penalties for violations of the statute, the RCRA regulations, and RCRA permits. If EPA determines that any person is violating "any requirement" of Subtitle C (the regulatory heart of RCRA), it may either issue an administrative order or commence a civil action against the alleged violator in federal district court. An administrative order issued by EPA under §3008 may impose penalties of up to $25,000 per day of noncompliance and is a final agency order, unless the person subject to it requests an administrative hearing before an EPA administrative law judge (ALJ) within 30 days "after the order is served" (§3008(b)). After a hearing is held, the ALJ may either reinstate or modify EPA's original order. Judicial review may then follow.

EPA may instead choose to file a lawsuit in federal district court. In such an event, the federal judge is very generally empowered to grant EPA "appropriate relief," including injunctions (i.e., the court can order the defendant to do or to refrain from doing a specific act) and penalties, again up to $25,000 per day of noncompliance. If EPA files a lawsuit, the normal rules of evidence and judicial procedure apply and the case will be heard *de novo* (i.e., the parties present their case from scratch, against a clean slate). However, judicial review of an appealed administrative order is generally limited to the administrative record the parties make before the ALJ; no new evidence (other than newly discovered evidence) can be presented to the court on appeal.

Normally, EPA, the states, and the courts do not assess the full range of civil penalties that could be levied under the statute. EPA has developed a civil penalty policy (which has, in turn, been adopted in various forms by many states) which assists the government in calculating the appropriate penalty for particular cases.

The penalty calculation usually consists of three elements: (1) the gravity of the infraction (an evaluation of the extent of the deviation from the requirements and the potential for harm to the environment and to EPA's program); (2) economic benefit from noncompliance (how much the company saved by not complying); and (3) any reasons, such as good faith or a history of compliance or noncompliance, to adjust the penalty upward or downward. Any person involved in penalty negotiations must review the applicable penalty policy and the government's specific penalty calculations in detail.

In addition to these civil penalties, §3008(d) establishes criminal penalties applicable to individuals and corporations. Any person who knowingly commits any of the following acts can be held criminally liable under RCRA:

1. transporting or causing transportation of hazardous waste to a TSD facility that does not have a permit
2. treating, storing, or disposing of hazardous waste without a permit or contrary to interim status regulations
3. omitting material information or making a false material statement in permit applications or records required to be kept under RCRA
4. destroying, altering, concealing, or failing to file records required to be kept under RCRA
5. transporting hazardous waste without a manifest
6. illegally shipping hazardous waste to a foreign country
7. storing, treating, transporting, disposing of, or otherwise handling hazardous waste in violation of a permit or regulations issued by EPA

A person convicted of any of these acts is subject to a fine of up to $50,000 per day or imprisonment for two years, or both. A person convicted of numbers 1 or 2, or both, can be sent to jail for up to five years. A repeat offender's penalties are doubled for subsequent convictions. Reasoning that a few prominent, well-publicized criminal convictions can do a lot to deter others from violating RCRA, the federal government and the states have been pursuing criminal sanctions more aggressively in the last few years. This trend is expected to continue.

Section 3008(e) creates a second category of criminal penalties for even more egregious conduct. Any person who knowingly commits one of the seven actions discussed above and "who knows at that time that he thereby places another person in imminent danger of death or serious bodily injury" is subject to a $250,000 fine and imprisonment for up to 15 years, or both. If the defendant is an organization, as opposed to a "natural person" (the law's definition of a human being), the fine can be up to $1,000,000. Special rules are detailed in §3008(f) for determining a knowing violation and whether someone knew he or she was placing someone else in danger.

Another important aspect of RCRA liability involves the dual federal/state nature of the RCRA program. As discussed elsewhere in this book, EPA can delegate enforcement of the federal RCRA program to a state which, upon final authorization, operates its program "in lieu of" the federal program (RCRA, §3006(b)).

At the time this chapter was written, no state has received final authorization for the 1984 Hazardous and Solid Waste Amendments (HSWA) portions of the RCRA program, due in large part to the fact that EPA has not yet finished developing all of the regulations mandated by HSWA.

To the extent that a person is dealing with a part of the RCRA program that has been fully delegated to a state, that person looks to that state, not EPA, for permits and enforcement. With respect to a fully authorized program compliance issue (for example, state approval of a closure plan), if a person buys its peace with the state through settlement or otherwise, a major issue arises: Is that person still subject to enforcement by EPA for the same violations or issues if EPA does not like the result reached by the state? This is the so-called overfiling issue. EPA believes that it always has the authority to ''overfile'' against a person who has violated a fully delegated RCRA program component even if that person has resolved that issue with the state. On the other hand, many people argue that since the state enforces the EPA-authorized federal program ''in lieu of'' EPA, EPA is divested of authority to pursue violations of fully delegated RCRA components. This issue will become even more pronounced when (and if) states begin receiving full delegation of HSWA authority. The important point for the reader is not to be lulled into a false sense of security upon settlement with the state. It is still possible that EPA—and citizens, under the citizen suit provisions in §7002 of RCRA—could bring another action seeking to compel a different result.[1]

Overfiling by EPA or by citizens is a very difficult issue. A corporate executive wants to know if he or she has bought peace. Companies concerned about the matter should consult with an attorney to try and tailor their situation or settlement to provide as much protection from subsequent EPA and citizen suits as possible.

CERCLA

As stated above, RCRA is not the only source of duties and liabilities associated with handling hazardous wastes. One of RCRA's major shortcomings when it was enacted in 1976 was that it principally governed activities that occurred after the effective date of the RCRA regulations (November 19, 1980). With the possible exception of a §7003 lawsuit to address imminent hazards (which some courts held could address prior practices, although other courts disagreed), RCRA did not address the problems of inactive sites or past disposal practices. To remedy

---

[1]Even if a company has reached consensus with EPA and the state, it is still potentially subject to citizen suits under §7002 of RCRA. Companies aware of (or defendants in) citizen suits under the Clean Water Act are all too familiar with the issues that can arise in this context.

these problems, Congress passed CERCLA in the waning days of the Carter administration in late 1980.[2]

The main thrust of Superfund initially was to create a $1.6 billion fund (increased to $9 billion with the 1986 Superfund Amendments and Reauthorization Act) to be used by EPA to clean up inactive sites posing a threat to public health or the environment. However, Superfund did much more. In addition to Superfund's revenue-creating aspects, it also created certain regulatory requirements, filling gaps in RCRA's prospective system. Three main aspects of CERCLA will be discussed in this chapter: (1) the liability-creating §107 and EPA's nonemergency response authority under §104; (2) EPA's emergency authority under §106; and (3) the reporting requirements for spills and other disposal activities under §103.

Section 107: Liability

The centerpiece of CERCLA liability is §107(a) (42 USC §9607(a)), which provides that certain persons (called "potentially responsible parties" or PRPs) are liable for certain things, including cleanup costs incurred by the government and others. There are four categories of PRPs under §107:

1. present owners or operators of a facility from which there is a release or threatened release of hazardous substances

---

[2]When Congress passed the HSWA amendments in 1984, it addressed the "retroactive gap" in RCRA as initially enacted in 1976. Under the 1976 statute and the 1980 implementing regulations, a person or organization could escape RCRA regulation (subject to the vagaries of §7003 litigation mentioned earlier) if, for example, placing of waste in a unit ceased before November 19, 1980. Although that person or organization could still be subject to RCRA for generator, transporter, and TSD activities that occurred after that date, the "abandoned" or "inactive" unit was not subject to RCRA.

Congress closed this gap with the enactment of three "corrective action" provisions in HSWA. Under §3004(u), any person or organization "seeking" a RCRA TSD permit must take appropriate corrective action for releases of hazardous waste or constituents from any solid waste management unit (SWMU) *regardless* of the time waste was placed in the SWMU. Thus, the onsite landfill that stopped receiving waste on November 1, 1980 was brought back into the RCRA system under §3004(u) *for those facilities currently seeking a TSD permit* for other, ongoing operations, and the permittee must now take appropriate corrective measures to address releases from the SWMU. In addition, §3004(v) requires the permittee to take corrective action beyond its facility's boundaries unless access cannot be obtained.

In addition, §3008(h) provides that EPA can issue an order (as opposed to requiring corrective action as part of a Part B TSD permit) requiring corrective action to remedy "a release of hazardous waste into the environment from a facility authorized to operate under" interim status. Although the language of §3004(u) is very different from §3008(h), and there is substantial dispute over the proper scope of §3008(h) vis-a-vis §3004(u), EPA takes the position that it can issue a §3008(h) order to anyone who has or ever had interim status to remedy a release from any part of the interim status facility (which EPA defines as the property boundaries, including nonregulated SWMUs), even those portions or areas that have been inactive since November 1980. Thus, despite the different language, EPA basically interprets the two sections the same way.

2. past owners or operators of a facility from which there is a release or threatened release of hazardous substances, provided that the person owned or operated the facility "at the time of disposal"
3. a person or organization (commonly called a "generator") arranging for waste to be treated or disposed of at another facility, if there is a release of hazardous substances from that facility
4. a person who transports waste to a facility from which there is a release or threatened release of hazardous substances, provided that the transporter actually selected that facility

These four categories of PRPs are liable for four things:

1. costs incurred by the federal or state government to clean up the facility
2. costs incurred by "any other person" to clean up the facility
3. damages to natural resources
4. costs of health assessments conducted by the federal Agency for Toxic Substances and Disease Registry to determine if people living around the Superfund site have been injured by the site

Thus, present owners or operators of a site; persons who owned or operated a site at the time of disposal; generators who arrange for their waste to go to the site; and transporters who actually select the site are all liable for government and private party cleanup costs and damages to natural resources.

The nature of this liability is very imposing. Liability is strict—it does not matter how carefully the PRP acted or if the PRP took all reasonable precautions against the release. Liability is retroactive—it does not matter that the PRP complied with the law in effect at the time. And liability can be "joint and several" in the case of indivisible injury. Joint and several means that one PRP may be liable for all the costs of cleanup even if there are other PRPs. Indivisible injury means that the environmental damage has become commingled and it is not possible to apportion discrete damage to a particular PRP, often the case at Superfund sites.

Not only is the liability very imposing, but there are also precious few defenses to CERCLA liability. To escape §107(a) liability, the PRP must prove that the release of hazardous substances was caused "solely" by (1) an act of God; (2) an act of war; (3) an act of a third party; or (4) some combination of these acts. These defenses to §107(a) liability are quite limited. Chances of proving that a release was caused solely by an act of God or war are slim. The third-party defense is similarly limited. To qualify, the PRP must prove by a preponderance of the evidence each of the following:

• The release was caused solely by the third party.
• The third party is not the PRP's employee or agent, or a person with whom the PRP has a direct or indirect contractual relationship.

- The PRP exercised due care in handling the hazardous substance.
- The PRP took precautions against foreseeable acts or omissions of the third party.

In most situations, it is very difficult to prove each of these elements. Since the statute says these are the only four defenses to CERCLA liability, it is very difficult to escape CERCLA liability once the person or organization fits into one or more of the PRP categories. It is also important to note that CERCLA liability attaches not only to sites to which you send waste, but also to your own property as well.

Because CERCLA liability is so all-pervasive, many believe it has done more than any other law to improve the way people handle their waste. Although there are a number of legal issues a PRP can advance (which are in addition to the defenses discussed above, but which are beyond the scope of this chapter), the best way to avoid CERCLA liability is to follow the steps outlined in Section III of this book: Conduct periodic audits of your own facility and those utilized to treat or dispose of your wastes; and, perhaps as important as anything else, follow exactly the Waste Management Hierarchy discussed in Chapters 15 and 18.

Section 107 liability is usually imposed as follows: EPA has authority under §104 to spend its CERCLA money to clean up a site and then to sue PRPs to recover the money expended. Usually, absent an emergency, EPA will write PRPs and tell them if they do not clean up the site, EPA will, and then sue the PRPs to recover its expenses. Since government-financed cleanups are almost always more expensive than privately-funded cleanups, it is in the PRPs' best interest to do the work themselves and to avoid paying EPA expenses (which include, according to some courts, government overhead, employee salaries, and the like) as well as the transaction and litigation costs of defending a §107 cost-recovery suit.

A final point: §107 liability is premised upon the existence of a "hazardous substance." A CERCLA "hazardous substance" covers a much broader range of chemicals and waste products than a RCRA "hazardous waste." In fact, a RCRA hazardous waste is only one of six subsets of CERCLA hazardous substances. A CERCLA hazardous substance includes not only RCRA hazardous wastes but also (1) a "hazardous substance" listed under §311(b) of the Clean Water Act; (2) a "toxic pollutant" listed under §307 of the Clean Water Act; (3) a "hazardous substance" for which EPA establishes a reportable quantity (RQ) under §102 of CERCLA; (4) a "hazardous air pollutant" listed under §112 of the Clean Air Act; and (5) an "imminently hazardous" material for which EPA has taken action under §7 of the Toxic Substances Control Act. An example helps clarify the distinction: Even if a waste is not RCRA hazardous for lead because the EP toxicity value is less than 5 ppm, the waste is still CERCLA hazardous because lead is on the §307 list of toxic pollutants, and because the courts have held that no minimum quantity or concentration of a hazardous substance such as lead is necessary under CERCLA.

Section 106: Emergency Authority

In addition to EPA's §104/107 authority, EPA has enormous powers under §106 to address emergency situations. If EPA finds there is an "imminent and substantial endangerment" to people or the environment, EPA may either issue an administrative order or file a lawsuit in federal court. Because some EPA officials foresee a time in the not-too-distant future when the Superfund will be gone, it has recently begun to use its §106 authority more aggressively, even in situations that many would not think rise to the level of an "imminent and substantial endangerment."

The receipt of a §106 order is one of the most significant events that can occur in environmental law. Typically, the order requires the respondent to take some immediate action to study and/or clean up a site. Failure to comply with a §106 order "without sufficient cause" can lead to draconian penalties of $25,000 per day *plus* treble damages (i.e., if it costs EPA $1 million to clean up a site, the PRP is liable for $3 million). As a result, the receipt of a §106 order requires an immediate, well-thought-out response and (almost always) the assistance of legal counsel to protect your interests and to create an appropriate administrative record, especially if you decide you have "sufficient cause" not to comply.

Section 103: Reporting Requirements

Section 103(c) of CERCLA established reporting requirements for hazardous waste sites not covered by RCRA. Within 180 days of December 11, 1980, past and present owners or operators of, or any person who accepted hazardous waste for transportation to, a facility where hazardous substances "are or have been stored, treated, or disposed of, shall, unless such facility has a permit" or interim status under RCRA, notify EPA of the existence of such a facility and specify "the amount and type of any hazardous substances to be found there, and any known, suspected, or likely release of such substances from the facility." Failure to notify can result in a $10,000 fine or imprisonment for one year, or both. EPA was also authorized to adopt regulations regarding the location of such facilities and the hazardous substances contained there, and to specify records that must be maintained by anyone required to notify under this section. It is unlawful for anyone required to notify under this section to destroy or otherwise render unavailable any records required to be kept under these regulations for a period of 50 years, unless a waiver is obtained from EPA. Although this section is primarily of historic significance, it raises an interesting issue for the corporate compliance manager who discovers that the required notice was not given in the early 1980s.

Sections 103(a) and (b) of CERCLA establish the "spill reporting" requirements. Anyone in charge of a vessel or onshore or offshore facility must immediately notify the National Response Center if hazardous substances are released into the environment from such vessel or facility. "Release" and "environment" are very broadly defined to cover nearly every type of spill, discharge, leaching, or disposal of hazardous waste into the air, land, or water. Notification does not need to occur unless there is a release of a "reportable quantity" of the hazardous substance, originally defined by statute as one pound, and since amended to provide variable quantities for a variety of substances. EPA's regulations at 40 CFR §302.4 contain the list of all the CERCLA hazardous substances and the reportable quantities (RQ) for each.

Although this reporting scheme seems straightforward enough (if you spill material on the list greater than its RQ, report it), a number of very troubling compliance issues arise. First, remember you have to spill more than the RQ in *any* 24-hour period. Second, mixtures of hazardous and other substances create problems in determining whether the RQ for a given substance has been exceeded. Third, if you spill a substance that is entirely contained within the factory or building, no report is required (50 FR 13462 [1985]).

Perhaps the most troubling §103 reporting issue involves the present discovery of a prior spill or release. For example, if it is discovered today that groundwater has been impacted by a release that occurred several years ago, does it need to be reported? Many questions arise: When did the spill occur? Was it after the effective date of the statute (December 1980)? Is this section of the statute retroactive? Most importantly, how much was spilled or released over what period of time, and what the RQ exceeded in any 24-hour period? These and other issues present nearly unanswerable questions that, again, probably create the need for legal counsel.

COMMON LAW

The federal statutes discussed above were not written in a void, but rather against the preexisting background of common law, which is not statutes or regulations, but "law" made (divined?) by judges. The issue addressed in this section is whether other obligations or liabilities attach to handling hazardous materials as a result of common law. There are two potential sources of additional responsibilities and liabilities: federal common law and state common law.

Federal Common Law

It must be emphasized that federal common law is very limited and applies only in the absence of an applicable act of Congress. Decisions of the United States Supreme Court in the early 1980s hold that when Congress enacts a comprehensive environmental statute containing detailed standards of conduct, authorizing comprehensive administrative regulations supervised by an expert agency, and providing elaborate provisions for enforcement of those standards, federal courts are not free to alter the terms of the statute by developing new standards of conduct or engrafting nonstatutory remedies onto the statute. In such an event, federal statutes preempt, or prevent, federal judges from developing additional requirements or liabilities under federal common law. Taken together, RCRA and CERCLA certainly establish a comprehensive congressional scheme addressing nearly every phase of hazardous waste management. Therefore, it is likely that federal judges will not be able to create new responsibilities or liabilities that are not already spelled out in the federal statutes and implementing regulations. As a result, federal judges' "lawmaking" authority will be limited to filling gaps in the statutes and regulations, and will not allow the federal judiciary to recognize new, nonstatutory federal common law responsibilities or causes of action.

State Common Law

It is fairly clear that comprehensive federal environmental statutes do not preempt state common law. State common law is not limited as is federal common law, and may be used not only to fill in gaps in statutory schemes, but also to recognize wholly separate and distinct liabilities and causes of action. As a result, state common law survives alongside or in addition to the comprehensive federal environmental programs. Although there is an argument that just as federal statutes can preempt federal common law, state statutes and regulations may preempt state common law, most state courts that have addressed this issue have concluded that, in the absence of strong state legislative intent in favor of such preemption, state common law remains. As a result, even if a person complies with all the federal and state environmental statutes and regulations, it is still possible that additional liabilities may be imposed under state common law.

Examples of common law doctrines that may apply in the area of handling hazardous substances include the law of nuisance, strict liability for ultrahazardous activities, trespass, and negligence. It is beyond the scope of this chapter to discuss fully the elements of these causes of action or the various permutations that exist under each state's common law. However, the following provides a flavor of these additional sources of liability.

Nuisance

Nuisance is an ancient doctrine inherited from the English common law that has long been used to address problems of an environmental nature. A nuisance encompasses any activity or situation that is injurious to health or offensive to the senses, or that interferes with the comfortable use and enjoyment of life and property. A landfill that leaches contaminants into groundwater, thereby polluting neighboring wells or a public body of water, could certainly be classified a nuisance. Most states hold that even a business lawfully operated in compliance with all applicable environmental laws and regulations may still constitute a nuisance. Therefore, compliance with statutes and regulations such as RCRA and Superfund would not be a defense to a nuisance action and additional responsibilities could be imposed by a court or jury. Damages could also be awarded to the injured party.

Strict Liability for Ultrahazardous Activities

The doctrine of strict liability for ultrahazardous activities traces its roots to a famous pollution case in England in the late 1800s, *Rylands v. Fletcher*. The modern doctrine essentially provides that one who carries on an "abnormally dangerous activity" that results in harm to another will be held liable for such harm, despite compliance with other laws or the exercise of utmost care and precaution. Existence of this strict liability tort depends on whether a given activity constitutes an "ultrahazardous activity." Historically, activities such as blasting and mining have been considered ultrahazardous. More recently, some state courts have held that handling hazardous waste constitutes an ultrahazardous activity, resulting in strict liability. However, it should also be noted that other states have held just the opposite. Many states have not addressed the issue at all. The law of each state must be reviewed to determine whether this far-reaching liability attaches to handling hazardous substances.

Negligence

Unlike strict liability, to be held liable under a negligence theory, a company or individual must be "at fault" or have failed to exercise a "reasonable standard of care" in handling hazardous substances. For example, if a generator of hazardous waste handles the waste haphazardly or fails to follow a RCRA standard, such as those relating to storage or containerization of wastes, and if such an action (or failure to act) causes damage to another party, that party could sue on a negligence theory, claiming the defendant failed to comply with a reasonable standard of careful conduct.

Trespass

Trespass is unauthorized entrance onto another's land, which could occur if hazardous substances were spilled from a truck or leached from a landfill onto another's property. Trespass, like nuisance and ultrahazardous activity, is a strict liability action. No amount of due care insulates one from liability; all that matters is that the unauthorized entry occurred.

Proximate Cause

In all of these common law theories, a plaintiff must overcome a substantial hurdle referred to in the law as "proximate cause." Essentially, proximate cause means that the injury must be "reasonably foreseeable" and the defendant's activity must be the direct and relatively immediate cause of plaintiff's injury. There cannot be intervening or superseding activities of others that "break the chain of causation" or become the "major," "superseding," or "more important" cause of plaintiff's injuries. This confusing doctrine can best be explained with an example. Say a generator contracts with a transporter to haul hazardous waste to a given disposal site, and, rather than transport the waste to the identified disposal site, the transporter instead dumps the waste along the road, thereby injuring a plaintiff. The defendant generator could argue that the transporter's intervening act was the proximate cause of plaintiff's injury and that it was not reasonable foreseeable that the transporter would illegally dump the waste. Such an argument could be successful for the generator. On the other hand, if the transporter took the waste to the identified landfill, and the landfill later leached and caused damage to someone, proximate cause may exist because, it would be argued, all landfills eventually leak, so the damage was foreseeable and there was no dramatic, intervening cause. The doctrine varies substantially from state to state. However, the lack of proximate cause could be a proper defense to the common law causes of action discussed above, and could exonerate the defendant from liability.

Regarding additional responsibilities and liabilities, in each of these common law theories, a court or jury can award damages (i.e., money) to the injured party. If the injury-causing activity is continuing, the court could also enter an injunction to prevent the activity from occurring or to restrict the activity to ameliorate injuries to people or the environment. As a result, it is possible that stricter controls than exist under RCRA, for example, could be imposed by common law.

CONCLUSION

As can be seen, there is a very good reason to comply with the requirements of the law of hazardous substances—noncompliance can hurt. The costs of developing and implementing a company-wide compliance program may seem infinitesimal when compared to the penalties associated with noncompliance, the bad publicity that may ensue, and the fact that, once caught, a company will be required to implement such a program anyway. Corporate planners must be aware of the fact that compliance with statutory and regulatory requirements may not be enough—additional burdens may be imposed through common law. Therefore, working knowledge of state common law will help companies further define and refine their duties and liabilities in this area.

CHAPTER 27

The Consultant's Role in Hazardous Waste Management

Freeman C. Cook, Jr.

As shown by the diverse topics discussed in the preceding chapters of this book, hazardous waste management is a dynamic and ever-expanding field of endeavor. This is certainly evidenced by the large body of current and pending hazardous waste regulation, the public attention focused on hazardous waste issues, the development of jargon and technology specific to the hazardous waste management field, and the ongoing efforts of the scientific and legal communities in a variety of hazardous waste-related areas. At the manufacturing plant level, hazardous waste management demands a significant commitment of manpower and resources for conducting routine tasks (e.g., compliance programs, waste collection/disposal, recordkeeping, etc.) as well as for extraordinary environmental circumstances such as spill events and remedial actions. The myriad of environmental issues and complexity of environmental problems often exceed in-house plant or corporate staff capabilities or resources. Time constraints can also become involved. When this occurs, the need for consulting services arises.

Like its associates in the civil engineering field, the hazardous waste management consulting firm must offer a wide array of engineering and technical services. Such firms typically employ engineers, scientists, regulatory analysts, chemists, industrial hygienists, and technicians, all possessing expertise and specialized training in hazardous waste management. Many full-service architectural/engineering (A/E) consulting firms offer hazardous waste management and engineering services including compliance assistance, waste characterization, permit preparation, groundwater monitoring, and facility/process design, to name a few. Some environmental specialty firms (i.e., TSD facilities, spill cleanup and remediation contractors, etc.) provide consulting assistance to their clients as needed. Sampling and analysis for performing required hazardous waste determinations and obtaining disposal approvals are examples of these types of services. Still other firms specialize only in hazardous waste management and other related projects such as RCRA compliance matters, hazardous materials training, employee and community right-to-know programs, and environmental litigation support.

Hazardous waste management consulting firms are typically multidisciplinary organizations featuring some or all of the following technical capabilities:

environmental engineering/design
regulatory analysis
sampling and analytical testing (see discussion below)
data collection, management, and evaluation
environmental chemistry
geology/hydrogeology
environmental auditing
environmental permitting

In addition, most qualified firms either own and operate analytical laboratories or maintain service agreements with contract laboratories for a wide range of testing services. Since decisionmaking (e.g., hazardous waste determinations) often relies upon analytical results, data accuracy, reliability, and reproducibility are essential. The rigorous sampling and analysis requirements imposed by RCRA may necessitate the use of an EPA-approved CLP laboratory. These labs strictly employ EPA methodologies (i.e., per EPA publication SW-846) and are required to develop and maintain a QA/QC program with established data acceptance/rejection criteria. Obviously, analytical needs can vary widely from project to project—from pH and metals determinations, which most labs routinely perform, to volatile organics and other nonconventional parameters requiring state-of-the-art instrumentation (e.g., GC/MS) and greater technical expertise. A qualified consultant should be able to provide valuable input with respect to procuring laboratory services or best utilizing those in-house or contract capabilities already available.

Two key factors in the effective use of consulting services are (1) adequate problem identification and (2) establishment of realistic goals and objectives for solving the problem. In many instances, consulting services may be required to make these initial determinations. For example, an environmental audit of an industrial facility typically generates a "laundry list" of environmental deficiencies or problems, some demanding immediate attention, others requiring more thorough evaluation and planning. Adequate problem identification is critical both to effective problem solving and client-consultant relations. Often, a given situation actually consists of several separate but interrelated problems. Ill-defined problems and vague goals (e.g., complete regulatory compliance) pose difficulties in identifying all project tasks, prioritizing among tasks, and in developing accurate cost estimates. The inevitable result is a difference in expectations on the part of client and consultant. Experienced consultants will take it upon themselves to assist their clients in better defining problem situations. This typically involves identifying multiple objectives, potential conflicts and constraints (regulatory, time, and financial), and the possible approaches to resolving the problem(s) at hand, along with the estimated costs associated with each approach. The consultant helps the client make a decision regarding project nature, scope, and timing, in light of the client's budgetary and other constraints. Regulatory enforcement actions sometimes impose overriding constraints that cause the project to proceed at a faster pace, and result in regulatory agency concerns being assigned higher priorities than would otherwise be the case.

The consultant's ability to perform certain tasks is critical to effective problem solving, and justifies the consultant's fees. The critical tasks include:

1. Assess the situation posed by the client.
2. Identify environmental and regulatory problems.
3. Distinguish between substantive and procedural aspects of the problems.
4. Identify internal and external constraints.
5. Develop viable approaches to resolving the problems identified, in consideration of the various constraints.
6. Adequately communicate assessment and recommendations to the client.

Once this initial project scoping is performed (and this can often take quite some time), the remainder of the project is, in many instances, more or less routine implementation of the plan of action decided upon. Proper project scoping allows for the identification of both project milestones and potential problem areas with respect to timely project completion and/or budgetary constraints.

Companies of all sizes and descriptions can usually benefit from establishing a relationship with a competent hazardous waste consulting firm. This is especially true for plants lacking full-time environmental engineering staff or those experiencing routine compliance difficulties. For these types of situations, many consulting firms offer on-demand problem-solving assistance under retainer agreements. One popular feature is a regulatory update service, providing industries with monthly summaries of existing and proposed regulations and other developments in hazardous waste management. Such retainers can also be a boon for environmental compliance personnel without sufficient time to review and digest reams of regulatory information.

For larger projects, companies may wish to engage a full-service environmental consulting firm whose capabilities include hazardous waste management and engineering. This is due, in part, to the inherently interrelated nature of environmental control regulations (i.e., air, water, and hazardous waste) and to technological overlap in the areas of industrial waste treatment and hazardous waste management. For example, a treatment plant design for wastewaters generated from an electroplating facility should include measures for water conservation and reuse as well as the evaluation of recovery of metal ions from plating baths or rinsewaters. These efforts can significantly reduce the size and expense of the necessary treatment equipment. The treated wastewater will be exempt from RCRA regulation, providing it is discharged through a control point subject to Clean Water Act requirements (i.e., an NPDES permit for receiving streams or a sewer permit). The F006 wastewater treatment sludge will be a listed hazardous waste subject to land disposal restrictions, and must meet specified treatment standards (or be delisted) prior to direct land disposal. Hazardous waste issues like these must be addressed early in the planning and design of such facilities in order to ensure full compliance with all appropriate environmental control regulations and to avoid often-costly future problems or modifications.

Once a determination has been made that a particular problem or project exceeds in-house capabilities, the consultant selection process ensues. Companies should consider and select consulting firms which demonstrate:

- a thorough knowledge of current environmental control regulations and regulatory initiatives (e.g., proposed rules)
- necessary engineering/technical staff capabilities
- an understanding of hazardous waste management principles and technologies (including new and innovative designs)
- successful completion of similar projects or hazardous waste management programs
- other project-specific capabilities

There are many factors to be considered in the selection of a hazardous waste management consultant for a particular problem or project, and selection criteria vary widely. However, the following list of suggestions provides some practical guidance for selecting a qualified firm.

1. Assemble a list of qualified environmental or hazardous waste management consulting firms before the need for services arises. Listings of such firms can be found in consultant directories which accompany numerous technical publications (e.g., *Pollution Engineering, Waste Age, Journal of The Water Pollution Control Federation*, etc.), in the journals themselves, in conference proceedings, and even in the Yellow Pages. Lists of qualified consultants may also be made available through industry trade associations and hazardous waste regulatory agencies, especially at the state level. Request Statement of Qualification (SOQ) packages, related project experience or summaries, and a list of government and/or private sector references to aid in your evaluation of each firm's capabilities. Pay particular attention to (1) the number and type of hazardous waste-related projects completed by the firm and (2) experience levels of key individuals identified in the qualification statement. This type of initial screening process should

enable you to better evaluate each firm's overall capabilities and to choose consultants wisely.

2. Unless your company is in an emergency response mode or is experiencing some other environmental crisis, utilize standard procurement procedures for selecting consultants. This usually includes the preparation of a detailed request for services from which consulting firms may submit proposals or quotations. The more detailed, plant-specific information is included in a bid request, the better. A loosely written scope of work will often result in proposals that lack specificity and are therefore difficult to evaluate. Some examples are provided below.

Request for Services—Bad Example:

Consultant shall provide all services necessary to characterize hazardous waste generated at the ACME Company plant. Consultant shall then assist ACME Company in disposing of hazardous waste generated at its plant.

Request for Services—Good Example:

The Consultant shall perform sampling and analysis of paint spray booth solids and filter sludge generated at the ACME Company plant to determine if these wastes meet current RCRA hazardous waste definitions. Upon making these determinations, the Consultant shall provide technical support in terms of evaluating potential disposal sites, filing requests for disposal and/or obtaining disposal permits, scheduling waste transportation and disposal, and preparing waste manifests. The Consultant shall utilize standard methods (per EPA Publication SW-846) in the collection and analysis of all waste samples. Chain-of-custody forms and other sample documentation will be provided to ACME Company within 5 working days after sample collection; all samples shall be held in storage by the Consultant for a minimum period of 90 days following receipt of analysis by the ACME Company. The Consultant shall submit a list of analytical test parameters and a cost estimate for performing the analytical work to ACME Company for authorization prior to submitting the samples for analysis. Upon completion of analytical testing, the Consultant shall provide ACME Company with a written summary of the data, raw laboratory data sheets including QA/QC reports, and recommendations relative to the disposal of the waste materials. The Consultant shall submit a proposed schedule for performing these project activities within 5 working days of receipt of a written notice to proceed and ACME Company purchase order authorization.

Unless you intend to be vague, avoid the bad example request format. Adding more detail to the request for services (as in the second example) helps to clarify the scope of work and should promote greater uniformity in the proposals submitted. One final note on this topic: If you lack experience in preparing bid requests, are unfamiliar with developing a scope of work for hazardous waste projects, or are faced with soliciting quotes for a large-scale project, you may wish to consider utilizing the services of a consultant to assist in (1) developing the necessary scope of work, (2) the preparation of the bid documents themselves, (3) evaluating the proposals received, and (4) selecting a qualified firm to provide the required services.

3. For some larger projects (e.g., closure of a RCRA surface impoundment) or any project with a poorly defined scope of work (e.g., site assessment of an abandoned disposal site), utilize the phased project approach. This involves dividing the project into manageable subunits of work activity as well as soliciting bids and awarding contracts at preestablished milestones. Table 27.1 presents an outline of a phased project for a contaminated site evaluation and cleanup. You may or may not wish to change consultants/contractors or assign separate purchase orders for different project phases. In

Table 27.1   Outline for Phased Project Contaminated Site Evaluation and Cleanup

Phase 1	• Scope of work and project objectives defined • Initial site visit • Regulatory evaluation and record search • Preparation of sampling/analysis plans • Preliminary findings and recommendations report
Phase 2	• Sampling and analysis (e.g., of waste, contaminated soil, groundwater) • Data evaluation • Site-specific cleanup criteria (e.g., health- or risk-based) established • Remedial action alternatives (e.g., no action, source removal, in situ treatment) screened; development of preliminary cost estimates and remedial plans • Regulatory agency coordination
Phase 3	• Project schedules and milestones established • Engineering/design of remedial measures • Plans and specifications prepared (as needed); bids solicited and reviewed; cleanup contractor(s) selected • Regulatory agency coordination
Phase 4	• Construction, start-up, operation of treatment systems • Implementation of remedial measures (e.g., source removal, groundwater recovery and treatment, etc.) • Verification sampling/analysis and continued monitoring (as necessary) • Evaluation of effectiveness of remedial measures; cleanup efforts augmented (as necessary) • Final report detailing cleanup project activities • Regulatory agency coordination

the outline provided in Table 27.1, a full-service environmental engineering firm should be able to perform or direct all of the anticipated project activities. Upon completion of the Phase 1 project, the consultant can prepare and submit proposals for the Phase 2 work activities. If you are satisfied with the work performed in Phase 1, the Phase 2 quotations may simply be a formality, but will certainly serve to reestablish (or modify, in some cases) project objectives and to broaden the financial outlook. This is of obvious benefit to project decisionmaking. As a practical matter and to avoid time-consuming and costly delays, you may wish to request preliminary estimates for subsequent project phases before Phase 1 work is completed, review and refine these estimates, and then simply issue a change order to the existing purchase agreement. Be aware that changing consultants in mid-project can often pose undesirable coordination and continuity problems.

4. In enforcement-sensitive matters, environmental litigation, or any other situation involving a significant potential for environmental risk or liability to the company, consultants should work in close association with competent environmental legal counsel. In the most serious cases, you may wish to request that a consultant be retained by counsel instead of being directly employed by the company. This will afford some degree of protection for any data or information generated under attorney-client privilege. This is not recommended as a means for sheltering or covering up findings which might be damaging to the company. This approach is intended to provide a buffer to allow companies an opportunity to fully define their problems and to consider appropriate corrective measures or strategies in advance of making public disclosures (e.g., to a regulatory agency). Please note that this manner of data/information management in no way supersedes the company's notification requirements under several environmental laws (e.g., CERCLA 103 (a) or (c) notice) in the event that a reportable incident has occurred or is discovered.

Though not typically needed for routine hazardous waste management activities, legal assistance can be helpful in the preparation and filing of permit applications and delisting petitions, in negotiating compliance schedules with regulatory agencies, investigating environmental impairment insurance claims, preparing environmental indemnification agreements for industrial real estate transactions, etc. As is obvious from some of the landmark Superfund cases, environmental problems and their associated liabilities can sometimes result in financial catastrophe for a company. In these types of situations, experienced legal assistance and technical support are absolutely necessary. Such services are also routinely sought in environmental policy matters and a variety of other related areas (e.g., health and safety, right-to-know, regulatory agency negotiations, etc.). Corporate attorneys acting alone or in concert with independent environmental counsel often serve as liaisons between their companies or clients and regulatory agencies.

A determination as to the type and degree of legal involvement required for a given situation is a key element of any environmental project decision tree. In enforcement negotiations and environmental litigation, all communication between the company and the agency flows through proper legal channels. Keep in mind that a lawyer who is anxious to litigate (i.e., "fight this thing if it takes every cent this client has") may not have the company's best interests in mind. Most experienced lawyers and consultants agree that, in most instances, litigation is a last-resort measure to be used only after other means of resolution (e.g., settlement conferences or hearings, administrative appeals, consent agreements, etc.) have been exhausted.

5. Fees for hazardous waste management or environmental consulting services vary widely according to project type and size. With early project planning (e.g., during the initial consultant screening process) you may be afforded the opportunity to "shop around" for a consulting firm by comparing hourly rate sheets, supply charges, and laboratory price lists. You should be aware that the cost of these types of consulting services can be significant. Charges of $100 per hour are not unusual and some professional services (e.g., expert testimony) can cost up to $1500 per day plus expenses. Fees are largely dependent upon a consultant's experience level—staff or junior level engineers/technicians may be billed at $40–$60 per hour, while senior staff rates typically range from $70 to $100 per hour. Locality is another important factor; unit rates for Washington, D.C.-based consultants typically exceed those for Midwestern firms, but not always. While some projects may be performed on a lump sum basis (e.g., RCRA plans or permit preparations), most hazardous waste projects are performed on a time and materials (T/M) basis due to inherent uncertainties relative to disposal approvals, sampling/analysis requirements, regulatory agency interaction, and many other factors. Estimates of project costs are typically available on T/M projects. It is important to remember in comparing hourly fees that knowledge, experience, and productivity largely determine the value of the consultant's efforts. If possible, try to include expenses associated with routine consulting services and analytical services in your operating budget. Large projects such as onsite remediation of abandoned disposal sites or closure of RCRA treatment, storage, or disposal facilities generally require special funding. Here again, the phased project approach offers certain advantages in that capital and operating expenses associated with subsequent project phases can be factored into company financial plans.

6. The hazardous waste management or environmental consultant should frequently interface with designated company employees (corporate and plant

level). The designated individuals should be familiar with all waste-generating processes, waste management practices, and the overall environmental condition of the facility. At the plant level, hazardous waste managers or wastewater treatment plant supervisors are good candidates for this role; purchasing agents who lack specific knowledge of plant processes and environmental control regulations are often not well suited. In addition to providing an important information link between the company and the consultant, the assigned individual can also serve as the company's representative to ensure that project schedules and objectives are achieved with the funds allocated. For larger projects, weekly or biweekly performance and budget evaluations are recommended.

7. In most instances, both the consultant and the company representatives should work to establish and maintain a good working relationship with regulatory agency officials. When an adversarial relationship exists, agency officials seldom make use of the regulatory discretion allowed them in many cases. Most typically encountered hazardous waste management enforcement or other compliance problems can be resolved through cooperation and positive (proactive) approaches. Those which cannot usually involve significant fines or penalties and require legal assistance.

The consultant's principal role in hazardous waste management is that of problem-solver and advisor. The impact of RCRA and other environmental control regulations, and the demands imposed by these regulations on industry, have created a permanent place for the hazardous waste consultant in the business community. The information presented in this chapter is intended to better define the consultant's role in hazardous waste management and to provide some practical guidance relative to the selection and use of consulting services. If your company is experiencing a hazardous waste problem or you have concerns about environmental conditions or health and safety issues at a particular plant site, and existing personnel have neither the time nor the expertise to address the situation, it may very well be advisable to engage the services of a qualified consultant.

APPENDIX A

EPA Form 8700-12:
Notification of Hazardous Waste Activity

Please print or type with ELITE type *(12 characters per inch)* in the unshaded areas only

Form Approved. OMB No. 2050-0028. Expires 9-30-88.
GSA No. 0246-EPA-OT

United States Environmental Protection Agency
Washington, DC 20460

♻EPA Notification of Hazardous Waste Activity

Please refer to the *Instructions for Filing Notification* before completing this form. The information requested here is required by law *(Section 3010 of the Resource Conservation and Recovery Act).*

For Official Use Only

Comments

C

C

Installation's EPA ID Number

T/A   C

Approved

Date Received
(yr.   mo.   day)

C

F

1

I. Name of Installation

II. Installation Mailing Address

Street or P.O. Box

C

3

City or Town

State

ZIP Code

C

4

III. Location of Installation

Street or Route Number

C

5

City or Town

State

ZIP Code

C

6

IV. Installation Contact

Name and Title *(last, first, and job title)*

Phone Number *(area code and number)*

C

2

V. Ownership

A. Name of Installation's Legal Owner

B. Type of Ownership *(enter code)*

C

R

VI. Type of Regulated Waste Activity *(Mark 'X' in the appropriate boxes. Refer to instructions.)*

A. Hazardous Waste Activity

☐ 1a. Generator   ☐ 1b. Less than 1,000 kg/mo.

☐ 2. Transporter

☐ 3. Treater/Storer/Disposer

☐ 4. Underground Injection

☐ 5. Market or Burn Hazardous Waste Fuel
(enter 'X' and mark appropriate boxes below)

   ☐ a. Generator Marketing to Burner

   ☐ b. Other Marketer

   ☐ c. Burner

B. Used Oil Fuel Activities

☐ 6. Off-Specification Used Oil Fuel
(enter 'X' and mark appropriate boxes below)

   ☐ a. Generator Marketing to Burner

   ☐ b. Other Marketer

   ☐ c. Burner

☐ 7. Specification Used Oil Fuel Marketer *(or On site Burner)*
Who First Claims the Oil Meets the Specification

VII. Waste Fuel Burning: Type of Combustion Device *(enter 'X' in all appropriate boxes to indicate type of combustion device(s) in which hazardous waste fuel or off-specification used oil fuel is burned. See instructions for definitions of combustion devices.)*

☐ A. Utility Boiler   ☐ B. Industrial Boiler   ☐ C. Industrial Furnace

VIII. Mode of Transportation *(transporters only — enter 'X' in the appropriate box(es)*

☐ A. Air   ☐ B. Rail   ☐ C. Highway   ☐ D. Water   ☐ E. Other *(specify)*

IX. First or Subsequent Notification

Mark 'X' in the appropriate box to indicate whether this is your installation's first notification of hazardous waste activity or a subsequent notification. If this is not your first notification, enter your installation's EPA ID Number in the space provided below.

☐ A. First Notification   ☐ B. Subsequent Notification *(complete item C)*

C. Installation's EPA ID Number

EPA Form 8700-12 (Rev. 11-85) Previous edition is obsolete.

Continue on reverse

Appendix A. EPA Form 8700-12: Notification of Hazardous Waste Activity (front).

		ID — For Official Use Only																
C																T/A	C	
W																	1	

X. Description of Hazardous Wastes *(continued from front)*

A. Hazardous Wastes from Nonspecific Sources. Enter the four-digit number from 40 *CFR* Part 261.31 for each listed hazardous waste from nonspecific sources your installation handles. Use additional sheets if necessary.

1	2	3	4	5	6
7	8	9	10	11	12

B. Hazardous Wastes from Specific Sources. Enter the four-digit number from 40 *CFR* Part 261.32 for each listed hazardous waste from specific sources your installation handles. Use additional sheets if necessary.

13	14	15	16	17	18
19	20	21	22	23	24
25	26	27	28	29	30

C. Commercial Chemical Product Hazardous Wastes. Enter the four-digit number from 40 *CFR* Part 261.33 for each chemical substance your installation handles which may be a hazardous waste. Use additional sheets if necessary.

31	32	33	34	35	36
37	38	39	40	41	42
43	44	45	46	47	48

D. Listed Infectious Wastes. Enter the four-digit number from 40 *CFR* Part 261.34 for each hazardous waste from hospitals, veterinary hospitals, or medical and research laboratories your installation handles. Use additional sheets if necessary.

49	50	51	52	53	54

E. Characteristics of Nonlisted Hazardous Wastes. Mark 'X' in the boxes corresponding to the characteristics of nonlisted hazardous wastes your installation handles. *(See 40 CFR Parts 261.21 — 261.24)*

☐ 1. Ignitable *(D001)*   ☐ 2. Corrosive *(D002)*   ☐ 3. Reactive *(D003)*   ☐ 4. Toxic *(D000)*

XI. Certification

I certify under penalty of law that I have personally examined and am familiar with the information submitted in this and all attached documents, and that based on my inquiry of those individuals immediately responsible for obtaining the information, I believe that the submitted information is true, accurate, and complete. I am aware that there are significant penalties for submitting false information, including the possibility of fine and imprisonment.

Signature	Name and Official Title *(type or print)*	Date Signed

EPA Form 8700-12 (Rev. 11-85) Reverse

Appendix A.   EPA Form 8700-12: Notification of Hazardous Waste Activity (back).

State Hazardous Waste Management Agencies

Alabama
Alabama Department of
  Environmental Management
Land Division
1751 Federal Drive
Montgomery, Alabama 36130
(205) 271-7730

Alaska
Department of Environmental
  Conservation
P.O. Box 0
Juneau, Alaska 99811
Program Manager: (907) 465-2666
Northern Regional Office
  (Fairbanks): (907) 452-1714
South-Central Regional Office
  (Anchorage): (907) 274-2533
Southeast Regional Office
  (Juneau): (907) 789-3151

Arizona
Arizona Department of Health
  Services
Office of Waste and Water Quality
2005 North Central Avenue,
  Room 304
Phoenix, Arizona 85004
Hazardous Waste Management:
(602) 255-2211

Arkansas
Department of Pollution Control
  and Ecology
Hazardous Waste Division
P.O. Box 9583
8001 National Drive
Little Rock, Arkansas 72219
(501) 562-7444

California
Department of Health Services
Toxic Substances Control Division
714 P Street, Room 1253
Sacramento, California 95814
(916) 324-1826

State Water Resources Control
  Board
Division of Water Quality
P.O. Box 100
Sacramento, California 95801
(916) 322-2867

Colorado
Colorado Department of Health
Waste Management Division
4210 E. 11th Avenue
Denver, Colorado 80220
(303) 320-8333 Ext. 4364

Connecticut
Department of Environmental
  Protection
Hazardous Waste Management
  Section
State Office Building
165 Capitol Avenue
Hartford, Connecticut 06106
(203) 566-8843, 8844

Connecticut Resource Recovery
  Authority
179 Allyn Street, Suite 603
Professional Building
Hartford, Connecticut 06103
(203) 549-6390

Delaware
Department of Natural Resources
  and Environmental Control
Waste Management Section
P.O. Box 1401
Dover, Delaware 19903
(302) 736-4781

Florida
Department of Environmental
  Regulation
Solid and Hazardous Waste Section
Twin Towers Office Building
2600 Blair Stone Road
Tallahassee, Florida 32301
(904) 488-0300

Georgia
Georgia Environmental Protection
  Division
Hazardous Waste Management
  Program
Land Protection Branch
Floyd Towers East, Suite 1154
205 Butler Street, S.E.
Atlanta, Georgia 30334
(404) 656-2833
Toll-Free: (800) 334-2373

Hawaii
Department of Health
  Environmental Health Division
P.O. Box 3378
Honolulu, Hawaii 96801
(808) 548-4383

Idaho
Department of Health and Welfare
Bureau of Hazardous Materials
450 West State Street
Boise, Idaho 83720
(208) 334-5879

Illinois
Environmental Protection Agency
Division of Land Pollution Control
2200 Churchill Road, #24
Springfield, Illinois 62706
(217) 782-6761

Indiana
Department of Environment
  Management
Office of Solid and Hazardous
  Waste
105 South Meridian
Indianapolis, Indiana 46225
(317) 232-4535

Iowa
U.S. EPA Region VII
Hazardous Materials Branch
726 Minnesota Avenue
Kansas City, Kansas 66101
(913) 236-2888
Iowa RCRA Toll-Free:
(800) 223-0425

Kansas
Department of Health and
  Environment
Bureau of Waste Management
Forbes Field, Building 321
Topeka, Kansas 66620
(913) 862-9360 Ext. 292

Kentucky
Natural Resources and
  Environmental Protection Cabinet
Division of Waste Management
18 Reilly Road
Frankfort, Kentucky 40601
(502) 564-6716

Louisiana
Department of Environmental Quality
Hazardous Waste Division
P.O. Box 44307
Baton Rouge, Louisiana 70804
(504) 342-1227

Maine
Department of Environmental
  Protection
Bureau of Oil and Hazardous
  Materials Control
State House Station #17
Augusta, Maine 04333
(207) 289-2651

Maryland
Department of Health and Mental
  Hygiene
Maryland Waste Management
  Administration
Office of Environmental Programs
201 West Preston Street, Room A3
Baltimore, Maryland 21201
(301) 225-5709

Massachusetts
Department of Environmental
  Quality Engineering
Division of Solid and Hazardous
  Waste
One Winter Street, 5th Floor
Boston, Massachusetts 02108
(617) 292-5589
(617) 292-5851

Michigan
Michigan Department of Natural
  Resources
Hazardous Waste Division
Waste Evaluation Unit
Box 30028
Lansing, Michigan 48909
(517) 373-2730

Minnesota
Pollution Control Agency
Solid and Hazardous Waste Division
1935 West County Road, B-2
Roseville, Minnesota 55113
(612) 296-7282

Mississippi
Department of Natural Resources
Division of Solid and Hazardous
  Waste Management
P.O. Box 10385
Jackson, Mississippi 39209
(601) 961-5062

Missouri
Department of Natural Resources
Waste Management Program
P.O. Box 176
Jefferson City, Missouri 65102
(314) 751-3176
Missouri Hotline: (800) 334-6946

Montana
Department of Health and
  Environmental Sciences
Solid and Hazardous Waste Bureau
Cogswell Building, Room B-201
Helena, Montana 59620
(406) 444-2821

Nebraska
Department of Environmental
  Control
Hazardous Waste Management
  Section
P.O. Box 94877
State House Station
Lincoln, Nebraska 68509
(402) 471-2186

Nevada
Division of Environmental Protection
Waste Management Program
Capitol Complex
Carson City, Nevada 89710
(702) 885-4670

New Hampshire
Department of Health and Human
  Services
Division of Public Health Services
Office of Waste Management
Health and Welfare Building
Hazen Drive
Concord, New Hampshire
03301-6527
(603) 271-4608

New Jersey
Department of Environmental
  Protection
Division of Waste Management
32 East Hanover Street, CN-028
Trenton, New Jersey 08625
Hazardous Waste Advisement
  Program: (609) 292-8341

New Mexico
Environmental Improvement
  Division
Ground Water and Hazardous Waste
  Bureau
Hazardous Waste Section
P.O. Box 968
Santa Fe, New Mexico 87504-0968
(505) 827-2922

New York
Department of Environmental
  Conservation
Bureau of Hazardous Waste
  Operations
50 Wolf Road, Room 209
Albany, New York 12233
(518) 457-0530
SQG Hotline: (800) 631-0666

North Carolina
Department of Human Resources
Solid and Hazardous Waste
  Management Branch
P.O. Box 2091
Raleigh, North Carolina 27602
(919) 733-2178

North Dakota
Department of Health
Division of Hazardous Waste
  Management and Special Studies
1200 Missouri Avenue
Bismarck, North Dakota 58502-5520
(701) 224-2366

Ohio
Ohio EPA
Division of Solid and Hazardous
  Waste Management
361 East Broad Street
Columbus, Ohio 43266-0558
(614) 466-7220

Oklahoma
Waste Management Service
Oklahoma State Department of
  Health
P.O. Box 53551
Oklahoma City, Oklahoma 73152
(405) 271-5338

Oregon
Hazardous and Solid Waste Division
P.O. Box 1760
Portland, Oregon 97207
(503) 229-6534
Toll-Free: (800) 452-4011

Pennsylvania
Bureau of Waste Management
Division of Compliance Monitoring
P.O. Box 2063
Harrisburg, Pennsylvania 17120
(717) 787-6239

Rhode Island
Department of Environmental
  Management
Division of Air and Hazardous
  Materials
Room 204, Cannon Building
75 Davis Street
Providence, Rhode Island 02908
(401) 277-2797

South Carolina
Department of Health and
  Environmental Control
Bureau of Solid and Hazardous
  Waste Management
2600 Bull Street
Columbia, South Carolina 29201
(803) 734-5200

South Dakota
Department of Waste and Natural
  Resources
Office of Air Quality and Solid
  Waste
Foss Building, Room 217
Pierre, South Dakota 57501
(605) 773-3153

Tennessee
Division of Solid Waste
  Management
Tennessee Department of Public
  Health
701 Broadway
Nashville, Tennessee 37219-5403
(615) 741-3424

Texas
Texas Water Commission
Hazardous and Solid Waste Division
Attn: Program Support Section
1700 North Congress
Austin, Texas 78711
(512) 463-7761

Utah
Department of Health
Bureau of Solid and Hazardous
  Waste Management
P.O. Box 16700
Salt Lake City, Utah 84116-0700
(801) 538-6170

Vermont
Agency of Environmental Conservation
103 South Main Street
Waterbury, Vermont 05676
(802) 244-8702

Virginia
Department of Health
Division of Solid and Hazardous
  Waste Management
Monroe Building, 11th Floor
101 North 14th Street
Richmond, Virginia 23219
(804) 225-2667
Hazardous Waste Hotline:
(800) 552-2075

Washington
Department of Ecology
Solid and Hazardous Waste Program
Mail Stop PV-11
Olympia, Washington 98504-8711
(206) 459-6322
In-State: (800) 633-7585

West Virginia
Division of Water Resources
Solid and Hazardous Waste/
  Ground Water Branch
1201 Greenbrier Street
Charleston, West Virginia 25311
(304) 348-5935

Wisconsin
Department of Natural Resources
Bureau of Solid Waste Management
P.O. Box 7921
Madison, Wisconsin 53707
(608) 266-1327

Wyoming
Department of Environmental
  Quality
Solid Waste Management Program
122 West 25th Street
Cheyenne, Wyoming 82002
(307) 777-7752

or

EPA Region VIII
Waste Management Division
  (8HWM-ON)
One Denver Place
999 18th Street, Suite 1300
Denver, Colorado 80202-2413
(303) 293-1502

APPENDIX C

EPA Regional Waste Management Offices

EPA Region I
State Waste Programs Branch
JFK Federal Building
Boston, Massachusetts 02203
(617) 223-3468
Connecticut, Massachusetts, Maine,
New Hampshire, Rhode Island,
Vermont

EPA Region II
Air and Waste Management Division
26 Federal Plaza
New York, New York 10278
(212) 264-5175
New Jersey, New York, Puerto
Rico, Virgin Islands

EPA Region III
Waste Management Branch
841 Chestnut Street
Philadelphia, Pennsylvania 19107
(215) 597-9336
Delaware, Maryland, Pennsylvania,
Virginia, West Virginia, District
of Columbia

EPA Region IV
Hazardous Waste Management
Division
345 Courtland Street, N.E.
Atlanta, Georgia 30365
(404) 347-3016
Alabama, Florida, Georgia, Kentucky,
Mississippi, North Carolina, South
Carolina, Tennessee

EPA Region V
RCRA Activities
230 South Dearborn Street
Chicago, Illinois 60604
(312) 353-2000
Illinois, Indiana, Michigan,
Minnesota, Ohio, Wisconsin

EPA Region VI
Air and Hazardous Materials Division
1201 Elm Street
Dallas, Texas 75270
(214) 767-2600
Arkansas, Louisiana, New Mexico,
Oklahoma, Texas

EPA Region VII
RCRA Branch
726 Minnesota Avenue
Kansas City, Kansas 66101
(913) 236-2800
Iowa, Kansas, Missouri, Nebraska

EPA Region VIII
Waste Management Division
  (8HWM-ON)
One Denver Place
999 18th Street, Suite 1300
Denver, Colorado 80202-2413
(303) 293-1502
Colorado, Montana, North Dakota,
  South Dakota, Utah, Wyoming

EPA Region IX
Toxics and Waste Management Di-
  vision
215 Fremont Street
San Francisco, California 94105
(415) 974-7472
Arizona, California, Hawaii,
  Nevada, American Samoa, Guam,
  Trust Territories of the Pacific

EPA Region X
Waste Management Branch—MS-530
1200 Sixth Avenue
Seattle, Washington 98101
(206) 442-2777
Alaska, Idaho, Oregon, Washington

Commonly Used RCRA/CERCLA Acronyms

Hazardous Material/Waste Acronyms and Recognized Abbreviations

ACM	asbestos-containing material
BDAT	Best Demonstrated Available Technology
CAS RN	Chemical Abstracts Service Registry Number
CERCLA	Comprehensive Environmental Response, Compensation and Liability Act ("Superfund")
CERCLIS	Comprehensive Environmental Response, Compensation and Liability Information System (inventory), formerly the ERRIS List
CFR	*Code of Federal Regulations*
40 CFR	Protection of the Environment—U.S. EPA
49 CFR	Transportation—U.S. DOT
29 CFR	General Industry Standards—OSHA
CHEMTREC	Chemical Transportation Emergency Center (CMA)
CMA	Chemical Manufacturers Association
DOT	U.S. Department of Transportation
EP	extraction procedure (toxicity)
EPA	U.S. Environmental Protection Agency
EPCRA	Emergency Planning and Community Right-To-Know Act (Title III of SARA)
FR	*Federal Register*
HMT	Hazardous Materials Table (49 CFR 172.101)
HMTA	Hazardous Materials Transportation Act
HSWA	Hazardous and Solid Waste Amendments of 1984 (to RCRA)

HWM	hazardous waste management
MSDS	Material Safety Data Sheet
MSWLF	municipal solid waste landfill
NCP	National Oil and Hazardous Substances Pollution Contingency Plan (National Contingency Plan—40 CFR Part 300)
NFPA	National Fire Protection Association
NIOSH	National Institute for Occupational Safety and Health
n.o.s.	not otherwise specified
NPDES	National Pollutant Discharge Elimination System
NPL	National Priorities List
NRC	National Response Center
O/O	owner/operator
ORM	other regulated material
OSHA	U.S. Occupational Safety and Health Administration
PCB	polychlorinated biphenyl
PEL	Permissible Exposure Limit
POTW	publicly owned treatment works (sewage treatment plant)
PRP	potentially responsible party (often "responsible party" [RP])
RCRA	Resource Conservation and Recovery Act of 1976
RFA/RFI	RCRA Facility Assessment/RCRA Facility Investigation
RQ	reportable quantity
RSPA	Research and Special Programs Administration (DOT)
SARA	Superfund Amendments and Reauthorization Act of 1986 (to CERCLA)
SIC	Standard Industrial Classification (Code)
SPCC	Spill Prevention Control and Countermeasures (Plan)
SQG	small quantity generator
SWDA	Solid Waste Disposal Act (RCRA predecessor)
SWMU	solid waste management unit (often SMU)
SW-846	"Test Methods for Solid Waste: Physical/Chemical Methods"
TCLP	Toxicity Characteristic Leaching Procedure
TLV	Threshold Limit Value
TSD	treatment, storage, or disposal facility (often TSDF)
UIC	underground injection control
UST	underground storage tank
VOC	volatile organic compound (chemical)
WWT	wastewater treatment

List of Acronyms and Recognized Abbreviations in the CERCLA Program and Select Other Programs

ACL	Alternate Concentration Limit
AHERA	Asbestos Hazard Emergency Response Act of 1986 (Title II of TSCA)
ARAR	Applicable or Relevant and Appropriate Requirement (cleanup standard)
ATSDR	Agency for Toxic Substances and Disease Registry
CAA	Clean Air Act
CAER	Community Awareness and Emergency Response Program (CMA)
CAS RN	Chemical Abstracts Service Registry Number
CEPP	Chemical Emergency Preparedness Program (EPA)
CERCLA	Comprehensive Environmental Response, Compensation and Liability Act ("Superfund")
CERCLIS	Comprehensive Environmental Response, Compensation and Liability Information System (inventory), formerly the ERRIS List
CHEMTREC	Chemical Transportation Emergency Center (CMA)
CMA	Chemical Manufacturers Association
CWA	Clean Water Act
EHS	extremely hazardous substance
EPA	U.S. Environmental Protection Agency
EPCRA	Emergency Planning and Community Right-to-Know Act (Title III of SARA)
FIFRA	Federal Insecticide, Fungicide, and Rodenticide Act
HCS	Hazard Communication Standard (OSHA 29 CFR 1910.1200)
HRS	hazard ranking system
HSWA	Hazardous and Solid Waste Amendments of 1984 (to RCRA)
LEPC	local emergency planning committee
MCL	maximum contaminant level
MCLG	maximum contaminant level goal
NBAR	Nonbinding Preliminary Allocation of Responsibility
NCP	National Oil and Hazardous Substances Pollution Contingency Plan (National Contingency Plan—40 CFR Part 300)
NCRIC	National Chemical Response and Information Center (CMA)
NESHAP	National Emission Standard for Hazardous Air Pollutants
NPL	National Priorities List

NRC	National Response Center
NRT	National Response Team
NTP	National Toxicology Program
OSC	on-scene coordinator
OTA	Office of Technology Assessment of the U.S. Congress
PA/SI	preliminary assessment/site inspection
PRP	potentially responsible party (often ''responsible party'' [RP])
RAP	remedial action plan
RCRA	Resource Conservation and Recovery Act
RA/RD	remedial action/remedial design
RI/FS	remedial investigation/feasibility study
ROD	record of decision
RP	responsible party (*see* PRP)
RQ	reportable quantity
RRT	Regional Response Team
RTECS	Registry of Toxic Effects of Chemical Substances
SARA	Superfund Amendments and Reauthorization Act of 1986 (to CERCLA)
SDWA	Safe Drinking Water Act
SERC	State Emergency Response Commission
SITE	Superfund Innovative Technology Evaluation (program)
SWDA	Solid Waste Disposal Act (RCRA predecessor)
TITLE III	*See* EPCRA
TPQ	Threshold Planning Quantity
TSCA	Toxic Substances Control Act
UIC	underground injection control
UST	underground storage tank
WQA	Water Quality Act (1987 CWA amendments)
WQC	water quality criteria

APPENDIX E

Definitions of Selected Regulatory Terms

California List wastes   A group of liquid hazardous wastes, including ones with PCBs, cyanides, heavy metals, and halogenated organic compounds, that EPA must have evaluated by July 8, 1987, to determine if they should be banned from land disposal or if restrictions should be placed on the land disposal of these wastes. See HSWA Section 201(d)(2).

CERCLA   The Comprehensive Environmental Response, Compensation and Liability Act of 1980, as amended by the Superfund Amendments and Reauthorization Act of 1986 (SARA). CERCLA gives the federal government the power to respond to releases, or threatened releases, of any hazardous substance into the environment, as well as to a release of a pollutant or contaminant that may present an imminent and substantial danger to public health or welfare. CERCLA established a Hazardous Substance Trust Fund (Superfund), available to finance responses taken by the federal government.

CFR   The *Code of Federal Regulations,* a document containing all finalized regulations.

closure   The act of securing a hazardous waste management facility pursuant to the requirements of 40 CFR Part 264 or 265. *See also* final closure.

commercial chemical product   Products and intermediates, off-specification variants, spill residues, and container residues listed in 40 CFR 261.33 (e) and (f).

container   Any portable device in which material is stored, transported, treated, disposed of, or otherwise handled.

designated facility   A hazardous waste treatment, storage, or disposal facility which has received an EPA or state permit (or has interim status) or that is regulated at 40 CFR 261.6(c)(2) or Subpart F of 40 CFR Part 266, and has been designated on the manifest by the generator as the facility to which the generator's waste should be delivered.

discharge or hazardous waste discharge The accidental or intentional spilling, leaking, pumping, pouring, emitting, emptying, or dumping of hazardous waste into or on any land or water.

disposal The discharge, deposit, injection, dumping, spilling, leaking, or placing of any solid waste or hazardous waste into or on any land or water so that such solid or hazardous waste or any constituent thereof may enter the environment or be emitted into the air or discharged into any waters, including ground water.

disposal facility A facility or part of a facility at which hazardous waste is intentionally placed into or on any land or water, and at which waste will remain after closure.

EPA identification number The unique number assigned by EPA to each generator or transporter of hazardous waste, and to each treatment, storage, or disposal facility.

existing facility A HWM facility which was in operation or for which construction commenced on or before November 19, 1980.

facility or HWM facility 40 CFR 260.10: all contiguous land, structures, other appurtenances, and improvements on the land, used for treating, storing, or disposing of hazardous waste. A facility may consist of several treatment, storage, or disposal operational units (e.g., one or more landfills, surface impoundments, or combinations of them). 40 CFR 270.2: any HWM facility or other facility or activity (including land or appurtenances thereto) that is subject to regulation under the RCRA program.

Federal Register **(FR)** A document published daily by the federal government that contains proposed and final regulations.

final closure The closure of all hazardous waste management units at the facility in accordance with all applicable closure requirements so that hazardous waste management activities regulated at 40 CFR Parts 264 or 265 and requiring a permit or interim status are no longer conducted at the facility.

generator Any person, by site, whose act or process produces hazardous waste or whose act first causes a hazardous waste to become subject to regulation.

hazardous waste management unit A contiguous area of land on or in which hazardous waste is placed, or the largest area in which there is significant likelihood of mixing hazardous waste constituents in the same area. Examples of hazardous waste management units include a surface impoundment, a waste pile, a land treatment area, a landfill cell, an incinerator, a tank and its associated piping and underlying containment system, and a container storage area. A container alone does not constitute a unit; the unit includes containers and the land or pad upon which they are placed.

HSWA The Hazardous and Solid Waste Amendments of 1984 (Public Law 98616) that significantly expanded both the scope and the coverage of RCRA.

interim status   Allows owners and operators of TSD facilities that were in existence, or for which construction had commenced, prior to November 19, 1980 to continue to operate without a permit after this date. Owners and operators of TSD facilities are eligible for interim status on an ongoing basis if the TSD facility is in existence on the effective date of regulatory changes under RCRA that cause the facility to be subject to Subtitle C regulation. Owners and operators in interim status are subject to and must comply with the applicable standards in 40 CFR Part 265. Interim status is gained through the notification process and by submitting Part A of the permit application.

land disposal   The placement of hazardous waste in or on the land. Includes, but is not limited to, placement in a landfill, surface impoundment, waste pile, injection well, land treatment facility, salt dome formation, salt bed formation, underground mine or cave, or concrete vault or bunker intended for disposal purposes.

landfill   A disposal facility or part of a facility where hazardous waste is placed in or on land and which is not a land treatment facility, a surface impoundment, or an injection well.

leachate   Any liquid, including any suspended components in the liquid, that has percolated through or drained from hazardous waste.

management or hazardous waste management   The systematic control of the collection, source separation, storage, transportation, processing, treatment, recovery, and disposal of hazardous waste.

offsite   The opposite of onsite (see next term).

onsite   On the same or geographically contiguous property which may be divided by public or private right(s)-of-way, provided the entrance and exit between the properties are at a crossroads intersection, i.e., access is by crossing as opposed to going along the right(s)-of-way. Noncontiguous property owned by the same person but connected by a right-of-way which the person controls and to which the public does not have access is also considered onsite property.

open dump   Any facility or site where solid waste is disposed of, which is not a sanitary landfill meeting the criteria listed in 40 CFR Part 257 and which is not a facility for the disposal of hazardous waste.

Part A   The first part of the two-part application that must be submitted by a TSD facility to receive a permit. Part A contains general facility information. There is a standard form for the Part A.

Part B   The second part of the permit application that includes detailed and highly technical information concerning the TSD facility in question. There is no standard form for the Part B; the facility must submit information based on the regulatory requirements.

permit   An authorization, license, or equivalent control document issued by EPA or an authorized state to implement the regulatory requirements of RCRA. Permit does not include interim status.

personnel or facility personnel   All persons who work at, or oversee the operations of, a hazardous waste facility, and whose actions or failure to act may result in noncompliance with regulatory requirements.

Regional Administrator   The highest ranking official in each of the 10 EPA regions.

regulation   The legal mechanism that spells out how a statute's broad policy directives are to be carried out. Regulations are published in the *Federal Register* and then codified in the *Code of Federal Regulations.*

representative sample   A sample of a universe or whole (e.g., waste pile, lagoon, ground water, or waste stream) which can be expected to exhibit the average properties of the universe or whole.

site   The land or water area where any facility or activity is physically located or conducted, including adjacent land used in connection with the facility or activity.

sludge   Any solid, semisolid, or liquid waste generated from a municipal, commercial, or industrial wastewater treatment plant, water supply treatment plant, or air pollution control facility, exclusive of the treated effluent from a wastewater treatment plant.

statute   The law as passed by Congress and signed by the President.

storage   The holding of hazardous waste for a temporary period, at the end of which the hazardous waste is treated, disposed of, or stored elsewhere.

Superfund   *See* CERCLA.

surface impoundment   A facility or part of a facility which is a natural topographic depression, manmade excavation, or diked area formed primarily of earthen materials (although it may be lined with manmade materials), which is designed to hold an accumulation of liquid wastes or wastes containing free liquids, and which is not an injection well. Examples of surface impoundments are holding, storage, settling, and aeration pits, ponds, and lagoons.

tank   A stationary device, designed to contain an accumulation of hazardous waste, which is constructed primarily of nonearthen materials (e.g., wood, concrete, steel, plastic) that provide structural support.

tank system   A hazardous waste storage or treatment tank and its associated ancillary equipment and containment system.

transfer facility   Any transportation-related facility, including loading docks, parking areas, storage areas, and other similar areas where shipments of hazardous waste are held during the normal course of transportation.

transporter   A person engaged in the offsite transportation of hazardous waste by air, rail, highway, or water, if such transportation requires a manifest under 40 CFR Part 262.

treatment   Any method, technique, or process, including neutralization, designed to change the physical, chemical, or biological character or composition of any hazardous waste so as to neutralize such waste; to recover energy or material resources from the waste; to render it nonhazardous or less hazardous; or to make it safer to transport, store or dispose of, amenable for recovery or storage, or reduced in volume.

TSD   Treatment, storage, or disposal (as in TSD facility).

wastewater treatment unit   A device which:

1. is part of a wastewater treatment facility which is subject to regulation under either section 402 or 307(b) of the Clean Water Act;

2. receives and treats or stores an influent wastewater which is a hazardous waste; generates and accumulates a wastewater treatment sludge which is a hazardous waste; or treats or stores a wastewater treatment sludge which is a hazardous waste; and

3. meets the definition of tank in 40 CFR 260.10.

EPA Form 7530-1: Notification for Underground Storage Tanks

Notification for Underground Storage Tanks

FORM APPROVED
OMB NO. 2050-0068
APPROVAL EXPIRES 9-30-91

EPA estimates public reporting burden for this form to average 30 minutes per response, including time for reviewing instructions, gathering and maintaining the data needed, and completing and reviewing the form. Send comments regarding this burden estimate to Chief, Information Policy Branch, PM-223, U.S. Environmental Protection Agency, 401 M St., S.W., Washington, D.C. 20460; and to the Office of Information and Regulatory Affairs, Office of Management and Budget, Washington, D.C. 20503, marked "Attention: Desk Officer for EPA."

STATE USE ONLY

I.D. Number

Date Received

GENERAL INFORMATION

Notification is required by Federal law for all underground tanks that have been used to store regulated substances since January 1, 1974, that are in the ground as of May 8, 1986, or that are brought into use after May 8, 1986. The information requested is required by Section 9002 of the Resource Conservation and Recovery Act, (RCRA), as amended.

The primary purpose of this notification program is to locate and evaluate underground tanks that store petroleum or hazardous substances. It is expected that the information you provide will be based on reasonably available records, or, in the absence of such records, your knowledge, belief, or recollection.

Who Must Notify? Section 9002 of RCRA, as amended, requires that, unless exempted, owners of underground tanks that store regulated substances must notify designated State or local agencies of the existence of their tanks. Owner means—
(a) in the case of an underground storage tank in use on November 8, 1984, or brought into use after that date, any person who owns an underground storage tank used for the storage, use, or dispensing of regulated substances, and
(b) in the case of any underground storage tank in use before November 8, 1984, but no longer in use on that date, any person who owned such tank immediately before the discontinuation of its use.

What Tanks Are Included? Underground storage tank is defined as any one or combination of tanks that (1) is used to contain an accumulation of "regulated substances," and (2) whose volume (including connected underground piping) is 10% or more beneath the surface of the ground. Some examples are underground tanks storing: 1. gasoline, used oil, or diesel fuel, and 2. industrial solvents, pesticides, herbicides or fumigants.

What Tanks Are Excluded? Tanks removed from the ground are not subject to notification. Other tanks excluded from notification are:
1. farm or residential tanks of 1,100 gallons or less capacity used for storing motor fuel for noncommercial purposes;
2. tanks used for storing heating oil for consumptive use on the premises where stored;
3. septic tanks;

4. pipeline facilities (including gathering lines) regulated under the Natural Gas Pipeline Safety Act of 1968, or the Hazardous Liquid Pipeline Safety Act of 1979, or which is an intrastate pipeline facility regulated under State laws;
5. surface impoundments, pits, ponds, or lagoons;
6. storm water or waste water collection systems;
7. flow-through process tanks;
8. liquid traps or associated gathering lines directly related to oil or gas production and gathering operations;
9. storage tanks situated in an underground area (such as a basement, cellar, mineworking, drift, shaft, or tunnel) if the storage tank is situated upon or above the surface of the floor.

What Substances Are Covered? The notification requirements apply to underground storage tanks that contain regulated substances. This includes any substance defined as hazardous in section 101 (14) of the Comprehensive Environmental Response, Compensation and Liability Act of 1980 (CERCLA), with the exception of those substances regulated as hazardous waste under Subtitle C of RCRA. It also includes petroleum, e.g., crude oil or any fraction thereof which is liquid at standard conditions of temperature and pressure (60 degrees Fahrenheit and 14.7 pounds per square inch absolute).

Where To Notify? Completed notification forms should be sent to the address given at the top of this page.

When To Notify? 1. Owners of underground storage tanks in use or that have been taken out of operation after January 1, 1974, but still in the ground, must notify by May 8, 1986. 2. Owners who bring underground storage tanks into use after May 8, 1986, must notify within 30 days of bringing the tanks into use.

Penalties: Any owner who knowingly fails to notify or submits false information shall be subject to a civil penalty not to exceed $10,000 for each tank for which notification is not given or for which false information is submitted.

INSTRUCTIONS

Please type or print in ink all items except "signature" in Section V. This form must be completed for each location containing underground storage tanks. If more than 5 tanks are owned at this location, photocopy the reverse side, and staple continuation sheets to this form.

Indicate number of continuation sheets attached

I. OWNERSHIP OF TANK(S)

Owner Name (Corporation, Individual, Public Agency, or Other Entity)

Street Address

County

City   State   ZIP Code

Area Code   Phone Number

Type of Owner (Mark all that apply ☒)

☐ Current   ☐ State or Local Gov't   ☐ Private or Corporate
☐ Former   ☐ Federal Gov't   ☐ Ownership uncertain
                (GSA facility I.D. no. _____)

II. LOCATION OF TANK(S)

(If same as Section 1, mark box here ☐)

Facility Name or Company Site Identifier, as applicable

Street Address or State Road, as applicable

County

City (nearest)   State   ZIP Code

Indicate number of tanks at this location

Mark box here if tank(s) are located on land within an Indian reservation or on other Indian trust lands ☐

III. CONTACT PERSON AT TANK LOCATION

Name (If same as Section I, mark box here ☐)   Job Title   Area Code   Phone Number

IV. TYPE OF NOTIFICATION

☐ Mark box here only if this is an amended or subsequent notification for this location.

V. CERTIFICATION (Read and sign after completing Section VI.)

I certify under penalty of law that I have personally examined and am familiar with the information submitted in this and all attached documents, and that based on my inquiry of those individuals immediately responsible for obtaining the information, I believe that the submitted information is true, accurate, and complete.

Name and official title of owner or owner's authorized representative   Signature   Date Signed

CONTINUE ON REVERSE SIDE

EPA Form 7530-1 (Revised 9-88)   Page 1

Appendix F. EPA Form 7530-1: Notification for Underground Storage Tanks (page 1).

Owner Name (from Section I) _____ Location (from Section II) _____ Page No. _____ of _____ Pages

VI. DESCRIPTION OF UNDERGROUND STORAGE TANKS (Complete for each tank at this location.)

Tank Identification No. (e.g., ABC-123), or Arbitrarily Assigned Sequential Number (e.g., 1,2,3...)	Tank No.	Tank No.	Tank No.	Tank No.	Tank No.
1. Status of Tank *(Mark all that apply ☒)* — Currently in Use / Temporarily Out of Use / Permanently Out of Use / Brought into Use after 5/8/86					
2. Estimated Age (Years)					
3. Estimated Total Capacity (Gallons)					
4. Material of Construction *(Mark one ☒)* — Steel / Concrete / Fiberglass Reinforced Plastic / Unknown // Other, Please Specify					
5. Internal Protection *(Mark all that apply ☒)* — Cathodic Protection / Interior Lining (e.g., epoxy resins) / None / Unknown // Other, Please Specify					
6. External Protection *(Mark all that apply ☒)* — Cathodic Protection / Painted (e.g., asphaltic) / Fiberglass Reinforced Plastic Coated / None / Unknown // Other, Please Specify					
7. Piping *(Mark all that apply ☒)* — Bare Steel / Galvanized Steel / Fiberglass Reinforced Plastic / Cathodically Protected / Unknown // Other, Please Specify					
8. Substance Currently or Last Stored in Greatest Quantity by Volume *(Mark all that apply ☒)* — **a. Empty**					
b. Petroleum — Diesel / Kerosene / Gasoline (including alcohol blends) / Used Oil / Other, Please Specify					
c. Hazardous Substance					
Please Indicate Name of Principal CERCLA Substance OR Chemical Abstract Service (CAS) No.					
Mark box ☒ if tank stores a mixture of substances					
d. Unknown					
9. Additional Information (for tanks permanently taken out of service)					
a. Estimated date last used (mo/yr)	/	/	/	/	/
b. Estimated quantity of substance remaining (gal.)					
c. Mark box ☒ if tank was filled with inert material (e.g., sand, concrete)					

EPA Form 7530-1 (Revised 9-88) Reverse

Page 2

Appendix F. EPA Form 7530-1: Notification for Underground Storage Tanks (page 2).

Owner Name (from Section I) _____ Location (from Section II) _____ Page No. _____ of _____ Pages

VII. CERTIFICATION OF COMPLIANCE (COMPLETE FOR ALL NEW TANKS AT THIS LOCATION)

10. Installation (mark all that apply):

☐ The installer has been certified by the tank and piping manufacturers.

☐ The installer has been certified or licensed by the implementing agency.

☐ The installation has been inspected and certified by a registered professional engineer.

☐ The installation has been inspected and approved by the implementing agency.

☐ All work listed on the manufacturer's installation checklists has been completed.

☐ Another method was used as allowed by the implementing agency. Please specify:

_____

11. Release Detection (mark all that apply):

☐ Manual tank gauging.

☐ Tank tightness testing with inventory controls.

☐ Automatic tank gauging.

☐ Vapor monitoring.

☐ Ground-water monitoring.

☐ Interstitial monitoring within a secondary barrier.

☐ Interstitial monitoring within secondary containment.

☐ Automatic line leak detectors.

☐ Line tightness testing.

☐ Another method allowed by the implementing agency. Please specify:

_____

12. Corrosion Protection (if applicable)

☐ As specified for coated steel tanks with cathodic protection.

☐ As specified for coated steel piping with cathodic protection.

☐ Another method allowed by the implementing agency. Please specify:

_____

13. I have financial responsibility in accordance with Subpart I. Please specify:

Method: _____

Insurer: _____

Policy Number: _____

14. OATH: I certify that the information concerning installation provided in Item 10 is true to the best of my belief and knowledge.

Installer: _____   _____

　　　　　　　　　　　　　Name　　　　　　　　　　　　　　　　　　　Date

　　　　　　_____

　　　　　　　　　　　　　　　　　　　Position

　　　　　　_____

　　　　　　　　　　　　　　　　　　　Company

Appendix F. EPA Form 7530-1: Notification for Underground Storage Tanks (page 3).

Examples of Potentially Incompatible Wastes

Many hazardous wastes, when mixed with other waste or materials at a hazardous waste facility, can produce effects which are harmful to human health and the environment, such as (1) heat or pressure, (2) fire or explosion, (3) violent reaction, (4) toxic dusts, mists, fumes, or gases, or (5) flammable fumes or gases.

Below are examples of potentially incompatible wastes, waste components, and materials, along with the harmful consequences which result from mixing materials in one group with materials in another group. The list is intended as a guide to owners or operators of treatment, storage, and disposal facilities, and to enforcement and permit granting officials, to indicate the need for special precautions when managing these potentially incompatible waste materials or components.

This list is not intended to be exhaustive. An owner or operator must, as the regulations require, adequately analyze his wastes so that he can avoid creating uncontrolled substances or reactions of the type listed below, whether the wastes are listed below or not.

It is possible for potentially incompatible wastes to be mixed in a way that precludes a reaction (e.g., adding acid to water rather than water to acid) or that neutralizes them (e.g., a strong acid mixed with a strong base), or that controls substances produced (e.g., by generating flammable gases in a closed tank equipped so that ignition cannot occur, and burning the gases in an incinerator).

In the lists below, the mixing of a Group A material with a Group B material may have the potential consequences as noted.

Group 1A	Group 1B
Acetylene sludge	Acid sludge
Alkaline caustic liquids	Acid and water
Alkaline cleaner	Battery acid
Alkaline corrosive liquids	Chemical cleaners
Alkaline corrosive battery fluid	Electrolyte, acid
Caustic wastewater	Etching acid liquid or solvent
Lime sludge and other corrosive alkalines	Pickling liquor and other corrosive acids
Lime wastewater	Spent acid
Lime and water	Spent mixed acid
Spent caustic	Spent sulfuric acid

Potential consequences: Heat generation; violent reaction.

Group 2A	Group 2B
Aluminum	Any waste in Group 1A or 1B
Beryllium	
Calcium	
Lithium	
Magnesium	
Potassium	
Sodium	
Zinc powder	
Other reactive metals and metal hydrides	

Potential consequences: Fire or explosion; generation of flammable hydrogen gas.

Group 3A	Group 3B
Alcohols	Any concentrated waste in Groups 1A or 1B
Water	Calcium
	Lithium
	Metal hydrides
	Potassium
	SO_2Cl_2, $SOCl_2$, PCl_3, CH_3SiCl_3
	Other water-reactive waste

Potential consequences: Fire, explosion, or heat generation; generation of flammable or toxic gases.

Group 4A	Group 4B
Alcohols	Concentrated Group 1A or 1B wastes
Aldehydes	Group 2A wastes
Halogenated hydrocarbons	
Nitrated hydrocarbons	
Unsaturated hydrocarbons	
Other reactive organic compounds and solvents	

Potential consequences: Fire, explosion, or violent reaction.

Group 5A	Group 5B
Spent cyanide and sulfide solutions	Group 1B wastes

Potential consequences: Generation of toxic hydrogen cyanide or hydrogen sulfide gas.

Group 6A	Group 6B
Chlorates	Acetic acid and other organic acids
Chlorine	Concentrated mineral acids
Chlorites	Group 2A wastes
Chromic acid	Group 4A wastes
Hypochlorites	Other flammable and combustible wastes
Nitrates	
Nitric acid, fuming	
Perchlorates	
Permanganates	
Peroxides	
Other strong oxidizers	

Potential consequences: Fire, explosion, or violent reaction.

Source: "Law, Regulations, and Guidelines for Handling of Hazardous Waste." California Department of Health, February 1975. Reprinted as Appendix V of 40 CFR Part 265.

APPENDIX H

Industrial Waste Exchanges in North America

Alabama Waste Exchange
P.O. Box 6373
University of Alabama
Tuscaloosa, AL 35487-6373
(205) 348-5889

Alberta Waste Materials Exchange
4445 Calgary Trail South
Edmonton, Alberta
Canada T6H 5R7
(403) 450-5461

California Waste Exchange
Department of Health Services
Toxic Substances Control Division
714 P Street
Sacramento, CA 95814
(916) 324-1807/1818

Canadian Waste Materials Exchange
Ontario Research Foundation
Sheridan Park Research Community
Mississauga, Ontario
Canada L5K 1B3
(416) 822-4111

Great Lakes Regional Waste
  Exchange
470 Market Street S.W., Suite 100
Grand Rapids, MI 49503
(616) 451-8992

Indiana Waste Exchange
1220 Wateway Boulevard
P.O. Box 1220
Indianapolis, IN 46206
(317) 634-2142

Industrial Material Exchange Service
P.O. Box 19276
Springfield, IL 62794-9276
(217) 782-0450

Industrial Waste Information
  Exchange
New Jersey Chamber of Commerce
5 Commerce Street
Newark, NJ 07102
(201) 623-7070

Manitoba Waste Exchange
c/o Biomass Energy Institute
1329 Niakwa Road
Winnipeg, Manitoba
Canada R2J 3T4
(204) 257-3891

Montana Industrial Waste Exchange
Montana Chamber of Commerce
P.O. Box 1730
Helena, MT 59624
(406) 442-2405

Northeast Industrial Waste Exchange
90 Presidential Plaza, Suite 122
Syracuse, NY 13202
(315) 422-6572, (800) 237-2481

Southeast Waste Exchange
Urban Institute
UNCC Station
Charlotte, NC 28223
(704) 547-2307

Southern Waste Information Exchange
P.O. Box 6487
Florida State University
Institute of Science and Public Affairs
Tallahassee, FL 32313
(904) 644-5516

Tennessee Waste Exchange
Tennessee Association of Business
226 Capitol Boulevard, Suite 800
Nashville, TN 37219
(615) 256-5141

Western Waste Exchange
Arizona State University Center for
    Environmental Studies
Krause Hall
Tempe, AZ 85287
(602) 965-1858

Zero Waste Systems, Inc.
2928 Poplar Street
Oakland, CA 94608
(415) 893-8257

Summaries of DOT Hazard Class Definitions

Note: All regulatory citations refer to Title 49 of the CFR.

Explosive   Any chemical compound, mixture, or device, the primary or common purpose of which is to function by explosion, i.e., with substantially instantaneous release of gas and heat, unless such compound, mixture, or device is otherwise specifically classified.

> **Class A Explosive:**   Detonating or otherwise of maximum hazard. The nine types of Class A explosives are defined in 173.53.
>
> **Class B Explosive:**   In general, function by rapid combustion rather than detonation and include some explosive devices such as special fireworks, flash powders, etc. *Flammable hazard.* (173.88)
>
> **Class C Explosive:**   Certain types of manufactured articles containing Class A or Class B explosives, or both, as components but in restricted quantities, and certain types of fireworks. *Minimum hazard.*

Blasting Agent   A material designed for blasting that has been tested and found to be so insensitive that there is very little probability of accidental initiation to explosion or of transition from deflagration to detonation.

Combustible Liquid   Any liquid having a flash point above 100°F and below 200°F as determined by tests listed in 173.115(d). Exceptions to this are found in 173.114(b).

Corrosive   Any liquid or solid that causes visible destruction of human skin tissue or a liquid that has a severe corrosion rate on steel. (See 173.240 (a) and (b) for details.)

Flammable Liquid   Any liquid having a flash point below 100°F as determined by tests listed in 173.115(d). Exceptions are listed in 173.115(a).

Pyrophoric Liquid   Any liquid that ignites spontaneously in dry or moist air at or below 130°F. (173.115(c))

Compressed Gas   Any material or mixture having in the container a pressure exceeding 40 psia at 70°F, or a pressure exceeding 104 psia at 130°F; or any liquid flammable material having a vapor pressure exceeding 40 psia at 100°F. (173.300(a))

    Flammable Gas: Any compressed gas meeting the requirements for lower flammability limit range, flame projection, or flame propagation criteria as specified in 173.300(b).

    Nonflammable Gas: Any compressed gas other than a flammable gas.

Flammable Solid   Any solid material, other than an explosive, which is liable to cause fires through friction or retained heat from manufacturing or processing, or which can be ignited readily and when ignited burns so vigorously and persistently as to create a serious transportation hazard. (173.150)

Organic Peroxide   An organic compound containing the bivalent -0-0 structure and which may be considered a derivative of hydrogen peroxide where one or more of the hydrogen atoms have been replaced by organic radicals (except under certain conditions). (See 173.151(a) for details.)

Oxidizer   A substance, such as chlorate, permanganate, inorganic peroxide, or a nitrate, that yields oxygen readily to stimulate the combustion of organic matter. (See 173.151.)

Poison A   *Extremely Dangerous Poisons*—poisonous gases or liquids of such nature that a very small amount of the gas or vapor of the liquid mixed with air is *dangerous to life.* (173.326)

Poison B   *Less Dangerous Poisons*—gases, liquids, or solids (including pastes and semi-solids), other than Class A Poisons or Irritating Materials, which are known to be so toxic to man as to afford a hazard to health during transportation; or which, in the absence of adequate data on human toxicity, are presumed to be *toxic to man.* (See 173.343.)

Irritating Material   A liquid or solid substance which upon contact with fire or when exposed to air gives off dangerous or intensely irritating fumes, but *not including any poisonous material, Class A.* (173.381)

Etiologic Agent   A viable microorganism, or its toxin, which causes or may cause human disease. (173.386)

Radioactive Material   Any material, or combination of materials, that spontaneously emits ionizing radiation and having a specific activity greater than 0.002 Ci/g. (173.389)

ORM (other regulated material)   Any material that (1) may pose an unreasonable risk to health and safety or property when transported in commerce; and (2) does not meet any of the definitions of the other hazard classes specified; or (3) has been reclassed an ORM. *Note:* A material with a flash point from 100°F to 200°F may not be classed as an ORM if it is a hazardous waste or is offered in a packaging having a rated capacity of more than 110 gallons.

ORM-A: A material which has an anesthetic, irritating, noxious, toxic, or other similar property and which can cause extreme annoyance or discomfort to passengers and crew in the event of leakage during transportation. (173.500(a)(1))

ORM-B: A material (including a solid when wet with water) capable of causing significant damage to a transport vehicle or vessel from leakage during transportation. Materials meeting one or both of the following criteria are ORM-B materials: (1) The material is a liquid substance that has a corrosion rate exceeding 0.250 inches per year (IPY) on aluminum (nonclad 7075-T6) at a test temperature of 130°F. (An acceptable test is described in NACE Standard TM-01-69.) (2) The material is specifically designated by name in 172.101. (173.500(a)(2))

ORM-C: A material, not described as an ORM-A or ORM-B, which has other inherent characteristics which make it unsuitable for shipment, unless properly identified and prepared for transportation. Each ORM-C material is specifically named in 172.101. (173.500)(a)(4))

ORM-D: A material such as a consumer commodity which, though otherwise subject to DOT regulations, presents a limited hazard during transportation due to its form, quantity, and packaging. A shipping description applicable to each ORM-D material is found in 172.101. (173.500(a)(4))

ORM-E: A material that is not included in any other hazard class, but is subject to the DOT regulations. Materials in this class include hazardous wastes and hazardous substances, as they are defined in 171.8.

However, a material with a flash point of 100°F–200°F may not be classed as an ORM-E if it is a hazardous waste or offered in a packaging having a rated capacity of more than 110 gallons. (173.500(b)(5))

APPENDIX J

Useful Telephone Numbers

EPA RCRA/Superfund Hotline	(800) 424-9346
	or (202) 382-3000
EPA Chemical Emergency Preparedness Hotline	(800) 535-0202
("Title III Hotline")	or (202) 479-2449
EPA Asbestos and Small Business Ombudsman Hotline	(800) 368-5888
	or (703) 557-1938
EPA TSCA Assistance Office	(202) 554-1404
EPA Asbestos Technical Information Service	(202) 554-1404
EPA Asbestos in Schools Hotline	(800) 835-6700
EPA Safe Drinking Water Hotline	(800) 426-4791
	or (202) 382-5533
EPA RCRA Ombudsman	(202) 475-9361
National Pesticide Telecommunications Network	(800) 858-7378
U.S. National Response Center (Emergency Only)	(800) 424-8802
CMA CHEMTREC (Emergency Only)	(800) 424-9300
CMA Chemical Referral Center	(800) 262-8200
CMA National Chemical Response and Information Center (NCRIC)	(202) 887-1216
DOT Office of Hazardous Materials Transportation	(202) 366-0656
U.S. Government Printing Office	(202) 783-3238
National Technical Information Service (NTIS)	(800) 336-4700
NIOSH Publications	(513) 533-8287
OSHA Publications	(202) 523-9667

APPENDIX K

Summary of Regulatory Requirements for Containerized Hazardous Wastes

1. Containers used for holding hazardous waste must be in good condition. If the container becomes damaged or deteriorated or begins to leak, the wastes should be transferred to a container that is in good condition.

2. Containers used for holding hazardous waste must not be deteriorated by the waste. The container or liner must be compatible with the wastes to be stored.

3. Each container must be labeled or marked clearly with the words "Hazardous Waste."

4. The accumulation start date for each container is to be marked clearly on each container. The accumulation start date marking must be visible for inspection.

5. Containers holding hazardous wastes must always be closed during storage. The only time containers can be opened is to add or remove waste.

6. Containers holding hazardous wastes are to be managed to avoid rupturing or damaging the container, or otherwise causing the container to leak.

7. Areas where ignitable or reactive wastes are stored should be located at least 50 feet from the facility property line.

8. Ignitable or reactive wastes are to be separated and protected from sources of ignition or reaction (e.g., open flames, smoking, cutting, welding, hot surfaces, frictional heat, sparks, and radiant heat).

9. "No Smoking" signs are to be posted wherever there is a hazard from ignitable or reactive wastes.

10. Incompatible wastes, or incompatible wastes and materials, must not be placed in the same container for storage purposes. Further, hazardous waste cannot be placed in an unwashed container that previously held an incompatible waste or material.

11. Incompatible hazardous wastes and hazardous wastes incompatible with nearby materials must be separated or protected from each other by means of a dike, berm, or wall, or separated by sufficient distance.

12. Emergency equipment is required to be available at each accumulation area.
    a. internal communications or alarm
    b. telephone or two-way radio
    c. portable fire extinguishers
    d. fire control equipment
    e. spill control equipment
    f. decontamination equipment
    g. water at adequate volume and pressure

13. Adequate aisle space in the container storage area is to be maintained to allow unobstructed movement in response to an emergency, as well as to perform weekly inspections.

14. Weekly inspections must be made of container storage areas, looking for leaks or other evidence of actual or pending releases.

15. Containerized wastes are to be shipped to offsite (commercial) hazardous waste management (HWM) facilities within 90 days of the accumulation start date.

Bibliography

"Administrative Procedures for RCRA Permits for Hazardous Waste Facilities," SW-934, U.S. EPA, U.S. Government Printing Office (1981).

"Alternatives to the Land Disposal of Hazardous Wastes—An Assessment for California," Toxic Waste Assessment Group, Governor's Office of Appropriate Technology, Sacramento, CA (1981).

Amstead, B.H., P. Ostwald, and M. Begeman. *Manufacturing Processes,* 7th ed. (New York: John Wiley and Sons, Inc., 1977).

"Application for a Hazardous Waste Permit—Consolidated Permits Program," Forms 3510-1 and 3510-3, U.S. EPA, U.S. Government Printing Office (1980).

"Assessment of Industrial Hazardous Waste Practices: Electronic Components Manufacturing Industry," SW-140c, U.S. EPA, U.S. Government Printing Office (1977).

"Assessment of Industrial Hazardous Waste Practices: Electroplating and Metal Finishing Industries," SW-136c, U.S. EPA, U.S. Government Printing Office (1977).

"Assessment of Industrial Hazardous Waste Practices: Inorganic Chemicals Industry," SW-104c, U.S. EPA, U.S. Government Printing Office (1975).

"Assessment of Industrial Hazardous Waste Practices: Leather Tanning and Finishing Industry," SW-113c, U.S. EPA, U.S. Government Printing Office (1976).

"Assessment of Industrial Hazardous Waste Practices in the Metal Smelting and Refining Industry," SW-145c.1, U.S. EPA, U.S. Government Printing Office (1977).

"Assessment of Industrial Hazardous Waste Practices: Organic Chemical, Pesticides, and Explosives Industries," SW-118c, U.S. EPA, U.S. Government Printing Office (1977).

"Assessment of Industrial Hazardous Waste Practices: Paint and Allied Products Industry, Contract Solvent Reclaiming Operations, and Factory Application of Coatings," SW-119c, U.S. EPA, U.S. Government Printing Office (1976).

"Assessment of Hazardous Waste Management Practices in the Petroleum Refining Industry," SW-129c, U.S. EPA, U.S. Government Printing Office (1976).

"Assessment of Industrial Hazardous Waste Management Practices: Petroleum Refining Industry," SW-144c, U.S. EPA, U.S. Government Printing Office (1977).

"Assessment of Industrial Hazardous Waste Practices: Rubber and Plastics Industry—Vol. II, Plastic Materials and Synthetics Industry," SW-163c.2, U.S. EPA, U.S. Government Printing Office (1978).

"Assessment of Industrial Hazardous Waste Practices: Special Machinery Manufacturing Industries," SW-141c, U.S. EPA, U.S. Government Printing Office (1977).

"Assessment of Industrial Hazardous Waste Practices: Storage and Primary Batteries Industries," SW-102c, U.S. EPA, U.S. Government Printing Office (1975).

"Assessment of Industrial Hazardous Waste Practices: Textiles Industry," SW125c, U.S. EPA, U.S. Government Printing Office (1976).

Cahill, L.B., and R.W. Kane, Eds. *Environmental Audits,* 4th ed. (Rockville, MD: Government Institutes, 1985).

"Chemical Emergency Preparedness Program Interim Guidance," U.S. EPA, U.S. Government Printing Office (1985).

"Choosing the Optimum Financial Strategy for Pollution Control Investments," EPA-625/3-76-005, U.S. EPA, U.S. Government Printing Office (1976).

"Choosing Optimum Management Strategies: Pollution Control Systems," EPA-625/3-77-008, U.S. EPA, U.S. Government Printing Office (1977).

Cleaning Our Environment (Homewood, IL: American Chemical Society, 1975).

"Closure and Postclosure: Interim Status Standards," SW-912, 40 CFR 265 (1981).

"Compendium of Technologies Used in the Treatment of Hazardous Wastes," EPA/625/8-87/014, U.S. EPA, U.S. Government Printing Office (1987).

Comprehensive Environmental Response, Compensation and Liability Act (CERCLA) (PL 96-510), 94 Stat. 2781 (1980), as amended by the Superfund Amendments and Reauthorization Act (SARA), PL 99-499 (1986), codified at 42 USC 9601 et seq.

"Controlling Pollution from the Manufacturing and Coating of Metal Products: Water Pollution Control," EPA-625/3-77-009, U.S. EPA, U.S. Government Printing Office (1977).

"Decision-Makers Guide in Solid Waste Management," 2nd ed. SW-500, U.S. EPA, U.S. Government Printing Office (1978).

"Driver's Pocket Guide to Hazardous Materials," (J.J. Keller and Associates, Inc., 1988).

Ehlers, V., and E. Steel. *Municipal and Rural Sanitation,* 6th ed. (New York: McGraw-Hill Book Company, 1965).

Elements of Toxicology and Chemical Risk Assessment—A Handbook for Non-Scientists, Attorneys and Decision Makers (Washington, DC: Environ, 1986).

"Emergency Response Guidebook: Guidebook for Initial Response to Hazardous Materials Incidents," DOT P 5800.4, U.S. DOT (1987).

"EPA's Final PCB Ban Rule: Over 100 Questions and Answers to Help You Meet These Requirements," U.S. EPA, U.S. Government Printing Office (1979).

"EPA/DOT: Hazardous Waste Transportation Interface," SW-935, U.S. EPA, U.S. Government Printing Office (1981).

Epstein, S., L. Brown, and C. Pope. *Hazardous Waste in America* (San Francisco: Sierra Club Books, 1982).

"Estimating Releases and Waste Treatment Efficiencies for the Toxic Chemical Release Inventory Form," EPA 560/4-88-002, U.S. EPA, U.S. Government Printing Office (1987).

"Everybody's Problem: Hazardous Waste," SW-826, U.S. EPA, U.S. Government Printing Office (1980).

"*Federal Register:* What It Is and How To Use It," Office of the *Federal Register,* U.S. Government Printing Office (1985).

"Financial Assurance and Liability Insurance: Requirements for Owners and Operators of Hazardous Waste Treatment, Storage, and Disposal Facilities Under RCRA, Subtitle C, Subpart H," SW-926, U.S. EPA, U.S. Government Printing Office (1981).

"Financial Requirements: Interim Status Standards," SW-913, 40 CFR 265, Subpart H (1981).

"Fire Protection Guide on Hazardous Materials," 8th ed. (Quincy, MA: National Fire Protection Association, 1984).

Greenwood, D.R., et al. "A Handbook of Key Federal Regulations and Criteria for Multimedia Environmental Control," EPA-600/7-79-175, U.S. EPA, U.S. Government Printing Office (1979).

"Guidance Manual for Electroplating and Metal Finishing Pretreatment Standards," U.S. EPA, U.S. Government Printing Office (1984).

"Hazardous Materials Emergency Planning Guide NRT-1," U.S. National Response Team (1987).

"Hazardous Materials Identification System Implementation Manual (Revised)," (Washington, DC: National Paint and Coatings Association, 1985).

"Hazardous Waste Disposal Methods: Major Problems with Their Use," CED-81-21, Report by the Comptroller General of the United States, General Accounting Office (1980).

"Hazardous Waste Generation and Commercial Hazardous Waste Management Capacity—An Assessment," SW-894, U.S. EPA, U.S. Government Printing Office (1980).

"The Hazardous Waste System," U.S. EPA, U.S. Government Printing Office (1987).

"Hazardous Waste Treatment, Storage, and Disposal Facilities (TSDF)—Air Emission Models," EPA-450/3-87-026, U.S. EPA, U.S. Government Printing Office (1987).

Henry, M. F., Ed. *Flammable and Combustible Liquids Code Handbook,* 2nd ed. (Quincy, MA: National Fire Protection Association, 1984).

"In-Process Pollution Abatement: Upgrading Metal-Finishing Facilities to Reduce Pollution," EPA 625/3-73-002, U.S. EPA, U.S. Government Printing Office (1973).

Krofchak, D. *Management and Engineering Guide to Economic Pollution Control* (Guelph, Ontario: Go Print G. T. & C. Ltd., 1972).

"Less is Better: Laboratory Chemical Management for Waste Reduction" (American Chemical Society, 1985).

Linville, J. L., Ed. *Industrial Fire Hazards Handbook,* 2nd ed. (Quincy, MA: National Fire Protection Association, 1984).

Lowry, G. G., and R. C. Lowry. *Lowrys' Handbook of Right-to-Know and Emergency Planning* (Chelsea, MI: Lewis Publishers, Inc., 1988).

Martin, E., and J. Johnson, Eds. *Hazardous Waste Management Engineering* (New York: Van Nostrand Reinhold, 1987).

"Meeting Hazardous Waste Requirements for Metal Finishers," EPA/625/4-87/018, U.S. EPA, U.S. Government Printing Office (1987).

Metcalf & Eddy, Inc. *Wastewater Engineering: Collection, Treatment, Disposal* (New York: McGraw-Hill Book Company, 1972).

Metcalf & Eddy, Inc. *Wastewater Engineering: Treatment/Disposal/Reuse,* 2nd ed. (New York: McGraw-Hill Book Company, 1979).

"Method for Determining the Compatibility of Hazardous Waste," EPA-600/2-80-076, U.S. EPA, U.S. Government Printing Office (1980).

"More About Leaking Underground Storage Tanks: A Background Booklet for the Chemical Advisory," U.S. EPA, U.S. Government Printing Office (1984).

Prudent Practices for Disposal of Chemicals from Laboratories (Washington, DC: National Research Council, 1983).

Prudent Practices for Handling Hazardous Chemicals in Laboratories (Washington, DC: National Research Council, 1981).

"NIOSH/OSHA/USCG/EPA Occupational Safety and Health Guidance Manual for Hazardous Waste Site Activities," NIOSH 85-115, U.S. Government Printing Office (1985).

"NIOSH Guide to Industrial Respiratory Protection," NIOSH 87-116, U.S. DHHS, U.S. Government Printing Office (1987).

"Notification of Hazardous Waste Activity," EPA Form 8700-12, U.S. EPA, U.S. Government Printing Office (Revised 11-85).

Nyer, E. *Groundwater Treatment Technology* (New York: Van Nostrand Reinhold, 1985).

"Occupational Health Guidelines for Chemical Hazards," NIOSH Publication No. 81-123, National Institute for Occupational Safety and Health, U.S. Government Printing Office (1981).

"From Pollution to Prevention—A Progress Report on Waste Reduction" Office of Technology Assessment, U.S. Congress, U.S. Government Printing Office (1987).

"Superfund Strategy," Office of Technology Assessment, U.S. Congress, U.S. Government Printing Office (1987).

"Mock RCRA Part B Permit Application," PEDCO Environmental, U.S. EPA Conference, April 6, 1982.

"Petitions to Delist Hazardous Wastes—A Guidance Manual," U.S. EPA, EPA/530-SW-003, Office of Solid Waste, (1985).

"Permit Applicants' Guidance Manual for the General Facility Standards of 40 CFR 264," SW-968, U.S. EPA, U.S. Government Printing Office (1983).

"Permitting Hazardous Waste Incinerators Workshop Proceedings," U.S. EPA, U.S. Government Printing Office (1986).

"Pharmaceutical Industry: Hazardous Waste Generation, Treatment, and Disposal," SW-508, U.S. EPA, U.S. Government Printing Office (1976).

Phifer, R. W., and W. McTigue. *Handbook of Hazardous Waste Management for Small Quantity Generators* (Chelsea, MI: Lewis Publishers, Inc., 1988).

Pipitone, D. A., Ed. *Safe Storage of Laboratory Chemicals* (New York: John Wiley and Sons, 1984).

"Plans, Recordkeeping, Variances, and Demonstrations for Hazardous Waste Treatment, Storage, and Disposal Facilities," SW-921, U.S. EPA, U.S.Government Printing Office (1981).

"Pocket Guide to Chemical Hazards," NIOSH Publication No. 85-114, National Institute for Occupational Safety and Health, U.S. Government Printing Office (1987).

"Practical Assessment of Industrial Waste Management Strategies," paper presented at the CESOS/CER Company Conference, February 16–18, 1981.

"Principles of Environmental Analyses," American Chemical Society (1983).

"Questions and Answers on Hazardous Waste Regulations," SW-853, U.S. EPA, U.S. Government Printing Office (1980).

"Questions and Answers Regarding the July 14, 1986 Hazardous Waste Tank System Regulatory Amendments," EPA/530-SW-87-012, U.S. EPA, U.S. Government Printing Office (1987).

"RCRA Inspection Manual," U.S. EPA, U.S. Government Printing Office (1980).

"RCRA Personnel Training Guidance Manual," SW-915, U.S. EPA, U.S. Government Printing Office (1980).

"RCRA Orientation Manual," EPA/530-SW-86-001, U.S. EPA, U.S. Government Printing Office (1980).

"Regional Guidance Manual for Selected Interim Status Requirements," U.S. EPA, U.S. Government Printing Office (1980).

"Regulation Information for Hazardous Waste Generators and Transporters," SW-906, U.S. EPA, U.S. Government Printing Office (1980).

"Report to Congress: How to Dispose of Hazardous Wastes," CED-79-13, U.S. General Accounting Office, U.S. Government Printing Office (1978).

"Report to Congress: Waste Minimization," U.S. EPA.

Resource Conservation and Recovery Act (RCRA) (PL 94-580), October 21, 1976, as amended by the Hazardous and Solid Waste Amendments of 1984 (HSWA) PL 98-616, codified at 42 USC 3251 et seq.

Robbins, R.L., Ed. *Proceedings of the Conference on Limiting Liability for Hazardous Waste* (Chicago, IL: Chicago-Kent College of Law, 1981).

Ruch, W., and B. Held. *Respiratory Protection—OSHA & The Small Business- man* (Ann Arbor, MI: Ann Arbor Science Publishers, 1975).

"Samplers and Sampling Procedures for Hazardous Waste Streams," EPA-600/2-80-018, U.S. EPA, U.S. Government Printing Office (1980).

Sarokin, D., W. Muir, C. Miller, and S. Sperber. *Cutting Chemical Wastes* (New York: Inform, 1985).

Schwendeman, T. G., and H. K. Wilcox. *Underground Storage Systems: Leak Detection and Monitoring* (Chelsea, MI: Lewis Publishers, Inc., 1987).

Serious Reduction of Hazardous Waste (Washington, DC: Office of Technology Assessment, U.S. Congress, 1986).

"State Decision-Maker's Guide for Hazardous Waste Management," SW-612, U.S. EPA, U.S. Government Printing Office (1977).

"Superfund Public Health Evaluation Manual," EPA/540/1-86/060, U.S. EPA, U.S. Government Printing Office (1986).

Tchobanoglous, G., H. Theisen, and R. Eliassen. *Solid Wastes: Engineering Principles and Management Issues* (New York: McGraw-Hill Book Company, 1977).

"Technical Resource Document for the Storage and Treatment of Hazardous Waste in Tank Systems," PB87-134391, U.S. EPA, U.S. Government Printing Office (1986).

Technologies and Management Strategies for Hazardous Waste Control, Office of Technology Assessment, U.S. Congress (1983).

"Test Methods for Evaluating Solid Waste: Physical/Chemical Methods," SW-846, U.S. EPA, U.S. Government Printing Office (1986).

Code of Federal Regulations, Title 40 Parts 190-399, U.S. EPA.

Code of Federal Regulations, Title 49 Parts 100-199, U.S. DOT.

Code of Federal Regulations, Title 29 Part 1910, U.S. OSHA.

"TLVs: Threshold Limit Valves and Biological Exposure Indices for 1987-1988" (Cincinnati, OH: American Conference of Governmental Industrial Hygienists, 1987).

"Training Requirements in OSHA Standards and Training Guidelines," U.S. Dept. of Labor–OSHA, OSHA 2254 (1985).

Tuck, C. A., Ed. *NFPA Inspection Manual,* 5th ed. (Quincy, MA: National Fire Protection Association, 1982).

"Underground Storage Tank Corrective Action Technologies," EPA/625/5-7-015 (1987).

"Understanding the Small Quantity Generator Hazardous Waste Rules: A Handbook for Small Business," EPA/530-SW-86-109 (1986).

"Waste Exchanges: Background Information," SW-887.1, U.S. EPA, U.S. Government Printing Office (1980).

"Waste Minimization: Environmental Quality with Economic Benefits," EPA/ 530-SW-87-026, U.S. EPA, U.S. Government Printing Office (1987).

Water Quality and Treatment, 3rd ed. (New York: American Water Works Association, 1971).

Whelan, E. *Toxic Terro* (Ottawa, IL: Jameson Books, 1985).

Young, R. A., Ed. *Proceedings of the HazPro '86 Professional Certification Symposium and Exposition* (Northbrook, IL: Pudvan Publishing, 1986).

Young, R. A., Ed. *Proceedings of International Congress on Hazardous Materials Management* (Northbrook, IL: Pudvan Publishing, 1987).

Index